D0142417

Helping and Communal Breeding in Birds

MONOGRAPHS IN BEHAVIOR AND ECOLOGY

Edited by John R. Krebs and Tim Clutton-Brock

Five New World Primates: A Study in Comparative Ecology, *by John Terborgh*

Reproductive Decisions: An Economic Analysis of Gelada Baboon Social Strategies, *by R. I. M. Dunbar*

Honeybee Ecology: A Study of Adaptation in Social Life, *by Thomas D. Seeley*

Foraging Theory, *by David W. Stephens and John R. Krebs*

Helping and Communal Breeding in Birds: Ecology and Evolution, *by Jerram L. Brown*

Helping and Communal Breeding in Birds

Ecology and Evolution

JERRAM L. BROWN

Princeton University Press
Princeton, New Jersey

Published by Princeton University Press, 41 William Street,
Princeton, New Jersey 08540
In the United Kingdom: Princeton University Press,
Guildford, Surrey

Library of Congress Cataloging in Publication Data will be
found on the last printed page of this book

ISBN 0-691-08447-5
0-691-08448-3 (pbk.)

This book has been composed in Lasercomp Times Roman
and Univers Med. 689

Clothbound editions of Princeton University Press books
are printed on acid-free paper, and binding materials are
chosen for strength and durability. Paperbacks, although
satisfactory for personal collections, are not usually
suitable for library rebinding

Printed in the United States of America
by Princeton University Press, Princeton, New Jersey

To my kin and their kin

Contents

Preface

In the early 1960s biologists began to realize that not only morphological structures but also social relationships and species-typical patterns of sociality were the product of natural selection. This was reflected in the first comparative studies of social organization in relation to environment (e.g., Orians, 1961; Crook, 1963) and in early cost-benefit natural-selection models for social behavior (e.g., Brown, 1964; Hamilton, 1964). The fundamental realization of the adaptiveness of social behavior was an impetus for the subsequent birth and growth of sociobiology and behavioral ecology in the 1970s. In surveying a small but active part of these largely overlapping fields this book reflects their recent history.

Acts of apparent altruism or cooperation by animals, such as helping behavior in nesting birds and in insects, have puzzled thoughtful people for ages and have stimulated recent advances in evolutionary theory. Outstanding among these are inclusive fitness theory (Hamilton, 1964) and, to a lesser extent, ESS (evolutionarily stable strategy) theory (Maynard Smith and Price, 1973). As observers of birds and mammals became aware in the early 1970s of the relevance of these theories to helping behavior, more and more of them undertook field studies to test predictions derived from the new theories.

The number of persons studying communally breeding birds has increased from perhaps 4 or 5 in the 1960s (Skutch, 1961a; Brown, 1963a; Carrick, 1963; Rowley, 1965a,b) to 50 or more by 1985. Accordingly, there has been an explosion in the number of published studies of helping behavior based on field work in nature. On the theoretical side there has also been a great increase in the number of papers and books on altruism and mutualism (reviewed in Michod, 1982; D. S. Wilson, 1983).

It is no longer possible for a textbook account or even a normal review paper to do justice to the literature devoted to these basic questions in evolutionary biology. This book is, therefore, intended as an overview of helping and communal social organizations for persons who require more coverage than can be given in a journal review article but who have not the time or inclination to read the extensive original sources. Although the main goal of this book is to integrate fact and theory from a wide array of sometimes obscure papers already in the literature, considerable new material has been included.

Why is such a lengthy overview needed? Firstly, reviews published through 1985 are incomplete or already out of date. Furthermore, there has been no thorough and detailed review. Secondly, major advances involving theory and field work which promise to alter the field significantly appeared in 1984 and 1985. Thirdly, where anarchy and uninformed emotion once prevailed, several unifying themes are now solidifying. Fourthly, there has been some misunderstanding of basic concepts and of the implications of particular findings, not to mention semantic confusion. This noise factor has been magnified by unstable terminology, and by the deleterious effects of the "method of advocacy" (explained in E. O. Wilson, 1975). In this book a special effort is made to employ the method of alternative working hypotheses (Chamberlin, 1897, a paper that every graduate student should read) and to avoid one-sided advocacy. Following Popperian methods, hypotheses should be capable of being falsified by empirical test. Controversies will be subjected to this approach.

The scientific terms used in this book are based wherever possible on original sources, such as Hamilton (1964) for altruism, Maynard Smith (1964) for kin selection, and Skutch (1935, 1961a) for helping. Since some writers and reviewers have not used these terms in their original sense, the original definitions may be unfamiliar to some readers. Consequently, I have been liberal in quoting sources and have supplied detailed justifications for my choices of controversial terms in the text and glossary.

Lastly, it is probably unnecessary for me to say that some of the issues discussed in this book have been controversial for a decade or more, especially that of the importance of what has been too loosely called kin selection and is more accurately termed indirect selection. It is important that *both sides* of these debates be presented. Since some of the recent literature has emphasized the anti-indirect-selection position, it is especially important that this issue be re-examined more critically and that a fair statement of alternative views be given. A consensus in this field is not yet in sight, but it is my hope that this book will restore a more balanced perspective.

The text uses the English names of animals. Scientific names of species with helpers are listed in Table 2.2. For other species, scientific names are provided when the species is first mentioned.

The mathematical symbols used are defined in each chapter as they appear. They are consistent and nonoverlapping within a chapter. Some symbols are not consistent between chapters because (1) they have been drawn from many different published works by various authors, (2) where possible I have retained the original symbolism, especially to conform with the original figures reproduced here, (3) some overlap is unavoidable because of the large number of symbols needed and the small number available on standard word processors, and (4) the symbols for a word are easier to remember when the first letter of that word is used. For example, in Chapter 8 it is convenient to use *D* for defense costs, since it is not used for donor; and in Chapter 14 it is convenient to employ *D* to symbolize donor, since defense costs are not mentioned.

J.L.B.
December 31, 1985

Acknowledgments

In twenty-seven years of studying communally breeding birds I have received aid from many people. One person has stood by me, always ready to help, for this entire period—my wife, Esther R. Brown. She has helped in every imaginable way, in field observations, in processing and analyzing data, in preparation of manuscripts, and in matters of home, family, and university life. In particular, thanks to her energy and efficiency in processing this manuscript, it proved possible to have it printed by computer and sent for private reviews (March 1985) within eleven months of the decision to write the book, April 1984.

My own understanding of communal birds has profited greatly from field experience around the world with many of the researchers whose work is featured in this book. For generously showing me their species under natural conditions on their study areas I thank J. Counsilman and D. Dow (Australia), R. Hegner, D. Ligon, S. Ligon, and D. Lewis (Africa), R. Ali (India), H. Bell (New Guinea), S. Yamagishi (Japan), and S. Strahl (Venezuela).

For discussion of relevant theoretical constructs I thank W. D. Hamilton, J. Maynard Smith, R. Michod, and D. S. Wilson. I have been fortunate in being able to draw on help from several biologists who have provided mathematical models or helped me with my own; these include T. Caraco, J. Gilliam, S. Pimm, and H. R. Pulliam. For discussion of the biology of communal breeding I am indebted to many, but I am particularly grateful to C. Barkan, J. Craig, S. Strahl, K. Rabenold, U. Reyer, S. Vehrencamp, R. Koford, and B. Bowen.

My research on Mexican Jays and Grey-crowned Babblers was made possible by research grants from the National Institute of Mental Health and the National Science Foundation, and by the cooperation of the Southwestern Research Station of the American Museum of Natural History.

I am grateful to the following persons for constructive comments covering all or most of this book: S. Austad, D. Siemens, W. Koenig, R. Mumme, K. Rabenold, S. Strahl. I also thank the following for their comments on selected subjects: J. Alcock, C. Barkan, B. Bowen, N. Davies, S. Emlen, A. Grafen, R. Koford, U. Reyer, I. Rowley, P. Stacey, S. Vehrencamp.

For help in arranging for the cover illustration I thank S. Yamagishi.

All errors are my responsibility.

Helping and Communal Breeding in Birds

1 Why Study Helping Behavior?

"A 'helper' is a bird which assists in the nesting of an individual other than its mate, or feeds or otherwise attends a bird of whatever age which is neither its mate nor its dependent offspring. Helpers may be of almost any age; they may be breeding or nonbreeding individuals; they may aid other birds of the most diverse relationships to themselves, including those of distinct species; and they may assist in various ways." (Skutch, 1961a)

Helping behavior has a special meaning in biology. It is not just any behavior that may appear to be helpful. Conventional usage of the term *helping* has been somewhat narrower than Skutch's definition above. *Helping* is parent-like behavior toward young that are not the genetic offspring of the helper. Social organizations in which individuals additional to a single male-female pair typically aid in the care of young at a single nest, not as a result of brood parasitism or brood mixing, are referred to as *communal breeding systems* (see glossary; the term *cooperative breeding* is also used). Paradoxically, helping is parental in a behavioral sense but nonparental in a genetic sense. The resolution of this paradox is the subject of this book. How has natural selection produced this exception to the rule that only parents exhibit "parental" behavior? And, in the case of breeding helpers, why do some parents devote much of their parental behavior to the young of other parents?

As a technical term, helping is derived from the ornithological term, helpers-at-the-nest, and has been in use for half a century (Skutch, 1935). *Alloparental behavior* (E. O. Wilson 1975) is a perfect synonym and is more precisely descriptive; however, in practice the term helper is so widely understood and established that it is usually the more useful of the two.

Helping behavior, especially as it is performed in the social insects by sterile workers, has been the impetus for one of the greatest revolutions in thinking about natural selection that has occurred in this century. This body of theory was first formally developed and applied by W. D. Hamilton (1963, 1964), and is known as *inclusive fitness theory*. It will be further elaborated in Chapter 4 and applied throughout the rest of this book. Helping behavior then is the *principal test case* for the ability of inclusive fitness theory to elucidate the evolution of aid-giving behaviors, some of which when carefully examined might qualify as altruism.

The principal organisms for the testing of predictions based on inclusive fitness theory have rightly been the social insects. The theory was developed by an entomologist (Hamilton, 1964), popularized by an entomologist (E. O. Wilson, 1975), and has been heavily studied by entomologists (e.g., Oster and Wilson, 1978). Yet insects are extremely different from vertebrates in their biology and they have limitations as models for vertebrates. One limitation is that many social insects are haplodiploid while the vertebrate genome is basically diploid. This has focused attention on the difference that haplodiploidy makes in the evolution of helping. For vertebrates this literature provides little insight.

This book, therefore, recognizes a need to focus on the vertebrates. In this group helping has been studied for fifty years in birds but has only just begun as a recognized subject for research by mammalogists (Bekoff et al., 1984; Gittleman, 1985; Moehlman, 1979; Riedman, 1982; Rood, 1978; Macdonald, 1983; Owens and Owens, 1984.) Most of this book, consequently, is about birds. Examples from the mammals will be used occasionally.

The history of the study of helping behavior has certainly been influenced at least qualitatively by inclusive fitness theory, which appeared in 1963 and 1964; however, the cumulative number of publications on helping behavior shown in Figure 1.1 does not show a dramatic increase in the rate of appearance of new papers on helping in the 5-year period, 1965–70. This may be because papers in the ornithological literature linking helping behavior with inclusive fitness theory did not start coming out until the end of this period (Brown, 1969a, 1970). The connection between helping in birds and inclusive fitness theory was developed more extensively in theoretical papers that appeared in the 5-year period ending in 1975 (Maynard Smith and Ridpath, 1972; Brown, 1974; Ricklefs, 1975). The period following these papers did show an increase in rate of appearance of publications on helping, as shown in Figure 1.1. In fact, the number of publications on helping was only 16 in the interval 1961–65, and 20 in 1966–70, but it doubled in 1971–75 ($N = 47$), and more than doubled again in 1976–80 ($N = 127$).

The curves reveal, however, that the literature on helping in birds was growing exponentially *before* the introduction of inclusive-fitness theory. The causes of this pattern are unclear. The curves might, for example, simply reflect the rate of growth in the number of ornithologists. The main surprise from Figure 1.1 is that the cumulative total of publications on avian helping has been growing exponentially over a 50-year period, with two thirds of the titles appearing in the last fifteen years.

Inclusive fitness theory, however, is not sufficient to explain the evolution of helping, nor is it necessary in all cases. Helping behavior exists in a social framework, or *social organization*; and social organizations

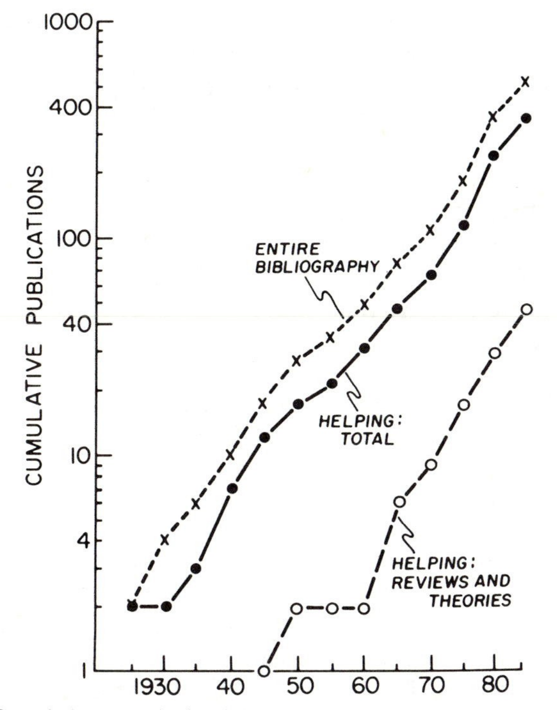

Figure 1.1 Cumulative record of publications on helping in birds. These curves are based on the bibliography in this book, which is as complete as I could make it. The curve for "helping: total" includes all papers on helping *per se*, as well as papers on relevant aspects of the biology of species with helpers. Papers that are mainly reviews, theories, and comments about avian helping are shown by the curve labeled "helping: reviews and theories." Theoretical papers on inclusive fitness or altruism are not included unless they specifically discuss helping in birds at length. Each point is for the cumulative total through the stated year. The last point on each curve is an underestimate because the curve was drawn in October 1985. Thus papers appearing later in 1985 as well as others in 1985 that were not immediately noticed by me were necessarily omitted.

differ greatly among species. To understand helping, we must understand the ecological determinants of the social organization within which it occurs. These are outlined in Chapter 2.

To learn why species differ in their social organizations and why helping is found in some species but not others, it is necessary to examine their ecologies. We begin by describing the eco-geographic distributions of helping in Chapter 3. Of particular interest is the *demographic context* of helping (Brown, 1969a, 1974; and many recent authors). This phrase refers to such phenomena from population ecology as survival rate, age structure, age-specific reproductive success, age at first breeding, fluctuations in population density, and dispersal. The demographic context "sets the stage" for selection to act on helping behavior and is the subject of Chapters 5 and 6. Studies on natural populations of species with helpers have made important contributions to our knowledge of the population ecology of birds and mammals. This constitutes an additional reason for studying helping.

Helping behavior, like all social behavior, involves at least two parties. Since the participants are typically not genetically identical, the possibility of *conflict of genetic interest* occurs. Donor and recipient may differ on how much help is to be given, on the cost to the donor and what is expected of the recipient as compensation. In many cases parent-offspring conflict (Trivers, 1974) might arise. Do parents "*manipulate*" their offspring to the parent's advantage (Alexander, 1974), or is it vice versa (Trivers and Hare, 1976). Or is this too simple a view? Perhaps a broader view expressed in the concepts of *variance utilization* and *variance enhancement* provides more insight. These problems are discussed in Chapter 15 along with a discussion of the role of dominance in cooperative social systems.

Behavioral conflict inevitably involves decisions by two or more participants. For such situations models that incorporate two or more decision makers are needed. Selection acting in such situations is best modeled using evolutionary game theory as embodied in the literature on *evolutionarily stable strategies*, ESS (Maynard Smith and Price, 1973; Maynard Smith, 1982) and exemplified by the adaptation of the prisoner's dilemma game to the study of animal mutualisms by Axelrod and Hamilton (1981). These approaches are developed and the relevant data on "*reciprocity*" evaluated in Chapter 14.

Because of the interest in helping and genetic relatedness stimulated by inclusive fitness theory, new ways of conceptualizing the *genetic structure of populations* have developed. It is now recognized that there is significant "structure" at levels below the deme. The concept of a structured deme (D. S. Wilson, 1975, 1977, 1980) has led to a modernization of "group selection" (Wade, 1978; D. S. Wilson, 1983), which I prefer to call either trait-group selection, using D. S. Wilson's term to distinguish it from interdemic group selection, or kin-group selection (Brown, 1974). The study of helping has led to new empirical approaches to the study of population genetic structure and processes, and these will be reviewed in Chapter 12. Kin recognition provides a behavioral genetic structure, especially when kin-group structure is diffuse rather then discrete; kin in a large group of non-kin may associate preferentially with other kin.

Lastly, the study of helping is in some ways a microcosm of the present state of naturalistic evolutionary biology with respect to the formulation and testing of selection models. This area of sociobiology is in a rapidly developing phase of its history. It has its full share of controversies over real and supposed issues. It is a fascinating exercise to examine how various workers have used (or not used):

(1) the method of alternative working hypotheses (Chamberlin, 1897),
(2) the method of weak inference or "natural experiment" and correlation,
(3) the method of strong inference by controlled experiment (Platt, 1964),
(4) the idea of parsimony, and, unfortunately,
(5) the method of advocacy (explained by E. O. Wilson, 1975).

Throughout this survey of helping behavior we shall keep in view the *scientific method* of resolving differences between theories. As explained in Chapter 18, the method of alternative working hypotheses has not always been used in strict Popperian manner.

2 The Discovery of Helping Behavior and a Classification of Avian Communal Breeding Systems

"In the great majority of bird species whose nesting has been carefully studied, each pair build their nest and rear their offspring without help from others of their kind. This, indeed, is almost a corollary of the theory of Territory, which teaches that each breeding pair occupy a definite nesting area from which they vigorously expel other individuals of their own species. While the concept of territory in bird life has done much to stimulate and give definite direction to bird study, as so often happens in the first enthusiasm of working out the details suggested by a fertile scientific theory, it has resulted in a tendency to neglect the opposite side of the story. There are many species in which the mated pair are not so exclusive in their territory, and as a result of this, coupled with other peculiar circumstances, receive more or less assistance in the duties of the nest. The number of recorded cases of helpers at the nest which have come to my notice is relatively small, but this appears to be, at least in part, because their discovery requires a more concentrated attention than is commonly devoted to studies of nesting birds." (Skutch, 1935)

In 1935 Alexander Skutch wrote a brief paper titled "Helpers at the nest." In it he described the feeding of young Brown Jays, Bushtits, and Banded Cactus Wrens by more than two unbanded birds of their own species. Some of the birds must have been feeding young that were not their own; hence some were helpers. For decades this paper was known to a few ornithologists, but it stimulated almost no further work except by Skutch himself. In the 1930s, '40s and '50s few ornithologists were interested in field studies of social relationships and few of them could visit the tropics where most of Skutch's helper species were found. There were a few exceptions. A bird-loving Japanese army officer took advantage of the Japanese occupation of the island of Taiwan (Formosa) to study "A sociable breeding habit among Timaliine birds" (Yamashina, 1938). David E. Davis carried out an extensive comparative study of the Crotophaginae (Davis, 1942).

A quarter-century later Skutch (1961a) reviewed the literature on helping that had accumulated. This review established the generality of helping by documenting its widespread taxonomic occurrence. Still ornithologists paid little attention—probably because helping had little relevance to the theoretical issues and controversies of those years, such as speciation, the subspecies concept, ecological niches, the motivation of displays, imprinting, and the phylogeny of behavior.

This chapter outlines the early history of the study of helping, then describes some studies of helping based upon color-banding, and closes with a survey of the types of social organizations in which helping is found. Some evolutionary hypotheses based on the comparative method are entertained briefly.

The Pre-color-banding Era

As pointed out by Skutch (1935) and Rowley (1968) the mind-set of temperate-zone naturalists assumed that parents exclude all others from their territory and take care of their own young exclusively. This had been observed so often that to many observers it seemed futile to look for exceptions. Since birds are usually not easily recognized as individuals, exceptions to the rule were rarely noticed and were apparently rare and confined for the most part to the tropics and Australia. Furthermore, there was virtually no incentive to employ techniques that might have revealed the presence of extra birds at a nest.

Nevertheless, studies on unbanded birds did allow some progress. Three patterns of group living were identified. In north temperate regions it is common for birds of many species to form *flocks during the winter* and to disband into pairs on large territories to breed. Sparrows, warblers, waterfowl, and many other taxa form such nonbreeding groups. Another common pattern is for birds to congregate in *colonies* to breed in groups, with each pair defending only the nest or a small space around it. In neither case do the groups stay together for the entire year. Birds in their first spring (at the age of one year) either attempt breeding or not, but they are not expected to be with their parents. Studies on unbanded birds by Skutch (1935, 1953, 1954, 1959, 1960) showed that some Central American species, such as the Brown Jay, departed from the temperate-zone pattern by the association of young with adults even during the breeding season and probably for a few years at least. These 1- and 2-year olds (Skutch's "innubiles") even helped in care of the nestlings. The resultant *groups existed all year.*

Davis (1940a,b, 1941) and Skutch (1959) showed that in tropical America anis live all year in social groups. Anis belong to a group of New World cuckoos known as the Crotophaginae. Davis showed by comparative studies on the four species of this group that they could be arranged in a loose series ranging from simple to complex in social organization. In the simpler forms, pairs tended to live by themselves, while in the more complex forms pairs tended to live together and share the same nest, which Skutch called joint nesting.

In California, Leach (1925) described communal nesting by groups of Acorn Woodpeckers; and Ritter (1938), in a rambling, philosophical work, provided more details.

In Australia, Robinson (1956) carried out detailed studies of the Australian "Magpie," a black and white, crow-sized bird of the endemic Australian family Cracticidae which is not taxonomically close to the Holarctic magpies. He also showed that groups of more than two birds persisted all year.

In Arizona, Gross (1949) showed that Mexican Jays breed in groups in which more than two birds frequent a nest.

The Advent of Color-banding

Colored leg-bands to enable individual recognition of wild birds under natural conditions were used by various workers as early as 1912. Butts (1930) used them to study the Black-capped Chickadee and traced the early history of their use. Burkitt (1924) and Lack (1953) used them to study European Robins, and they were used later to study dominance hierarchies in various titmice (Colquhoun, 1942) and sparrows (Sabine, 1949, 1959). In recent years color-banding has become invaluable not only for the study of helping behavior *per se*, but also for looking at associated demography, dispersal, and related phenomena. Recent studies on communal birds have exploited this method to the fullest.

With the use of color-banding it is possible to observe how many individuals share in building at a given nest, how frequently each of them feeds the nestlings, and how much the parents feed each other's young if there is more than one nest with young, as well as whether any of the helpers have preferences for a given nest. The construction of time budgets is a possibility that has just begun to be utilized.

As an illustration of how color-banding has provided insights into the study of communal birds, Figure 2.1 shows the results of one of the earlier studies of this sort. The feeding of nestlings was studied in a population of the Mexican Jay living in Arizona (Brown, 1970, 1972). All the members of one social unit in 1969 and 1970 are listed. In each year there were two active nests in this territory. The same two mothers were present each year (XMY-YMY and XRR-RRY) and confined their feedings mainly to their own nestlings. All the birds in this rather large flock participated in feeding the young. Most individuals fed young at both nests each year, though a

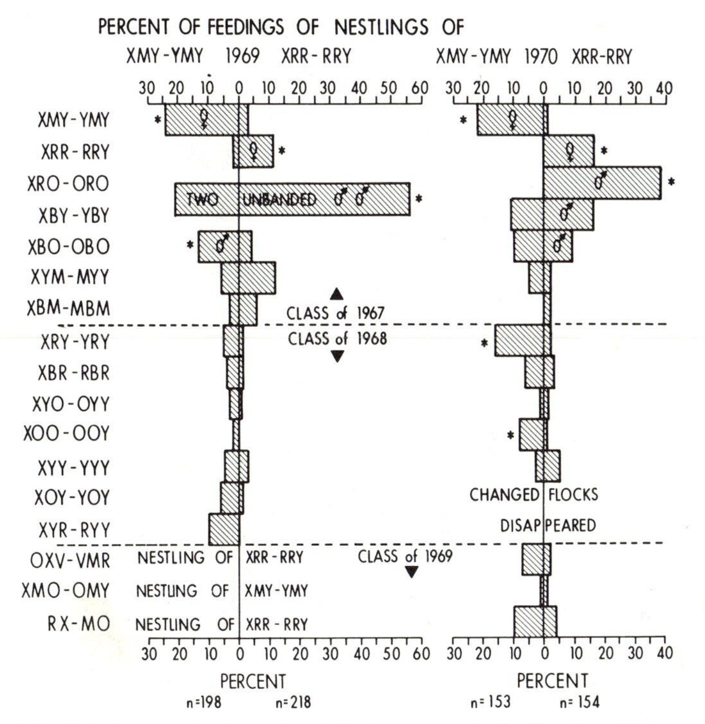

Figure 2.1 Relative contribution of each flock member to the feeding of the young in each of two nests in a social unit of Mexican Jays in 1969 and 1970. An asterisk designates statistically significant ($P < 0.01$) preference for delivering food to one of the nests. Year of hatching is indicated where known; jays in class of 1967 were hatched in 1967 or earlier and are of uncertain age. Sample size is indicated at the bottom for each nest. The "name" of each jay in this unit is shown at left. Each letter of the name stands for a color, and the name specifies the bird's unique combination of color bands. Note the retention in 1970 of two offspring from the previous year (1969) and the "reciprocal" helping by the two mothers (XMY-YMY and XRR-RRY). (From Brown, 1972.)

few showed a preference for a particular nest. Note that some of the nestlings from 1969 were helpers for their parents and nonparents in 1970, an example of generational mutualism (discussed in Chapter 14).

Collectively, the helpers contributed roughly half of the feedings of nestlings. The proportion of feedings by all helpers at the 1969 nests is shown in Figure 2.2. For fledglings the helpers contributed an even larger portion. This increase is not because the nonbreeding helpers worked harder; it is because the breeders increased their feeding of each other's young, thus becoming helpers for each other's broods in a kind of non-score-keeping mutualism (see Chapter 14). This study and that of Rowley (1965a) were the first to employ color-banding to reveal such striking

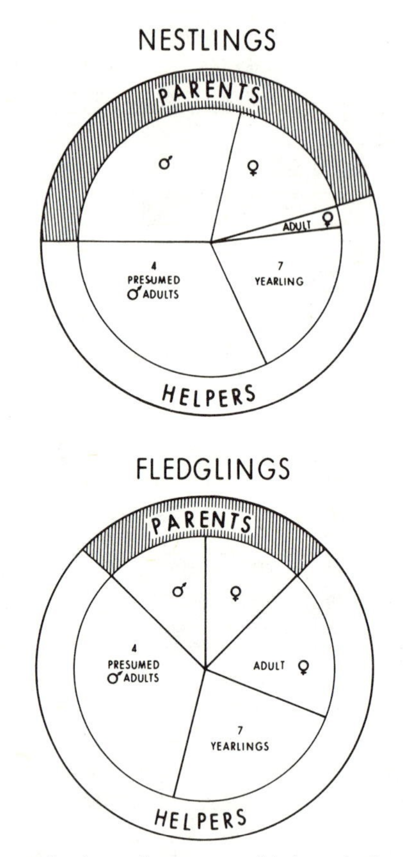

Figure 2.2 Relative contributions of parents and helpers in feeding the nestlings and fledglings of the color-banded unit of Mexican Jays in Figure 2.1. The increase in the proportion of feedings to fledglings by helpers was caused by an increase in reciprocal feeding of each other's young by parents after fledging. (From Brown, 1970.)

social complexity in the feeding of young. When the jay study was first published, two noted ornithologists were incredulous and told me that they did not believe it. This incident reveals the climate of opinion in 1970 with regard to helping. Opinion has changed, largely as a result of several extensive studies of color-banded birds.

In an earlier study with color-banded individuals Brown (1963a) plotted the home ranges of every member of two flocks of Mexican Jays in Arizona and demonstrated that all the members of a flock shared the same home range and had the same territorial boundaries. There were no subterritories within the flock territory. Thus the flock behaved as a *social unit.* Its members tended to forage together and rest together in a loose aggregation. Carrick (1963) showed that certain color-banded flocks of Australian magpies behave in a similar manner, and Rowley (1965a) did the same for the Superb Blue Wren in Australia. Such species are said to be *group-territorial.* The territory is held by a social unit of 3 or more individuals. Many mammals are also group-territorial (Macdonald, 1983). Group territoriality is a common but not universal attribute of species that have regular helpers, and is an integral part of certain theories for their evolution.

Communal versus Colonial Social Systems

Without color-banding it is difficult to reliably distinguish consistent social units, such as in group-territorial species, from groups that form casually and irregularly by aggregation of different individuals at various places and times. For example, in the Steller's Jay (*Cyanocitta stelleri*) color banding revealed that groups are formed only when pairs whose ranges overlap partially come together at a food source, to mob a predator, or for some other reason (Brown, 1963b).

The differences between these two origins of social groups are illustrated in Figure 2.3. Both types of sociality are thought to have their origins in a system of pairs with territories, as in western populations of the Scrub Jay. This is probably the commonest social system among temperate-zone passerine birds. A step toward coloniality is found in the Steller's Jay. Here defense of the boundaries has been relaxed so that intruders are tolerated if they behave submissively. These areas of dominance are called *dominions* rather than territories because rivals are not excluded from the area of dominance. This allows groups of jays to form, but the composition of the groups is not constant over a particular home range. Instead the composition varies according to the location, and the groups do not have constant membership as they do in the Mexican Jay. Helping is absent. The culmination of this trend toward sociality is found in the colonial Pinyon Jay (Balda and Bateman, 1971). Here again rivals are not excluded and the home range is shared by all members of a colony. Unlike the Mexican Jay system there are no group territories. The main component of sociality in a colonial species is the presence at the breeding site of many independent pairs. In the vast majority of colonial species most pairs lack helpers altogether, although a few pairs may have a helper or two, as in the Pinyon Jay. The present trend is for the discovery of more colonial species that have some helpers, particularly in the tropics.

Quite a different route toward sociality is shown in the lower part of Figure 2.3. The pattern here is for the young to stay with their parents on the parental territory at least through the next breeding season. The result

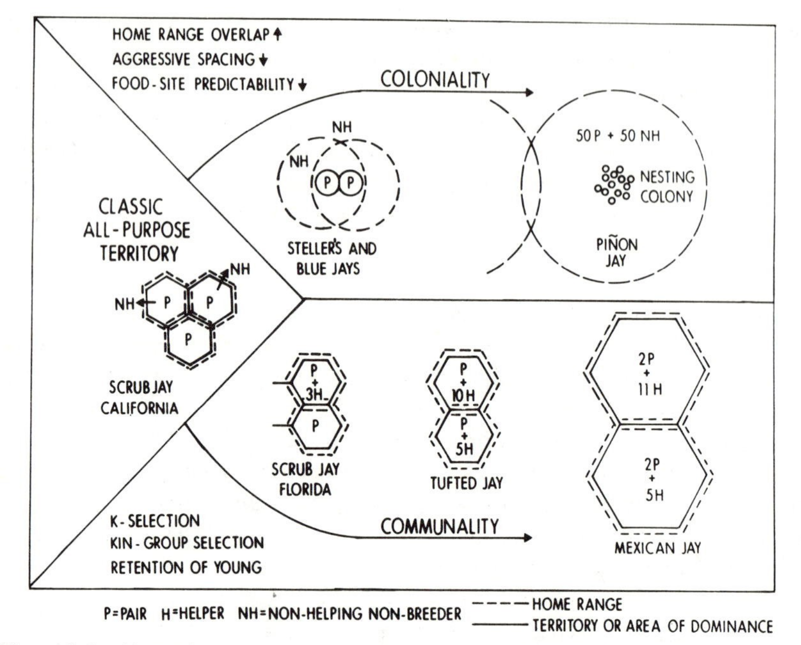

Figure 2.3 Alternative routes to sociality in jays. Sociality may be achieved by progressive overlapping of home ranges together with reduction of aggressive spacing, which results ultimately in a colonial social structure. Or it may be achieved through retention of the young in the territory for progressively longer periods, which results in a communal social structure via the subsocial route. An additional route to sociality in birds not known in jays is shown in Figure 2.4. (From Brown, 1974.)

is group territoriality. In some species, such as the Florida population of the Scrub Jay, the young stay about a year and then leave to breed elsewhere (Woolfenden, 1975). In other species, such as the Tufted Jay of western Mexico, the young appear to stay longer, perhaps two years, thus forming larger groups; but they still leave to breed (Crossin, 1967). Finally, some young may even breed on the home territory in the presence of the parents; or the parents may allow some outsiders to join and breed, as in the Mexican Jay (Brown and Brown, 1984).

The Florida Scrub Jay and the Tufted Jay, in which only one pair breeds per territory, are known as *singular-breeding species.* The Mexican Jay, in which two or more pairs commonly breed in a territory, is a *plural-breeding species.*

These two routes to sociality (Fig. 2.3) may be compared with the findings for the evolution of sociality in insects. The upper, or colonial, route corresponds in some respects to the semisocial route in insects. The lower, or group-territorial, route for birds corresponds to the matrifilial or subsocial route in insects (see E. O. Wilson, 1971, 1975; Michener, 1974).

A Classification of Avian Communal Breeding Systems

Helping occurs in a wide range of social systems. It is convenient as a basis for further discussion to recognize the major types briefly. These are classified in Table 2.1, and a speculative synopsis of their evolutionary relationships is shown in Figure 2.4. It is important to maintain a clear distinction between description of phenomena and explanations of them. Phenomena should not be described in terms of their presumed explanation (e.g., cooperative breeding). They should ideally be designated by unambiguous, observable features. I have tried to follow this convention here. This is, therefore, not a classification of theories of origin.

The major division in Table 2.1 is between group-territorial and colonial species. Note, however, that one colonial species, the White-fronted Bee-eater, has group territories in which *nesting does not occur* (Hegner et al., 1982). Helpers in colonial species are typically male and few in number. Pairs without helpers always occur and are usually in the majority.

Another important division is between singular- and plural-breeding species (see glossary for definitions). Singular-breeding, group-territorial species are usually simpler in social structure than plural-breeding species. Singular-breeding species vary importantly in the sex of the helpers. In

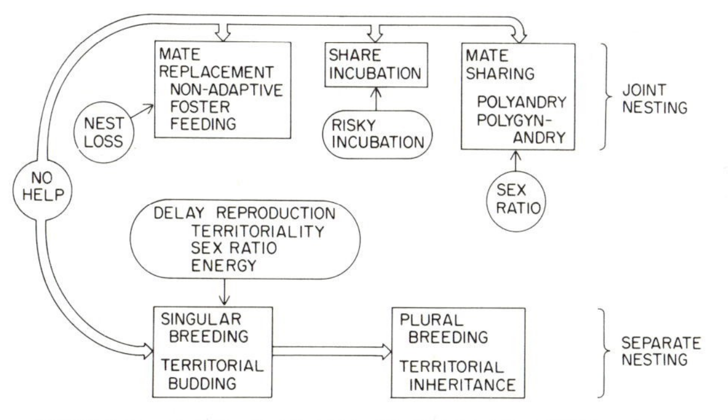

Figure 2.4 Multiple routes to regular communal breeding in birds. The upper arm shows origins that involve cooperation between rival breeders and that are not based on the nuclear family. Joint-nesting refers to those cases in which two or more females lay eggs in the same nest. The lower arm shows origins that are based upon the nuclear family and in which the helpers are nonbreeders at first (as in singular-breeders). Plural-breeding systems that are derived from singular-breeding systems may additionally involve helping by breeders. Plural-breeding also occurs in the parasocial route. The lower arm here corresponds to the lower arm in Figure 2.3.

Table 2.1

Major types of avian communal breeding systems. For names of species in each category see Table 2.2. This classification is based on observable features. It is not a classification of origins.

Categories	*Number of Species*
All-purpose territories defended by pairs and larger groups	
Singular-breeding. No more than one female breeds in unit	
Helpers mainly males	
Monogamous	10
Polyandrous	5
Mating system unknown	10
Helpers of both sexes	40
Sex of helpers unknown	34
Plural-breeding. Two or more females breed in same units	
Joint-nesting. Two or more females lay in same nest	
Monogamous	5
Polygynous	1
Polygynandrous in some units	4
Mating system unknown	2
Separate-nesting	10
Colonial in nesting. No all-purpose territory	20
Neither colonial nor group-territorial	5
Unclassified and miscellaneous	76
Total species with intraspecific helping	222

some, the helpers are usually males. An important factor in this system is the sex ratio. Rowley (1965a) showed for the Superb Blue Wren that adult helpers were nearly always males and that these males could breed if a female became available to them. If sex ratio is important, it follows that the proportion of helpers in the population must vary with the population sex ratio, as Emlen (1978) later illustrated. In some of these species with an extra male, both males copulate with the breeding female. In others, one male seems to be able to prevent the other from copulating, as in *Malurus*. It is not known why these types differ, but some hypotheses are discussed in Chapter 9.

In another important class of group-territorial, singular-breeding species, both sexes are represented among the helpers. Since sex ratio does not seem to be an important factor for the presence of potential helpers in these species, breeding is delayed for other reasons. Group size tends to be larger in such species and appears to depend mainly on the length of time that offspring stay on their natal territory. This category includes many of the better known cases of communal breeding.

The plural-breeding, group-territorial species are considerably more complex. In the jays, at least, they seem to be derived from singular-breeding forms, often with plural- and singular-breeding species in the same genus (e.g., *Cyanocorax*) or even in the same species (*Aphelocoma ultramarina*, Strahl and Brown, unpubl. ms.). The origins of plural breeding are discussed in Chapters 6 and 11 in relation to the difficulties of finding a territory. Group size tends to be large and quite variable in plural-breeding species, such as the Mexican Jay (Brown and Brown, 1985).

Joint nesting (Skutch, 1959:302), in which females pool their eggs in one nest, has the consequence that the risks of incubation are shared. The selection pressures leading to joint-nesting might, therefore, be hypothesized to involve higher mortality of incubating birds, at least in some species. Many of these species are weak fliers.

Few colonial species have been studied, but they are of special interest because in most of them the hypothesis of habitat saturation through territoriality can be rejected immediately. In addition, the opportunities for selection for kin recognition are much stronger than in the group-territorial species. This is because breeding pairs and their helpers have the possibility of close social relations with all members of the colony, which can be very large and is typically much larger than a group-territorial social unit. For further discussion of kin recognition in these contexts, see Chapter 12.

The Number of Communally Breeding Species

Intraspecific helping not due to adoption, brood parasitism, or mixing of fledged broods has been recorded in at least 222 of the 9,016 species of birds in the world listed by Morony et al. (1975; Table 2.1). Species strongly suspected to have helpers are not included in Table 2.1, for example many species listed by Grimes (1976b) as "unproved." Emlen (1984) did not list them but reported "over 300" communally breeding species, a figure that is much higher but perhaps includes some "unproved" species. Table 2.2 assigns the known species of birds with intraspecific helpers to the categories in Table 2.1. Roughly 146 can be assigned at least tentatively on present information. Another 76 are too poorly known, and for some of these, new categories may be needed. It can be quite difficult to assign species to categories, as the amount of detailed information available on each species varies enormously. Some species have been studied for decades, such as the Arizona population of the Mexican Jay; and some, such as Zavattariornis of Ethiopia, have rarely even been seen by an ornithologist, let alone studied.

Table 2.2

List of species of birds with intraspecific helpers. Each species has been entered only once. If a species may be put in more than one category, it has been put in the more complex or advanced one unless that occurrence is quite rare. This list is intended to include all species of birds in the world for which a published record of intraspecific helping is available.

Group territorial. Singular breeding. Male helpers. Monogamous.		
Red-cockaded Woodpecker	*Dendrocopos borealis*	Ligon 1970
Galapagos Mockingbird (Isla Genovesa)	*Nesomimus parvulus*	Grant & Grant 1979; Kinnaird & Grant 1982
Superb Blue Wren	*Malurus cyaneus*	Rowley 1965a
Splendid Fairy-wren	*M. splendens*	Rowley 1981a
Striated Thornbill	*Acanthiza lineata*	Bell 1983b, 1985
Buff-rumped Thornbill	*A. reguloides*	Bell 1983b, 1985
Pygmy Nuthatch	*Sitta pygmaea*	Norris 1958
Brown-headed Nuthatch	*S. pusilla*	Norris 1958
Cactus Ground-finch	*Geospiza scandens*	Price et al. 1983
Medium Ground-finch	*G. fortis*	Price et al. 1983
Group territorial. Singular breeding. Male helpers. Polyandrous.		
Galapagos Hawk	*Buteo galapagoensis*	de Vries 1973, 1975; Faaborg et al. 1980
Native Hen	*Tribonyx mortieri*	Ridpath 1972
Lonnberg Skua	*Catharacta mackormacki lonnbergi*	Young 1978
Dunnock	*Prunella modularis*	Birkhead 1981
Black Tit	*Parus niger*	Tarboton 1981
Group territorial. Singular breeding. Male helpers. Mating system unknown.		
Harris' Hawk	*Parabuteo unicinctus*	Mader 1975a,b, 1977, 1979; Bednarz 1985; Dawson & Mannan 1985
Spotted Wren	*Campylorhynchus jocosus*	Selander 1964
Logrunner	*Orthonyx temminckii*	Dow 1980a
Rufous Rockjumper	*Chaetops frenatus*	4 authors in Grimes 1976a
Abyssinian Black Wheatear	*Oenanthe lugens*	Reynolds in Grimes 1976a
Hooded Robin	*Petroica cucullata*	Rowley 1976; Bell 1984
Blue-breasted Fairy-wren	*Malurus pulcherrimus*	Chapman in Rowley 1976; Rowley 1981b
White-browed Scrub-wren	*Sericornis frontalis*	Harris & Newman 1974; Bell 1982b
Varied Sitella	*Daphoenositta chrysoptera*	Noske 1980a; Dow 1980a
Brown Treecreeper	*Climacteris picumnus*	Noske 1980b

(continued)

Table 2.2 *(continued)*

Group territorial. Singular breeding. Both sexes help. Plural breeding may occur rarely.		
Hoatzin	*Opisthocomus hoazin*	Strahl 1985
Purple gallinule	*Gallinula martinica*	Krekorian 1978; Hunter 1985
Laughing Kookaburra	*Dacelo novaeguineae (gigas)*	Parry 1973
Blue-winged Kookaburra	*D. leachii*	Parry in Dow 1980a
Green Woodhoopoe	*Phoeniculus purpureus*	Ligon & Ligon 1978
Ground Hornbill	*Bucorvus leadbeateri*	Kemp & Kemp 1980
Yellow-billed Shrike	*Corvinella corvina*	Grimes 1980
Bicolored Wren	*Campylorhynchus griseus*	Austad & Rabenold 1985
Stripe-backed Wren	*C. nuchalis*	Rabenold 1984, 1985
Band-backed Wren	*C. zonatus*	Skutch 1935, 1960, 1961; Selander 1964
Fasciated Wren	*C. fasciatus*	Rabenold pers. comm.
Gray-barred Wren	*C. megalopterus*	Selander 1964
Black-capped Donacobius	*Donacobius atricapillus*	Kiltie & Fitzpatrick 1984
Anteater Chat	*Myrmecocichla aethiops*	Grimes 1976a; Haas pers. comm.
Rufous Babbler	*Pomatostomus isidori*	Bell 1982a, 1983a
Grey-crowned Babbler	*P. temporalis*	King 1980
White-browed Babbler	*P. superciliosus*	Counsilman & Gardner in Anon. 1976
Hall's Babbler	*P. halli*	Balda & Brown 1977; Whitmore and Dow pers. comm.
Common Babbler	*Turdoides caudatus*	Gaston 1976, 1978a
Striated Babbler	*T. earlei*	Gaston 1977b
Large Grey Babbler	*T. malcolmi*	Gaston 1976
Arabian Babbler	*T. squamiceps*	Zahavi 1974, 1976
Jungle Babbler	*T. striatus*	Gaston 1976, 1977b, 1978b
White-headed Babbler	*T. affinis*	Zacharias unpubl. ms.; John-singh & Paramanandham 1982; Jeyasingh 1976
Arrowmarked Babbler	*T. jardinei*	Vernon 1976
White-winged Fairy-wren	*Malurus leucopterus*	Tidemann 1980
Long-tailed Tit	*Aegithalos caudatus*	Lack & Lack 1958; Gaston 1973; Nakamura 1972
Black-tailed Treecreeper	*Climacteris melanura*	Noske 1980b
Bell Miner	*Manorina melanophrys*	Swainson 1970; Smith & Robertson 1978; Clarke 1984
Bolivian Blackbird	*Oreospar bolivianus*	Orians et al. 1977a
White-browed Sparrow-weaver	*Plocepasser mahali*	Collias & Collias 1978a,b; Lewis 1981, 1982a,b
Golden-breasted Starling	*Cosmopsarus regius*	Huels 1981
Apostlebird	*Struthidea cinerea*	Baldwin 1974, 1975

(continued)

Table 2.2 *(continued)*

Group territorial. Singular breeding. Both sexes help. Plural breeding may occur rarely.		
Scrub Jay (Florida)	*Aphelocoma coerulescens*	Amadon 1944; Woolfenden 1975
Beechey Jay	*Cyanocorax beecheii*	Raitt & Hardy 1979; Winterstein & Raitt 1983
Tufted Jay	*C. dickeyi*	Crossin 1967
Green Jay (Colombia)	*C. y. yncas*	Alvarez 1975
Brown Jay (Guatemala)	*Psilorhinus morio*	Skutch 1935, 1960
Black-throated Magpie-jay	*Calocitta formosa colliei*	Winterstein 1985
Azure-winged Magpie	*Cyanopica cyana*	Hosono 1983; Yamagishi (pers. comm.)
Group territorial. Singular breeding. Sex of helpers unknown.		
White-fronted Nunbird	*Monasa morphoeus*	Skutch 1972
Red-and-yellow Barbet	*Trachyphonus erythrocephalus*	Short & Horne 1980
D'Arnaud's Barbet	*T. darnaudi*	Short & Horne 1980
Pied Barbet	*Lybius albicauda*	vanSomeren in Skutch 1961
Rufous-fronted Thornbird	*Phacellodomus rufifrons*	Skutch 1969a; Thomas 1983
White-bearded Flycatcher	*Conopias inornata*	Thomas 1979
Ground Cuckoo-shrike	*Coracina maxima*	Hindwood in Rowley 1976
Straight-crested Helmet-shrike	*Prionops plumata*	several authors in Grimes 1976a; Greig-Smith 1976
Retz's Redbilled Shrike	*P. retzii*	3 authors in Grimes 1976a
Chestnut-fronted Shrike	*P. scopifrons*	Sheppard & Vernon in Grimes 1976a; Britton & Britton 1977
Gray-backed Fiscal Shrike	*Lanius excubitorius*	Zack & Ligon 1985a,b
Galapagos Mockingbird	*Nesomimus macdonaldi*	R. Curry unpubl. data
Galapagos Mockingbird	*N. trifasciatus*	R. Curry unpubl. data
Alpine Accentor	*Prunella collaris*	Dyrcz 1977
Cinnamon Quail-thrush	*Cinclosoma cinnamomeum*	Brooker 1969
Brown Gerygone	*Gerygone mouki*	Mackness 1976
Yellow Thornbill	*Acanthiza nana*	Dow 1980a
Chestnut-rumped Thornbill	*A. uropygialis*	White 1950
Yellow-rumped Thornbill	*A. chrysorrhoa*	Immelman 1960; Ford 1963; Brown & Brown 1982
Large-billed Scrub-wren	*Sericornis magnirostris*	Ferrier in Dow 1980a
Eastern Yellow Robin	*Eopsaltria australis*	2 authors in Dow 1980a
Western Yellow Robin	*E. griseogularis*	4 authors in Dow 1980a
White-breasted Robin	*E. georgiana*	Brown, R. J. & M. N. 1980
White-spotted Wattle-eye	*Platysteira tonsa*	Erard in Grimes 1976a
Chestnut-cap Flycatcher	*Erythrocerus mccalli*	Erard in Grimes 1976a
Blue Flycatcher	*Trochocercus longicauda*	Erard in Grimes 1976a
Bushtit (Guatemala)	*Psaltriparus minimus*	Skutch 1935, 1960; Ervin 1979

(continued)

Table 2.2 *(continued)*

Group territorial. Singular breeding. Sex of helpers unknown.		
Red-browed Treecreeper	*Climacteris erythrops*	Noske 1980b
Seychelles White eye	*Zosterops modesta*	Grieg-Smith 1979
Rufous-throated Honeyeater	*Conopophila rufogularis*	Immelman 1961
Plain-colored Tanager	*Tangara inornata*	Skutch 1954
Pied Butcherbird	*Cracticus nigrogularis*	4 authors in Rowley 1976
Grey Butcherbird	*C. torquatus*	3 authors in Rowley 1976
Dusky Woodswallow	*Artamus cyanopterus*	Rowley 1976; Dow 1980a
Group territorial. Plural breeding. Joint nesting. Monogamous.		
Smooth-billed Ani	*Crotophaga ani*	Davis 1940a; Koester 1971
Groove-billed Ani	*C. sulcirostris*	Vehrencamp 1977, 1978
Greater Ani	*C. major*	Davis 1941, 1942
Guira	*Guira guira*	Davis 1940b
Speckled Mousebird (rare)	*Colius striatus*	Decoux 1982; Rowan 1967
Group territorial. Plural breeding. Joint-nesting. Polygynous.		
Magpie Goose	*Anseranas semipalmata*	Frith & Davies 1961
Group territorial. Plural breeding. Joint-nesting. Polygynandrous in part.		
Dusky Moorhen	*Gallinula tenebrosa*	Garnet 1980
Pukeko	*Porphyrio porphyrio*	Craig 1976, 1977, 1979, 1980a,b
Acorn Woodpecker	*Melanerpes formicivorus*	Leach 1925; Skutch 1943, 1969b; MacRoberts & MacRoberts 1976
White-winged Chough	*Corcorax melanorhamphus*	Rowley 1965, 1977
Group territorial. Plural breeding. Joint nesting. Mating system unknown.		
White-eared Barbet	*Buccanodon leucotis*	Moreau in Skutch 1961
Formosan Yuhina	*Yuhina brunneiceps*	Yamashina 1938
Group territorial. Plural breeding. Separate nesting.		
Red-fronted Woodpecker	*Melanerpes cruentatus*	Short 1970
Black-collared Barbet	*Lybius torquatus*	Skead & Vernon in Grimes 1976a
Chestnut-bellied Starling	*Spreo pulcher*	Wilkinson 1978, 1982; Wilkinson & Brown 1984
Superb Starling	*S. superbus*	Huels in Grimes 1976
Australian Magpie	*Gymnorhina tibicen*	Robinson 1956; Carrick 1963, 1972

(continued)

Table 2.2 *(continued)*

Group territorial. Plural breeding. Separate nesting.		
Mexican Jay (Arizona)	*Aphelocoma ultramarina*	Gross 1949; Brown 1963a, 1970
Bushy-crested Jay	*Cyanocorax melanocyanea*	Hardy 1976
San Blas Jay	*C. sanblasiana*	Hardy et al. 1981
Yucatan Jay	*C. yucatanica*	Raitt & Hardy 1976; Wagner 1955
White-throated Magpie-jay	*Calocitta f. formosa*	Skutch 1935, 1960; Wagner 1955; Innes, pers. comm.
Colonial in nesting. No all-purpose territory.		
Mississippi Kite	*Ictinia mississippiensis*	Parker & Ports 1982
Arctic Tern	*Sterna paradisea*	Cullen 1957; Skutch 1961
Pied Kingfisher	*Ceryle rudis*	Douthwaite 1973; Reyer 1980, 1984
Red-throated Bee-eater	*Merops bullocki*	Fry 1972; Dyer & Fry 1980
Carmine Bee-eater	*M. nubicus*	Dyer & Fry 1980
White-throated Bee-eater	*M. albicollis*	Dyer & Fry 1980
Rainbow Bee-eater	*M. ornatus*	Filewood et al. 1978
European Bee-eater	*M. apiaster*	Dyer & Fry 1980
White-fronted Bee-eater	*M. bullockoides*	Emlen 1981; Hegner et al. 1982; Emlen & Demong 1980
Barn Swallow	*Hirundo rustica*	Myers & Waller 1977; Ball 1982
New Holland Honeyeater	*Phylidonyris novaehollandiae*	Recher 1977
Little Wattlebird	*Anthochaera chrysoptera*	Sharland in Dow 1980a
Brown-and-yellow Marshbird	*Pseudoleistes virescens*	Orians et al. 1977b
Grey-capped Social Weaver	*Pseudonigrita arnaudi*	Collias & Collias 1980
Sociable Weaver	*Philetairus socius*	Maclean 1973; Collias & Collias 1978
House Sparrow	*Passer domesticus*	Sappington 1977
Pied Starling	*Spreo bicolor*	Craig, A. 1983
White-breasted Wood-swallow	*Artamus leucorhynchus*	Clunie 1973, 1976; Immelman 1966
Black-faced Wood-swallow	*A. cinereus*	Immelman 1966; Rowley 1976
Pinyon Jay	*Gymnorhinus cyanocephalus*	Balda & Bateman 1971
Neither group-territorial nor colonial in the usual sense.		
Ostrich	*Struthio camelus*	Sauer & Sauer 1959, 1966; Bertram 1980b
Striated Pardalote	*Pardalotus striatus*	Slater in Dow 1980a
Noisy Miner	*Manorina melanocephala*	see Chapter 9
Yellow-throated Miner	*M. flavigula*	Dow 1980a
Bay-winged Cowbird	*Molothrus badius*	Fraga 1972; Orians et al. 1977b

(continued)

Table 2.2 *(continued)*

Unclassified and miscellaneous.		
Horned Grebe	*Podiceps auritus*	Fjeldsa 1973
Australasian Grebe	*Tachybaptus novaehollandiae*	Lane 1978
Bateleur	*Terathopius ecaudatus*	Grimes 1976a
Peregrine Falcon	*Falco peregrinus*	Monnert 1983
Black Crake	*Limnocorax flavirostris*	Brooke 1975
Moorhen	*Gallinula chloropus*	Grimes 1976a
Red-knobbed Coot	*Fulica cristata*	Grimes 1976a
Southern Lapwing	*Vanellus chilensis*	Walters & Walters 1980
Chimney Swift	*Chaetura pelagica*	Dexter 1952, 1981
Striped Kingfisher	*Halcyon chelicuti*	Reyer 1980b
Forest Kingfisher	*H. macleayii*	Slater in Dow 1980a
Buff-breasted Paradise-kingfisher	*Tanysiptera sylvia*	Slater in Dow 1980a
Puerto Rican Tody	*Todus mexicanus*	Kepler 1977
African Hoopoe	*Upupa epops*	Zahavi 1976; Skead in Grimes 1976a
Casqued Hornbill	*Bycanistes subcylindricus*	Kilham 1956
White-crested Hornbill	*Berenicornis comatus*	Leighton 1982; Kemp 1978
Bushy-crested Hornbill	*Anorrhinus galeritus*	Leighton 1982
Ashy-tailed Swift	*Chaetura andrei*	Sick 1959
Short-tailed Swift	*C. brachyura*	Woolfenden 1976
Vaux's Swift	*C. vauxi*	Baldwin & Hunter 1963
Collared Aracari	*Pteroglossus torquatus*	Skutch 1958
Purple-throated Fruit-crow	*Querula purpurata*	Snow 1971
Rufous-margined Flycatcher	*Myiozetetes cayanensis*	Ricklefs 1980
Cape Wagtail	*Motacilla capensis*	Skead in Grimes 1976a
Spotted Greenbul	*Ixonotus guttatus*	Grimes 1976a
White-tailed Greenbul	*Thescelocichla leucopleura*	Grimes 1976a
White-crowned Shrike	*Eurocephalus anguitimens*	Moreau in Grimes 1976a
Magpie Shrike	*Corvinella melanoleuca*	Winterbottom in Grimes 1976a
Chabert Vanga	*Leptopterus chabert*	Appert in Grimes 1976a
Banded Wren	*Thryothorus pleurostictus*	Wagner 1955
European Robin	*Erithacus rubecula*	Harper 1985
Yellow-eyed Babbler	*Chrysomma sinensis*	Gaston 1978d
Black-lored Babbler	*Turdoides melanops*	Reynolds 1965
Blackcap Babbler	*T. reinwardii*	Grimes 1976a
Brown Babbler	*T. plebejus*	Grimes 1976a
Short-tailed Bush Warbler	*Cettia squameiceps*	Ohara & Yamagishi 1984, 1985
Seychelles Brush Warbler	*Acrocephalus sechellensis*	Diamond 1980
Tit Hylia	*Pholidornis rushiae*	Grimes 1976a
Green-backed Eremomela	*Eremomela pusilla*	Grimes 1976a
Dusky-faced Warbler	*E. scotops*	Vernon & Vernon 1978
Variegated Fairy-wren	*Malurus lamberti*	Dow 1980a
Red-winged Fairy-wren	*M. elegans*	Chapman & Rowley 1978

(continued)

Table 2.2 *(continued)*

Unclassified and miscellaneous.		
Yellow-bellied Gerygone	*Gerygone chrysogaster*	Bell 1983
Weebill	*Smicrornis brevirostris*	Dow 1980a
Pale Flycatcher	*Bradornis pallidus*	Beesley in Grimes 1976a
Abyssinian Slaty Flycatcher	*Melaenornis chocolatinus*	Beesley in Grimes 1976a
Forest Flycatcher	*Fraseria ocreata*	Erard in Grimes 1976a
Crested Shrike-tit	*Falcunculus frontatus*	Howe & Noske 1980
Tufted Titmouse	*Parus bicolor*	Brackbill 1970; Davis 1978; Tarbell 1983
Rufous Treecreeper	*Climacteris rufa*	Noske 1980b
White-plumed Honeyeater	*Lichenostomus penicillatus*	Stewart in Dow 1980a
Yellow-tufted Honeyeater	*L. melanops*	Morris in Dow 1980a
Brown-headed Honeyeater	*Melithreptus brevirostris*	Noske 1983
White-naped Honeyeater	*M. lunatus*	Dow 1980
White-throated Honeyeater	*M. albogularis*	Noske in Dow 1980a
Little Friarbird	*Philemon citreogularis*	Dow 1980a
Striped Honeyeater	*Plectorhyncha lanceolata*	Moffat et al. 1983
Gray-headed Bunting	*Emberiza fucata*	Nakamura 1973
Stripe-headed Sparrow	*Aimophila ruficauda*	Wagner 1955
Northern Cardinal	*Cardinalis cardinalis*	Brackbill 1944
Mexican Tanager	*Tangara mexicana*	Snow & Collins 1962
Speckled Tanager	*T. chrysophrys*	Skutch 1961a
Golden-masked Tanager	*T. larvata*	Skutch 1954, 1961a
Austral Blackbird	*Curaeus curaeus*	Orians et al. 1977b
Bobolink	*Dolichonyx oryzivorus*	Beason & Trout 1984
Cape Glossy Starling	*Lamprotornis nitens*	Craig, A. 1983
Fischer's Starling	*Spreo fischeri*	Miskell 1977
King Glossy Starling	*Cosmopsarus regius*	Van Someren in Craig 1983
Yellow-billed Oxpecker	*Buphagus africanus*	Dowsett and Reynolds in Grimes 1976a
Red-billed Oxpecker	*B. erythrorhynchus*	Henderson, Reynolds, and Stutterheim in Grimes 1976a
Black Drongo	*Dicrurus adsimilis*	Thangamani et al. 1981
Little Wood-swallow	*Artamus minor*	Estberg in Rowley 1976
Zavattariornis	*Zavattariornis stresemanni*	Benson 1946
Piapiac	*Ptilostomus afer*	Grimes 1976a
American Crow	*Corvus brachyrhynchos*	Kilham 1984
Northwestern Crow	*C. caurinus*	Verbeek and Butler 1981; Butler et al. 1984

The compilation of this list was greatly aided by the symposium on communal breeding organized in 1974 by Ian Rowley, in which each author listed the communal species in his geographical region. In addition to the papers in this symposium (Rowley, 1976; Grimes, 1976b; Woolfenden, 1976; Zahavi, 1976), two authors have provided updated or expanded lists (Dow, 1980a; Grimes, 1976a). I have added relatively few myself, some from the recent literature and some that were missed by the above reviewers.

Several authors of such lists have commented that the number of known communal species is still increasing, and some have named candidate species that occur in closeknit groups during the breeding season but which lack only an observation of helping to qualify. I too could suggest a few jays and babblers that probably have helpers; however, I would like to point out instead that not all the species listed should be considered as regular communal breeders. A list of regular communal breeders would be considerably shorter. Eliminated would be those species in which helping is rare or sporadic. A good example is the Northwestern Crow, which has supernumeraries around the nest that only occasionally help in reproduction and then only trivially (Verbeek and Butler, 1981).

It would have been desirable to subdivide the species in Table 2.2 further on the basis of regularity and number of nonbreeding and breeding helpers or perhaps by group size. For now, however, this will have to wait. There is no uniformity in reporting such data, and in most cases they are simply nonexistent. It would thus be very difficult at present to estimate what fraction of the species in Table 2.2 actually qualify as regular communal breeders.

Multiple Routes to Avian Communal Breeding

Using the information supplied in Tables 2.1 and 2.2, the comparative method can be employed to reduce the information on a large number of species to a small number of basic questions, theories, and hypotheses. The overall problem is to construct explanations for the evolution of helping behavior and associated characteristics of avian communal breeding systems for all species. Great diversity exists among species of communal birds in group size, mating system, age and sex composition of the helper corps, and behavior of unit members. Using the descriptive classification in Table 2.1, I have sorted the species into categories in Table 2.2. From this classification the scheme in Figure 2.4 has been constructed. Figure 2.4 is intended to represent some of the possible origins of avian communal breeding systems. The lower part represents the subsocial group-territory origin that we first recognized in Figure 2.3. In these cases helpers tend to include both sexes and to be immature in some sense, although immaturity is not a prerequisite. The young often have distinctive markings, such as color of bill or iris. Colonial species are not illustrated, but in many cases they follow the subsocial route, since offspring commonly help parents or other kin. Thus the breeding territory is commonly associated with the subsocial route but is not always necessary.

The upper group represents cases in which the helpers are breeders themselves or potential breeders in the same season. They may or may not be with their parents. In all cases their chances to obtain a mate and full possession of a breeding territory have been reduced and they have responded by some compromise. For example, a shortage of females leads some males to share a territory and sometimes paternity too. The sharing of a nest and incubation by females presumably is selected for by high costs associated with nest-building and incubation, and probably by other factors. These hypotheses are only mentioned here to indicate the scope of the problem. They will be developed more fully in later chapters. Figure 2.4 is intended to convey broad outlines. It does not cover every case, nor does it mention every important factor.

The three most important results of our comparative survey may be stated as follows.

(1) Helping behavior exists in a wide variety of social contexts.
(2) Helping behavior probably has multiple origins involving different selection pressures and requiring different explanations.
(3) Although these origins can probably be reduced to a small number, each of them is likely to involve several factors, and none of them is likely to be very simple.

Nontraditional or Nominal Helping

Certain phenomena are traditionally omitted from most discussions of avian helping but nevertheless deserve mention.

Alloparenting via Interspecific Brood Parasitism. The laying of eggs in the nest of another species resulting in interspecific alloparenting is the only means of reproduction by some birds, such as some cuckoos and cowbirds. Traditionally the evolution of brood parasitism and communal breeding are separate problems; however, they are related. Several groups of brood parasites have close taxonomic allies that are communal breeders. Parasitic cowbirds have the communal Bay-winged Cowbird (Fraga, 1972). Parasitic cuckoos have the communal anis (Davis, 1942). Parasitic honey-guides are related to barbets, which include communal species (Short and Horne, 1980). That interspecific brood parasitism may have originated from a communal social system is a plausible theory (Brown, 1975a, but see Payne, 1977).

Alloparenting via Intraspecific Brood Parasitism. Intraspecific brood parasitism is much harder to detect and little is known about it (Yom-Tov,

1980; Andersson, 1984). It has been reported for a variety of avian social systems, including the colonial Rook (*Corvus frugilegus*, Yom-Tov et al., 1974), Cliff Swallow (*Hirundo pyrrhonota*, C. Brown, 1984), the noncommunal Bluebird (*Sialia sialis*, Gowaty and Karlin, 1984), Shelduck (*T. tadorna*, Patterson, 1982), Starling (Yom-Tov et al., 1974), the communal Groove-billed Ani (Vehrencamp, 1978) and White-fronted Bee-eater (Emlen and Wrege, 1986). It is only a short step from intraspecific brood parasitism to joint nesting; indeed both are found in one species, the Groove-billed Ani.

Alloparenting via Mixing of Broods. Females of some waterfowl drive off mothers of conspecific broods and amalgamate the broods. Alloparenting results. During aggressive encounters between mothers in the Shelduck (Patterson, 1982) and Common Eider (Munro and Bedard, 1977) ducklings may become attached to the wrong parent. Some mothers may lose all their young and others may accumulate large creches by such brood capture. Patterson has suggested that such mixing is not selected *per se* but may be a byproduct of other factors (Patterson, 1982). His extensive data show no demonstrable benefit to chicks or parents. Some mouth-brooding cichlid fishes capture and brood the young of *other species* (McKaye, 1981), and this is thought to be adaptive.

Alloparenting by Adoption of Orphans. Adoptive suckling of orphans is well known in a variety of mammals (Riedman, 1982), but its counterpart in birds has received little attention. Although gulls appear to develop an ability to recognize their own young at an age that corresponds to the onset of chick mobility or potential mixing, cases of adoption have, nevertheless, been reported in the Herring Gull (*Larus argentatus*, Graves and Whiten, 1980). European Robins also rarely adopt strange chicks (Harper, 1985).

Interspecific Alloparenting by Mistake. Many cases of birds feeding nestlings of another species have been recorded (Skutch, 1961a; Shy, 1982). The most extreme example of interspecific feeding is that of a Cardinal that fed a goldfish (photo in Welty, 1975). Shy summarized 140 cases of interspecific feeding in birds. Adopting species belonged to 65 species in 22 families. In 95 cases nestlings were fed; in 30 cases, fledglings; in 11 cases, both; and in 4 cases no age was given. Shy has classified the cases according to probable cause, as shown in Table 2.3. Mixed clutches were frequent among hole-nesting species, such as Redstarts (*Phoenicurus phoenicurus*) laying in nests of Great (*Parus major*) and Coal (*Parus ater*) tits, and seemed to be caused by a shortage of holes rather than being an adaptive strategy of brood parasitism *per se*. Most categories are self-evident from the table and involve inappropriate responses to somewhat unusual conditions.

Table 2.3

Incidence of proximate reasons for interspecific feeding. (From Shy, 1982.)

Reason	*No. of Cases*	*Percent*[a]
Mixed clutch	32	21.2
Original nest destroyed	7	4.6
Close nest of another species	36	23.8
Calling by young	15	9.9
Orphaned birds	6	4.0
Mateless birds	7	4.6
Male, mate incubating	8	5.3
Miscellaneous	40	26.5

[a] Percentages reflect reasons cited; in some cases more than one reason was given.

In a similar intraspecific situation, a parent that encounters a neighbor chick, especially if it has lost its own young, may consequently allofeed young of a neighboring pair of its own species (e.g., Harper, 1985). As discussed in Brown (1983a:19), such cases are at present best viewed as accidents rather than altruistic adaptations. In some plural-breeding species, such as Mexican Jays, parents that lose their own nestlings may then begin to feed young from other nests in their *own social unit.* In such cases the capacity to feed another's young may be adaptive.

Compound Nests: Lodge and Mound Builders. The conventional definition of helping as alloparental care traditionally applies to the *feeding* of young, but does it apply to cases in which breeding pairs share in the *building* of a compound nest or lodge, such as that of the Sociable Weaver? The lodges of this species consist of 5 to 50 separate chambers constructed of grass, all covered by a common roof and superstructure of twigs and coarse stems (Maclean, 1973). The massive structure requires a sturdy tree for support, and the nest has often been compared to an apartment house. Lodges that are similar in principle but not so large as that of the Sociable Weaver have also been described for the Palm Chat (*Dulus dominicus,* Wetmore and Swales, 1931), Red-billed Buffalo-weaver (*Bubalornis albirostris,* Crook, 1958; Collias and Collias, 1964; Kemp and Kemp, 1974), and the Monk Parakeet (*Myiopsitta monachus,* Crook, 1965). In other species, which commonly nest in separate structures, nesting pairs sometimes overlap their nests or share a wall, thus resembling a small lodge but without a common superstructure. These include a Ploceine weaver (*Malimbus rubricollis,* Crook, 1960, 1964, 1965), the White-headed Buffalo-weaver (*Dinemellia dinemelli,* Collias and Collias, 1964), Shining Starling (*Aplonis metallica,* Rowley, 1976), the Grey-capped Social Weaver

(Collias and Collias, 1977), the House Sparrow (McGillivray, 1980), Spanish Sparrow (*Passer hispaniolensis*, Gavrilov, 1963) and some swallows. Especially in the latter category of species, each nest apartment is constructed by a pair which is not dependent on others for essential parts of the nest structure. In the Red-billed Buffalo-weaver a male makes several adjacent nests for a harem of females. Since nest building in all these cases is largely self-serving, I have not included lodge builders in the list of species with helpers unless allofeeding of young has also been reported. Only for the Sociable Weaver and House Sparrow have allofeeders been found, but helpers are likely to be discovered in some of the other species.

Building is also shared by more than two birds in many species that construct and sleep in structures called dormitories (Skutch, 1961b). Many of these are known to have allofeeders, such as the *Pomatostomus* babblers and the Rufous-fronted Thornbird (Table 2.2). In others such as the Bronze Mannikin (*Lonchura cucullata*), and often the above thornbird, the young remain and sleep with the family and participate in building and maintenance of the dormitory during the nonbreeding period, but are driven away by the parents before they can become allofeeders (Woodall, 1975). These species provide an intermediate step between families that immediately shed their independent offspring and those that retain them as helpers.

A different kind of sharing in nest building and maintenance is found in the Megapodidae. Generally, each incubation mound is owned by a pair, as in the Orange-footed Scrub-fowl (*Megapodius reinwardt* or *M. freycinet*), but occasionally two pairs may lay in the same mound (Crome and Brown, 1979). Thus it might be said that in caring for the mound each female benefits the other; but in my estimation the degree of cooperation described in this instance is slight, aberrant, and not enough to deserve to be termed helping. The females in the one recorded example of mound sharing held separate territories. One female had no mound in her territory and sneaked into the neighboring territory to lay and tend her eggs. Mixed broods from different mothers have not been reported.

Is Helping Adaptive?

Yes and no. Helping is probably *not adaptive* for the helper in certain cases. Interspecific helping, for example, is unlikely to be adaptive. Of course, one could postulate some future advantage to the interspecific helper such as interspecific reciprocation by the recipient young or improvement through learning by the helper; but such explanations are far-fetched, especially since interspecific helping is often done by mature breeders who have lost a brood or mate through predation rather than by immatures who might benefit more from learning. Furthermore, brood parasitism, which generates interspecific helping by deception, has usually been interpreted as being detrimental to the genetic interest of the alloparents.

It is a small step from accidental or deceptive interspecific helping to accidental or incidental intraspecific helping. It seems likely that many cases of intraspecific helping, particularly those of a sporadic or accidental nature, are not adaptive for the alloparent, while cases of regular helping may indeed be adaptive. This position is elaborated in the following section.

Grades and Contexts of Helping

It is possible to consider an array of species ranging from those with only rare and accidental helping to species that almost invariably have numerous helpers. It is the latter end of the spectrum with which this book is primarily concerned. For these latter species ample evidence exists of genetic benefit to the helpers, as will be shown in later chapters. Thus, in answer to the question raised above, helping is unlikely to be adaptive in some species, particularly those in which it is rare or irregular; but it is likely to be adaptive in other species, especially those in which beneficial effects on inclusive fitness have been measured. To document this position some of the grades and contexts of helping are briefly considered.

Helping by Failed Breeders in Noncommunal Species. In addition to the interspecific cases noted above, several cases of intraspecific helping by adults seem to have resulted from loss of a brood or in some cases from lack of a mate in the first place. In Barn Swallows (Myers and Waller, 1977; Crook and Shields, 1986) and in many Australian species (Harrison, 1969) helpers are typically adults and are unlikely to be young of the recipient parents. A few of these "helper" individuals may do some good for the recipients, but it seems more likely that their main interest is to enhance their chances of mating in the future, either by hindering a present nesting effort in hope of parenting a replacement brood, or by establishing a social bond with the female (males only).

In other cases, such as two species of Galapagos finch (Price et al., 1983), irregular helping may be simply neutral, neither beneficial nor harmful. Helping in these species was normally rare, but was frequent the year after a severe drought killed more females than males, leaving many males with territories but without females. Some of these males responded to the begging of nestlings in adjacent territories by feeding them. Such cases probably have negligible costs or benefits compared to the overriding benefit of flexibility *per se*.

Helping during Mate Replacement after Hatching. This may be considered a special case of the preceding one. When a female with nestlings loses her mate, she may need both aid in feeding her young and a male for a possible next brood. Helping by the replacement mate has been described in Savanna Sparrows (*Passerculus sandwichensis*, Weatherhead and Robertson, 1980), a junco (*Junco hyemalis*, Allan, 1979), and a sapsucker (*Sphyrapicus varius*, Kilham, 1977), among others. Such males qualify as helpers by the definition; however, I believe that Weatherhead and Robertson are correct in interpreting such behavior as a byproduct of a flexible mating strategy.

Since these mate replacements are probably no more than randomly related to the young they feed, one asks why they feed the young. The normal behavior of a male in these species, which are basically monogamous, is to feed his mate's young. The situation in which nestlings are not the father's is rare for males. To suppress feeding in such cases would require a special mechanism that seems largely unnecessary and possibly counterproductive; it might impair a male's chance of obtaining the female or impair the feeding of his own young in the next brood.

It follows from this interpretation that this helper situation is an unusual byproduct and not a suitable test case for studies of evolved aid-giving behavior.

In some populations unmated males attempt to become involved in the nesting of intact pairs. In a New York population of the Barn Swallow, unmated adult males frequented active nests but showed little alloparental behavior and frequent detrimental behavior, such as attempting copulation with the mother (Crook and Shields, 1986). The fact that these males were unrelated to the attended young is in agreement with the lack of benefit to the parents or young.

Helping Incidental. Helping has been described as an infrequent but perhaps regular occurrence by juveniles in a few populations of House Sparrow (*Passer domesticus*, Sappington, 1977), by a few yearlings in a population of Northwestern Crows (*Corvus caurinus*, Verbeek and Butler, 1981), and in a flock of Pinyon Jays (Balda and Bateman, 1971). The nonbreeders in these species usually do not act as helpers; and when they do, their effect on the breeders is typically scarcely measurable, perhaps even slightly negative in terms of the immediate nesting attempt. By the same token there has been no demonstration of a measurable benefit for the helper due solely to caring for young (as opposed to other effects of associating with the recipient breeders). There may be other benefits to the nonbreeder causing it to continue associating with its parents, such as learning of skills, use of resources in the parental territory, or exercise of behaviors that normally would appear at a later stage of ontogeny but which need not be suppressed at an earlier developmental stage. The occurrence of occasional helping in a few young individuals of these species appears to be incidental to other more important considerations in their development. Helping in these cases is probably not selected for its own sake. The importance of these cases is that they demonstrate a *substrate for helping* upon which selection could act should helping *per se* increase in value. They are not comparable with later stages discussed below in regard to selection pressures.

Helping Behavior Regular in a Moderate Fraction of the Species. In many species, helpers can be found regularly in most populations, although many pairs in the same population *lack helpers* and very few pairs have many. This category includes Pied Kingfishers (Douthwaite, 1973; Reyer, 1980a), Superb Blue Wrens (Rowley, 1965a), the Dunnock (Birkhead, 1981), some populations of Acorn Woodpecker (Stacey, 1979a), a skua (Young, 1978), some hawks (Mader, 1975a,b, 1979; de Vries, 1973; Faaborg et al., 1980), some jays (Woolfenden, 1975; Alvarez, 1975; Raitt and Hardy, 1979; Hardy, 1976), Kookaburras (Parry, 1973), Hoatzins (Strahl, 1985), and many others.

In many of these cases, the helpers are predominantly males that have been unable to mate primarily because of a shortage of females. This, on top of the usual variation, increases the variance in reproductive potential for males. Among the options for such males are (1) sharing a female in a polyandrous or polygynandrous unit, (2) bonding with a pair in a subordinate relationship as a way of being second in line for the female should she become available either through the disappearance of her mate or the behavior of the female, or (3) becoming an unbonded floater.

In this stage the probable value of bonding and helping would seem to consist mainly in raising the probability of reaching or improving breeding status. However, in some of these species the helpers seem to prefer to help their mother rather than a potential mate. Mother-son matings are rare and may be actively avoided (Chapter 12).

Nonbreeding Helpers Regular and Numerous. In many avian species social units without helpers are nearly unknown, and the fraction of helpers in an average unit may commonly exceed 50%, as in the White-winged Chough (Rowley, 1977), Mexican Jay (Brown and Brown, 1981a), Yellow-billed Shrike (Grimes, 1980), White-browed Sparrow-weaver (Lewis, 1982b), and Southern Ground Hornbill (Kemp and Kemp, 1980). These species often have some long-lived individuals that do not breed, but there are no clear castes. Yearlings, however, show signs of adaptation to the helper status, for in some species, at least, their gonads are underdeveloped (Band-backed and Gray-barred Wrens, Selander, 1964; Brown Jay, Selander, 1959; Grey-crowned Babbler, J. L. Brown, pers. observ.; Counsilman, 1979).

Breeding Helpers Regular. A product of many field studies carried out in the 1970s is the realization that helping by breeders feeding each other's young is widespread. As shown in Chapters 9 and 10, there is a clear benefit to the helper in feeding the young in its nest because some of them are likely to be its own offspring. Because the feeder cannot distinguish its own offspring from others, it functions as both parent and alloparent.

Conclusions

This chapter introduces helping behavior with some examples from early studies, and proceeds to provide a descriptive classification of communal breeding systems. Helping occurs in a great variety of situations and social systems. Without recognizing these and treating each separately it would be impossible to make progress on the general problem posed by helping behavior. Clearly, it is difficult to make useful statements about helping in general. On the other hand, when the different types of helper systems are recognized, some of the controversies that have arisen out of semantic confusion are dissipated and real progress on substantive issues is made possible.

Helping is considered to be nonadaptive in several situations, but when it is typical of a species and when benefits are measurable in natural populations (as shown in later chapters), helping is probably adaptive for helpers as well as recipients. Two theoretical "routes to communal breeding" are identified (Fig. 2.4), which will be examined in detail in subsequent chapters.

3 Climate, Geography, and Taxonomy

Is the frequency of avian species that have communal breeding systems correlated with climate, geography, or taxonomy? Yes, there are rough patterns of correlation with each of these factors, but their meaning is not entirely clear. We hope to gain some insight into the origins of communal breeding by studying these patterns. And, turning the problem around, we hope to be able to answer why these patterns exist through detailed studies of particular species. This chapter will review patterns of correlation with climate, geography, and taxonomy, and consider some of their possible causes. We shall return to the questions raised in this chapter later in the book to see what light has been shed on them by the more detailed field studies of recent years.

Taxonomy

The principal data available for inferences about taxonomy and climate are the geographic ranges of the various species exhibiting communal breeding. These are summarized in relation to taxonomic group in Table 3.1. The taxonomic representation of communal breeding is bewildering. No really clear patterns emerge, since communally breeding species are found in many orders and families. Communal breeding is clearly not a trait whose phylogeny can be usefully analyzed along phylogenetic lines, with the possible exception of certain genera (*Aphelocoma* jays) and subfamilies (Crotophaginae, anis). It is, of necessity, restricted to groups whose young require *more than minimal parental care*, and this factor may explain its rarity in taxa with precocial young (e.g., Charadriidae, Scolopacidae). It is absent from the hummingbirds and is rare or possibly absent in predominantly nectar-feeding birds (certain Meliphagidae being potential exceptions; see Chapter 17). It is rare in fruit specialists though it occurs in the frugivorous Purple-throated Fruit-crow, and some hornbills and toucans. Pelagic seabirds are notably deficient in helpers. In the end these curiosities in the taxonomic distribution tell us little.

Table 3.1

Taxonomic and geographic distribution of species with reported intraspecific helping. Taxonomy follows Peters as interpreted by Morony et al. (1975) purely as a convenience, not as an expression of the author's opinions on classification. All orders are listed but families and subfamilies lacking helpers are not listed. Numbers of species in taxa are from Bock and Farrand (1980). Many families do not occur on every continent. If a species occurs in more than one continent it is counted only where it was studied. Thus the entries for continents are underestimated. H/T = number of species with helpers/total species in taxon; NA = North America and Central America; SA = South America; A = North, Central, and South America; AS = Asia; EU = Europe; EA = Eurasia; AF = Africa; AU = Australia.

Order *Family* *Subfamily*	*H/T*	*Continent (number of species with helpers)*
Ratites	1/58	
Struthionidae		AF(1)
Sphenisciformes	0/18	
Gaviiformes	0/5	
Podicipediformes	2/20	
Podicipedidae		EU(1), AU(1)
Procellariiformes	0/104	
Pelecaniformes	0/62	
Ciconiiformes	0/114	
Phoenicopteriformes	0/6	
Anseriformes	1/150	
Anatidae		AU(1)
Falconiformes	5/288	
Accipitridae		AF(1), SA(1), NA(1), A(1)
Falconidae		EU(1)
Galliformes	1/269	
Opisthocomidae		SA(1)
Gruiformes	7/210	
Rallidae		AF(3), A(1), AU(3)
Charadriiformes	3/329	
Charadriidae		SA(1)
Stercorariidae		AU(1)
Laridae		EU(1)
Columbiformes	0/322	
Psittaciformes	0/340	
Cuculiformes	4/147	
Cuculidae		SA(2), A(2)
Strigiformes	0/146	
Caprimulgiformes	0/105	

(*continued*)

Table 3.1 (*continued*)

Order *Family* *Subfamily*	*H/T*	*Continent* *(number of species* *with helpers)*
Apodiformes	4/428	
Apodidae		SA(2), NA(2)
Coliiformes	1/6	
Coliidae		AF(1)
Trogoniformes	0/37	
Coraciiformes	19/200	
Alcedinidae		AF(2), AU(4)
Todidae		NA(1)
Meropidae		AF(4), EU(1), AU(1)
Upupidae		AF(1)
Phoeniculidae		AF(1)
Bucerotidae		AF(2), AS(2)
Piciformes	10/383	
Bucconidae		A(1)
Capitonidae		AF(5)
Ramphastidae		A(1)
Picidae		SA(1), NA(1), A(1)
Passeriformes	158/5274	
Furnariidae	1/218	SA(1)
Cotingidae	1/79	SA(1)
Tyrannidae	2/375	SA(1), A(1)
Hirundinidae	1/80	NA(1)
Motacillidae	1/54	AF(1)
Campephagidae	1/70	AU(1)
Pycnonotidae	2/123	AF(2)
Laniidae	7/74	AF(7)
Vangidae	1/13	AF(1)
Troglodytidae	7/60	SA(3), NA(3), A(1)
Mimidae	4/31	SA(4)
Prunellidae	2/12	EU(1), EA(1)
Muscicapidae	54/1427	
Turdinae		AF(3), EU(1)

(*continued*)

Geographic Distribution of Communal Breeding

The geographic distribution of communal breeding has stimulated much speculation and some systematic analyses. As noted by several early reviewers (Skutch, 1961a; Rowley, 1968; Harrison, 1969; Fry, 1972), communally breeding species are rare in the polar and north temperate land

Table 3.1 (*continued*)

Order / *Family* / *Subfamily*	*H/T*	*Continent (number of species with helpers)*
Orthonychinae		AU(2)
Timaliinae		AF(3), AS(9), AU(4)
Sylviinae		AF(4), AS(1)
Malurinae		AU(16)
Muscicapinae		AF(3), AU(4)
Platysteirinae		AF(1)
Monarchinae		AF(2)
Pachycephalinae		AU(1)
Aegithalidae	2/7	NA(1), EA(1)
Paridae	2/47	AF(1), NA(1)
Sittidae	3/22	NA(2), AU(1)
Climacteridae	4/6	AU(4)
Dicaeidae	1/58	AU(1)
Zosteropidae	1/83	AF(1)
Meliphagidae	13/172	AU(13)
Emberizidae	9/560	
Emberizinae		SA(2), NA(1), AS(1)
Cardinalinae		NA(1)
Thraupinae		SA(1), A(3)
Iteridae	5/95	SA(4), NA(1)
Ploceidae	4/144	AF(3), NA(1)
Sturnidae	9/111	AF(9)
Dicruridae	1/20	AS(1)
Grallinidae	2/4	AU(2)
Artamidae	4/10	AU(4)
Cracticidae	3/10	AU(3)
Corvidae	17/106	AF(3), SA(1), NA(12), AS(1)
Totals AF(65), SA(26), NA(29), A(12), AS(15), EU(6), EA(2), AU(67)		
Nonpasserines	58/3747 = .015	
Passerines	164/5274 = .031	
Combined	222/9021 = .025	

masses. They reach their highest abundance in Australia and the tropics of Africa, Asia, and the Americas.

The abundance of avian communal breeding systems in low latitudes and Australia contrasts dramatically with their rarity in north temperate regions. For example, on our study area of 13.5 km^2 in Meandarra, Australia (27°S), the most conspicuous species were those that typically breed in moderate to large communal groups. These included the White-winged Chough, Apostle Bird, Noisy and White-rumped Miners, Grey-crowned Babbler, and Australian Magpie. Among the less conspicuous residents that were also communal were the White-browed Babbler, Ground Cuckoo-shrike, Yellow Thornbill, Yellow-rumped Thornbill, two butcher birds, two fairy wrens, wood swallows, and others for a total of at least 18 species. In contrast, the region around Albany, New York, (43°N) has no species that compare in terms of communal breeding to those listed above. It does have several species in which a few helpers can be found if a diligent effort is made, such as the Chimney Swift and Barn Swallow, but the frequency of pairs with helpers in these species is probably lower than for any of the species listed above for Meandarra. Not only does Meandarra have many more communally breeding species; the degree of specialization for communal breeding is far higher in Meandarra birds than in those Albany birds that may have helpers. What then does this geographic difference mean?

Correlations with Climate

Several correlations with the observed pattern of geographic distribution have been suggested. Rowley (1965a) regarded the social system of the Superb Blue Wren in southeastern Australia as an adaptation to a "widely varying climate." In a review of communal breeding Rowley (1968) suggested that dispersal of the annual crop of young is not so urgent where the winter (or nonbreeding season) is not so rigorous as that in the Northern Hemisphere. Harrison (1969) considered the likeliest common factor underlying the high frequency of communal breeding among species of Australian birds to be climate, mentioning aridity and unpredictability of rainfall. Fry (1972) pointed out the prevalence of communal breeding in "hot climates," as compared with the cool temperate zones of the world.

None of these perceived correlations with avian communal breeding systems—climatic variability, aridity, unpredictability of rainfall, hot climates—is without numerous exceptions. Communal breeding occurs in tropical rainforest and cloud forest species too, but the birds are more easily studied in arid climates because vegetation does not obscure the view of the observer, and field work is easier. Dense tropical rain and cloud forests that are easily accessible are few in number compared to semiarid habitats, which retain much of their character and fauna after civilization. It is no accident that nearly all the more detailed recent studies have been done where rainfall is light and unpredictable. The vegetation in these regions is much more conducive to field studies of color-banded birds and to the persistence of the birds under civilization.

Avian Communal Breeding in Australia

The best known continental avifauna with regard to avian communal breeding systems, and the only one that has been studied systematically with respect to communal breeding, is that of Australia. Dow (1980a) has carried out a detailed survey of the occurrence of avian communal breeding systems in the various regions of Australia. These include an ecological range from wet and dry tropics in the north to wet and dry temperate zones in the south.

Dow found that the frequency of species with communal breeding systems is high not only in the hot, semiarid regions of erratic rainfall in central Australia, but also in coastal rainforests even in temperate regions. As shown in Figure 3.1, the highest number of species of communal breeders sympatric in an area was found in the east. This area is at 30°S latitude, and lies in the temperate zone 7° south of the edge of the tropical zone. It does not coincide with the hottest part of the continent, nor does

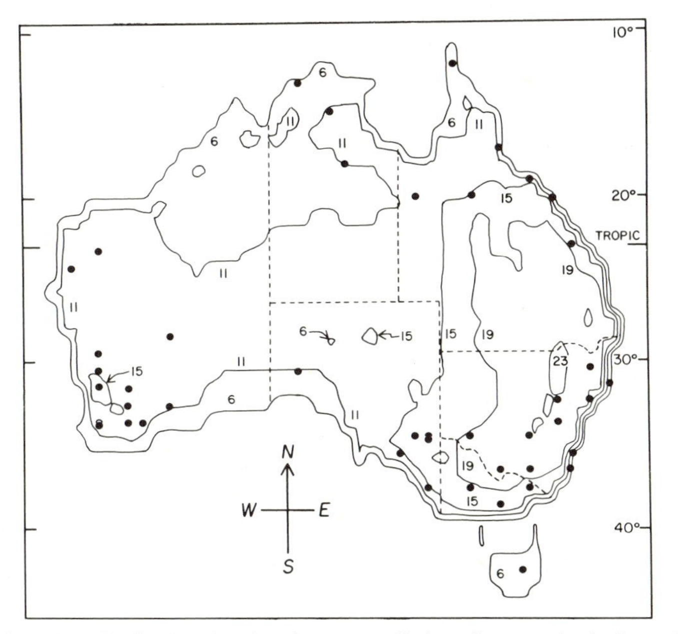

Figure 3.1 The distribution density of communally breeding species in Australia. The map was generated from a composite of the maps of distribution of thirty-one species. The number of sympatric communally breeding species varies from six to twenty-four. Isograms connect points with equal numbers of species (6, 11, 15, 19 and 23). Solid circles indicate sites mentioned by Dow. (From Dow, 1980a.)

it lie in the driest part or the part with the most variable or least variable climate. Statistical analysis of these variables also failed to provide much support for earlier theories. In Dow's words: "Thus, no overwhelmingly clearcut result emerges from this systematic analysis of possible effects of environmental factors. . . . I think it can be fairly stated that the present analysis does not strongly favour or exclude any one of these hypotheses."

The number of communally breeding species is part of the number of all species and may be expected to be influenced by the latter. Since each communal species is included in the total of all species, these variables are not independent. The correlation between these two variables was significant but not large for a part-whole relationship ($r = +0.55$).

Permanent Residency and the Surplus

As noted by Brown (1974), the species in which communal breeding systems occur are basically nonmigratory permanent residents. This is especially true for the group-territorial types. The principal exceptions to this rule are a few colonial breeders that do not defend feeding territories and do not construct lodges. Residence on the same territory all year depends on the availability of food all year, which in turn is related to diet (see Chapter 17). In the Northern Hemisphere the proportion of territorial species whose diet enables them to survive winters on their breeding territories diminishes rapidly going northward from the tropics. This factor alone may account for the correlations with climate described above, but it is only a permissive effect and does not explain why group-territoriality tends to evolve in permanent residents.

To understand why territorial groups form we need to know why some individuals delay breeding and stay with relatives, usually on the home territory. This is the subject of Chapters 5 and 6. For the moment, however, we simply assume that some nonbreeders do persist in their natal territory into the breeding season. If these persisters are numerous enough they can become involved in sharing various costs of territorial life (Chapter 8). The question now is under what *demographic* conditions do they accumulate.

What is required for accumulation of potential nonbreeding helpers is that the annual rate of production of potential breeders, b, exceed the annual rate at which "breeding spots" become available by death, d. Thus, the rate of accumulation is $a = b - d$.

The equilibrium fraction of surplus potential breeders of all ages, above and beyond the number of spots available, N_F, also depends on the annual survival of the surplus individuals, s. Thus, as shown in Brown (1969a) for this simple case,

$$\text{Nonbreeding surplus} = N_F = a/(1 - s).$$

The value of N_F is strongly dependent on s, as shown in Figure 3.2. For the same value of a, species with higher s will have higher N_F. This relationship suggests that the annual rates of survival of the nonbreeders and breeders are important determinants of the availability of potential helpers. This is especially likely around $s = 0.8$ and higher, a range that includes many communally breeding species (Table 3.2).

There is some support for the hypothesis that survival and longevity in regions where avian communal breeding systems are frequent, such as Australia, the tropics, and South Africa, are higher than in the north temperate zone for birds comparable in size. Small passerines can live to a remarkable age in tropical regions (e.g., Snow and Lill, 1974). There may be a trend to higher annual survival in tropical than temperate regions (Fry, 1980).

From such demographic considerations, Brown (1974) predicted that in congeneric pairs of communal versus noncommunal species the *communal species would have the higher survival rate.* Known annual survival rates for communally breeding species appear to be rather high in comparison to noncommunal species, but much variation occurs among species, with a few having rather low rates. This can be appreciated by comparing the data for communal species in Table 3.2 with the rough guidelines for passerines and nonpasserines in Figure 3.2.

A more controlled test of this prediction was performed by Zack and Ligon (1985a). They compared a communally breeding species with a congeneric, noncommunal species living on neighboring and partially overlapping areas, thus experiencing nearly identical climatic conditions. Confirming the prediction, they found that the communal species, the Grey-backed Fiscal Shrike, had a significantly higher annual rate of nondisappearance (0.64 and 0.67 for adults in two years) than the noncommunal one, the Common Fiscal Shrike (*Lanius collaris*; 0.39). Dispersing juveniles were not considered. Opportunities for controlled tests of this sort are rare and more tests are needed.

The conditions under which permanent residency and a pool of surplus potential breeders may lead to actual helping will be considered in later chapters. We note in passing, however, that the circumstances described above are consistent with two popular theories.

(1) Permanent residency facilitates aid-giving among relatives by making it possible or at least easier for relatives to stay together continuously through

Table 3.2

Adult annual survival rates (s) in some communally breeding species. Strictly speaking these are rates of nondisappearance from the study area. Representative ranges of annual survival rates for noncommunal passerines and nonpasserines are shown in Figure 3.2. M = male; F = female; Br = breeder; N = nonbreeding; G = unit size.

Species	*Location*	*Sex or Status*	*s*	*Reference*
Hoatzin	Venezuela		.85	Strahl 1985
Native Hen	Tasmania		.93	Ridpath 1972
Groove-billed Ani	Costa Rica		.69	Vehrencamp 1978
Pied Kingfisher	Kenya	M Br	.61	Reyer 1984
		F Br	.56	
		M N	.66	
White-fronted Bee-eater	Kenya		.80	Emlen in Ligon 1983
Green Woodhoopoe	Kenya		.65	Ligon 1983
Acorn Woodpecker	New Mexico		.57	Stacey in Ligon 1983
Acorn Woodpecker	California	M Br	.82	Mumme & Koenig unpubl.
		F Br	.71	
Yellow-billed Shrike	Ghana	M	.68	Grimes 1980
		F	.71	
Grey-backed Fiscal Shrike	Kenya		.66	Zack & Ligon 1985a
Stripe-backed Wren	Venezuela	M&F Br	.64	Wiley & Rabenold 1984
		M N	.78	
		F N	.47	
Bicolored Wren	Venezuela	G = 2	.72	Austad & Rabenold 1985, Rabenold 1985
		G = 3	.82	
		G = 4	.80	
Fasciated Wren	Peru		.70	Rabenold unpubl.
Galapagos Mockingbird	Galapagos Is.		.91	Kinnaird & Grant 1982
Dunnock	U.K.	M	.60	Snow & Snow 1982
		F	.61	
Grey-crowned Babbler	Australia	1 yr	.69	Brown & Brown 1981b
		2 yr	.76	
		3 yr	.67	
		4 + yr	.90	
Common Babbler	India	M Br	.88	Gaston 1978a
		M N	.64	
		F Br	.63	
		F N	.65	
Jungle Babbler	India	Br	.93	Gaston 1978b
		N	.69	
Superb Blue Wren	Australia	M	.62	Rowley 1981a
		F	.62	
Splendid Fairy Wren	Australia	M	.71	Rowley 1981a
		F	.43	
White-browed Sparrow-weaver	Zambia		.71	Lewis 1982b
Scrub Jay	Florida	M Br	.82	Woolfenden & Fitzpatrick 1984
		F Br	.82	
		M N	.80	
		F N	.65	
Mexican Jay	Arizona	M	.86	Brown unpubl.
		F	.81	
Beechey Jay	Sinaloa	1 yr	.48+	Raitt et al. 1984
		2 yr	.60+	
		3 yr	.70+	

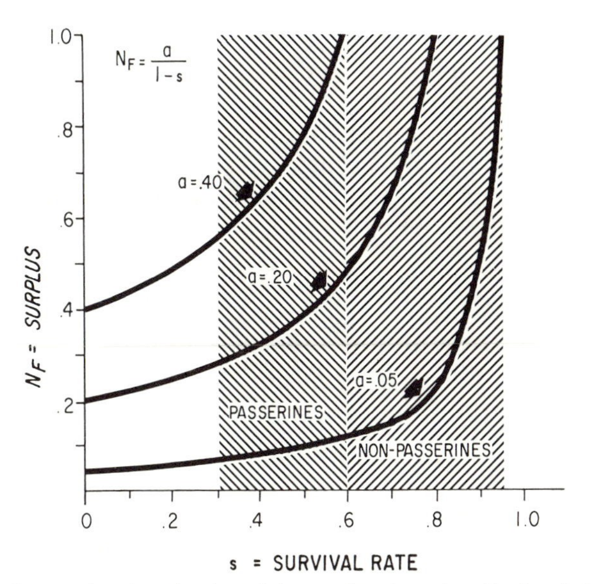

Figure 3.2 Curves showing the size of the predicted surplus, N_F, in relation to mean survival rate, *s*, of the surplus for various amounts of excess of population gains over losses, *a*. Both N_F and *a* are expressed as proportions of mean breeding density, which is taken as 1.0. The approximate ranges for adult survival rates of temperate zone passerines and nonpasserines are from Lack (1954:91–92), and are merely suggestive; game birds are excluded. Estimates of annual survival rates for communally breeding species are shown in Table 3.2. (From Brown, 1969a.)

the nonbreeding period, into the next breeding season, and beyond. In other words, the presence of kin together facilitates kin selection of both direct and indirect types (see next chapter).

(2) The presence of any individuals together for a long period of time in the same territory, whether they are kin or not, facilitates cooperative or mutualistic behavior.

Conclusions

Surveys of the taxonomic distribution of avian communal breeding systems have revealed numerous independent origins but little scope for interpretation along phylogenetic lines except within genera or subfamilies. The geographic distribution of avian communal breeding systems has suggested the existence of a pattern in which communal breeding systems are more numerous in low latitudes, hot climates, and possibly in regions of unpredictable rainfall.

The position taken here is that the primary pattern is that of a higher frequency of avian communal breeding systems in regions *where permanent residency on the same territory is common.* Demographically this may be mediated by higher survival rates in such regions, giving rise to a greater accumulation of a surplus population of nonbreeders, which are potential helpers. Such conditions are favorable for mutualistic interactions, particularly among kin.

4 Elements of Inclusive Fitness Theory for Field Studies

The Predominance of Individual Selection

In 1954 David Lack published his influential book, *The Natural Regulation of Animal Numbers*. Through this book and his many papers, Lack convinced many ecologists of the overpowering importance of selection at the level of the individual and of the weakness of the argument for reproductive restraint. The latter alternative refers to the idea that animals reproduce less than they are able, often with the implication that this restraint benefits their deme, population, or species. The test case in this issue was the control of clutch size. By supplementing conventional, natural-history approaches with strong-inference experiments, Lack convinced biologists that the reason birds do not lay larger clutches than the norm is that larger clutches result in lower fitness for the parents, for various reasons (Lack, 1947, 1948a,b, 1954a,b). "Lack was the first to realize the potential contribution of the study of life-history phenomena to understanding the nature of evolutionary adaptation" (Ricklefs, 1983). Lack summarized this pivotal work in the "group selection" debate along with a compendium on avian sociobiology (Lack, 1968).

Neither Lack (1968) nor Crook (1965) in their global surveys of avian social organization referred to Hamilton's (1964) theory of inclusive fitness or, of course, to Trivers's (1971) later theory of reciprocal altruism. Nevertheless, both concepts are included in Lack's concept of "natural selection," which was sufficient to explain much of avian social behavior and social organization. Lack's legacy was the establishment of the primacy of individual selection, and this forms the historical background into which newer developments must be integrated. Williams (1966a) later reviewed this debate, applying selection thinking to a wide range of phenomena.

The appearance of reproductive restraint is especially prominent in avian communal breeding systems. Some helpers are nonbreeders but have mature gonads and appear to be able to breed (Grey-crowned Babbler, Counsilman, 1979). Mutual helpers, or breeding helpers, appear to tolerate compromises such as sharing a mate, a nest, or a territory, with risks to paternity and tolerance of some abuse from others in the social unit. Even dominants must pay the cost of sharing resources.

Avian communal breeding systems, therefore, offer a temptation to suggest explanations based on interdemic selection, despite the historical background of rejection of such theories established by Lack. Skutch (1967) and V. Parry (1973) have invoked interdemic selection. Opposition to V. Parry's interpretation was supplied by Morton and G. D. Parry (1974). In general, communally breeding birds are a diverse lot; few of them appear to have the population structure and biology required by interdemic selection—many small, very isolated populations frequently subjected to extinction by overconsumption of their food supply. The latter condition might be difficult to satisfy because many communally breeding birds are omnivores and food generalists. Historically, the concept of interdemic selection stimulated only a few studies on avian communal breeding. In contrast, the concept of inclusive fitness has been the primary motivator of recent studies, even those done by authors who subsequently went on to espouse theories based more on mutualism than on kin selection.

The Difference between Classical and Inclusive Fitness

In order to appreciate how inclusive fitness theory enables new hypotheses to be generated, it is necessary to compare it with classical fitness theory. Therefore, we must first identify clearly and precisely the *difference* between these two theories.

Classical individual fitness, W, is conventionally conceived as being determined by the number of offspring that survive to maturity. If the average fitness of individuals in the population is 1.0, then we may express the fitness of an individual relative to the population as the basic unit, 1, plus all effects that either add to or subtract from it:

$$\text{Classical fitness} = W = \text{basic unit} + \text{fitness effects.}$$

When considering the evolution of aid-giving behavior using the classical model it is convenient to divide the fitness effects into two types, (1) effects due to parental care, grandparental care, etc. that influence the quality of their offspring, and (2) all other classical effects. The latter type includes effects on the number and quality of offspring produced, and adaptations for survival of the offspring exclusive of the effects of the behavior of the parents or helpers. These we call personal effects. They include effects on clutch size, egg size, and ability to survive until maturity, to obtain mates, and to reproduce. Therefore,

$$\begin{array}{ccccccc} W & = & 1 & + & a & + & P. \\ \text{Classical} & = & \text{Basic} & + & \text{Personal} & + & \text{Parental} \\ \text{fitness} & & \text{unit} & & \text{effects} & & \text{care effects} \\ & & & & & & \text{etc.} \end{array}$$

P includes the effects that an animal has on the fitness of its offspring and further descendants. All kinds of parental care are included under P.

Since we are considering only diploid species, each offspring should receive roughly half its genes from each of its parents, half in the sperm and half in the egg. Clearly, offspring are only one class of kin. Grasping the significance of this for the evolution of other kinds of aid-giving before anybody else, Hamilton (1964) generalized this category of effects to include not just effects of parents on their offspring, but also effects on nondescendent kin. I have included effects on grandoffspring and other descendent kin in P. Rewriting the above expression to include the more inclusive R instead of P, we obtain the equivalent of Hamilton's original expression for inclusive fitness, E:

$$\begin{array}{ccccccc} E & = & 1 & + & a & + & R. \\ \text{Inclusive} & = & \text{Basic} & + & \text{Personal} & + & \text{Effects on} \\ \text{fitness} & & \text{unit} & & \text{effects} & & \text{all kin} \end{array}$$

Selection acting on this broader class of relatives was termed kin selection by Maynard Smith (1964), who wrote as follows:

> By kin selection I mean the evolution of characteristics which favour the survival of close relatives of the affected individual, by processes which do not require any discontinuities in population breeding structure. In this sense, the evolution of placentae and of parental care (including "self-sacrificing" behaviour such as injury-feigning) are due to kin selection, the favoured relatives being the children of the affected individual.

With this background it is possible to identify the difference between inclusive fitness and classical fitness with respect to relatives affected. The difference is precisely $R - P$, but does this correspond to a familiar entity? In words it corresponds to "effects on all kin" minus "effects on descendent kin," or, simply to "effects on nondescendent kin." Some authors use "collateral kin" instead of nondescendent kin, but this is incorrect if the nondescendent kin is a lineal relative such as a parent being helped by its offspring or a grandparent helped by grandoffspring. Parents are lineal kin of their offspring, not collateral kin. These two distinctions are contrasted in Figure 4.1. Descendent kin receive their genes in common by

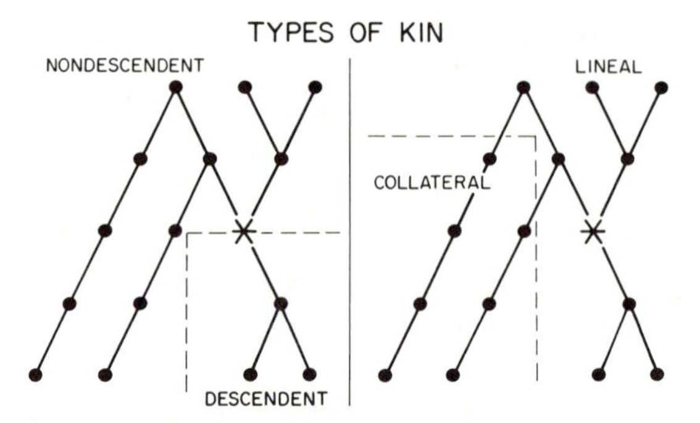

Figure 4.1 Two dichotomous classifications of kin. In the left pedigree the dashed line separates descendent and nondescendent kin of the starred individual. In the right pedigree the dashed lines separate lineal and collateral kin of the starred individual. The distinction between classical and modern components of inclusive fitness corresponds to the descendent-nondescendent dichotomy.

descent with X, the individual whose fitness is being described, *directly* through *gametes produced by* X. Nondescendent kin do not receive any of their genes in common with X from gametes of X. They receive them *indirectly* via one or more common ancestors of X, and X *need not produce any gametes at all.*

The distinction between direct and indirect effects on inclusive fitness is not an extension or alteration of Hamilton's theory in any way. Its importance is that it focuses attention on the *new element* of inclusive fitness and enables us to give a *concise and accurate verbal designation* of the new and old elements of selection theory. We may, for example, refer to direct and indirect selection or to direct and indirect fitness. The dichotomy is fundamental primarily in this sense, but the biological basis of the dichotomy also seems fundamental. The classical theory is built around the concept of reproduction and physical transmission of DNA in gametes, as outlined in Table 4.1. Indirect effects operate differently. In short, a *biological basis* exists for the dichotomy. It is not simply an arbitrary convention or historical accident.

To evaluate the importance of inclusive fitness theory as an improvement over classical fitness theory, the question we must ask is this: How important is indirect selection? We must know the importance of indirect selection for many different behaviors in a wide variety of species before its importance can be evaluated as a component of selection in nature. If indirect selection provides no additional explanatory power, then it must be rejected as a significant component of natural selection. On the other hand, if indirect selection has widespread applicability and utility, and

Table 4.1

Comparison of direct and indirect effects on inclusive fitness.

	Direct	*Indirect*
1.	Effects are on *descendent* kin.	Effects are on *nondescendent* kin.
2.	Effects are included in *classical theory*.	Effects are *not included* in classical theory.
3.	If X is the individual whose inclusive fitness is being considered with respect to its behavior toward Y, then the genetic similarity of X to Y depends on *production of gametes* by X,	Gametes need not be produced by X for the effect on inclusive fitness to occur.
4.	and on *physical transmission* of gametes from X to Y (without an intermediate in the parent-offspring case),	Physical transmission of gametes from X to Y does not occur.
5.	so that DNA from X becomes *integrated into the genome of* Y.	DNA integration is not involved.
6.	On a pedigree all links connecting X to Y run downward.	At least one link must run upward.
7.	For a given coefficient of relatedness the pedigree path is shorter and more direct; for example, for $r = 1/2$ only one link is involved (parent to offspring).	Pedigree path is longer; for example, for $r = 1/2$ four links are needed (sib to each parent or sib).

especially if there are some phenomena that have no other explanation, then inclusive fitness theory is correspondingly strengthened. Helping behavior is the prime candidate for this test. The widespread interest in helping in all animals results primarily from this fact.

Of course, in raising these questions we are not testing Hamilton's theory of inclusive fitness *per se*. The theory remains true as a mathematical statement regardless of whether or not it provides useful explanatory power.

The direct-indirect dichotomy was created to permit precise framing of this critical question (Brown, 1980; Brown and Brown, 1981b). Previously this question, "How important is kin selection?" had been asked; but as shown above, this wording does not distinguish between parental care and alloparental care. It is necessary to word the question so as to exclude that part of kin selection (or kinship component) that pertains to parental care.

Lacking a term for indirect fitness, some authors have substituted inclusive fitness. The reasons why this is incorrect or at least imprecise should now be evident. For further discussion of the origin of the direct-indirect dichotomy please see the glossary under indirect fitness.

It is convenient to point out now, following Hamilton (1964), that the direct and indirect components may be viewed in two ways, as shown in Figure 4.2: (1) in terms of their effects on *X* (upstream or incoming effects) from its parents (direct) and other kin (indirect), and (2) in terms of the effects of *X* (downstream or outgoing) on its descendent kin (direct) and nondescendent kin (indirect).

Strictly speaking, in population genetic models it is legitimate to count only upstream effects, since such models are based upon numbers of offspring. To count both upstream and downstream effects is referred to as double-accounting. This caveat has often been overlooked (Grafen, 1984).

The following diagram summarizes the meaning of the basic elements of inclusive fitness theory that we have described.

			Direct Components			Indirect Component
Inclusive Fitness	=	Basic Unit	+ Personal Component	+	Effects on Descendent Kin	+	Effects on Nondescendent Kin
					Kin Component	

The term "kin" as used here is short for a more complicated concept that is discussed in Chapter 12.

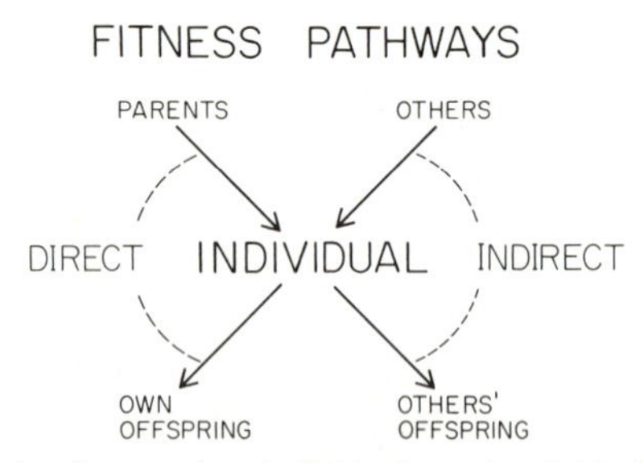

Figure 4.2 The inclusive fitness of an individual may be divided into two components, direct fitness, which is mediated through the genes in an individual's own offspring (and their descendants), and indirect fitness, which is mediated through copies of the same genes in the offspring of other individuals. Similarly, an individual's inclusive fitness is affected by influences from its parents (direct path) and from others (indirect path). These paths converge on the individual (upstream effects), and are blended and expressed as the individual's direct and indirect fitness (downstream effects). (From Brown and Brown, 1981b).

The Relative Fitness of Strategies

In the preceding treatment we have dealt with effects on fitness as departures from the norm, which is expressed as the basic unit. In asking questions about how natural selection works it is often useful to compare the fitnesses of two behavioral strategies with each other. For example, if helping and not helping are alternative strategies, then helping will be selected if inclusive fitness when helping is greater than inclusive fitness when not helping.

The Direct-Indirect Tradeoff in One Year. Since inclusive fitness has both direct and indirect components which are additive, it is possible that a loss in one can be made up by a gain in the other, and vice versa. Consider the effects of helping (and not breeding) versus nonhelping (and breeding) on inclusive fitness in a single year for a species such as the Grey-crowned Babbler, in which the effects of helping have been measured.

Our criterion for helping and not breeding to be selected can be stated in its general form as:

$$D_H + I_H > D_B + I_B,$$

where D_H = direct component when helping, I_H = indirect component when helping, D_B = direct component when breeding, I_B = indirect component when breeding. Since in this babbler $D_H = 0$ and $I_B = 0$, we simplify to

$$I_H > D_B.$$

I_H is the number of young attributable to a single helper, h, weighted by the relatedness (see Box 4.1) of the helper to those young, r_H. The quantity h can be estimated in field studies by subtracting the number raised without a helper from the number with a helper, or by estimating the slope of the regression of young raised per pair on number of nonbreeding helpers per breeding pair. See Chapter 11 for details.

D_B is the number of young the bird could expect if it left its parents and attempted to breed, m_B, weighted by its relatedness to those young, r_B. (Note: m_B includes the probability of acquiring a mate and territory as well as the reproductive success if these are obtained.)

Therefore, the criterion may be stated as:

$$r_H h > r_B m_B.$$

In contrast to the classical approach, in which the unit of measurement is offspring, the unit now is the gene-equivalent. The product, rm or rh, weights or corrects the number of offspring for their probability of carrying a gene of the decision-maker. It is more convenient to express such

BOX 4.1

r = COEFFICIENT OF RELATEDNESS

Definitions of r

(1) r_{AB} = the *probability* that a gene picked at random from the genome of individual A is identical by descent (IBD) with a gene in individual B. Note that because of the phrase IBD, r is *independent of the frequency* of the gene that is picked. This r is a measure of genealogical affinity and is intended to be independent of gene frequency, genetic polymorphism, and mutation.

(2) r_{AB} = the *expected fraction* of genes in the genome of B that is IBD with genes in the genome of A.

(3) r_{AB} = the *correlation* of genes in A and B.

Note that in noninbred diploid species $r_{AB} = r_{BA}$ for all three definitions with the exception of sex-linked genes. It is sometimes convenient to omit the designation of the donor, for example, using r_B instead of r_{AB}.

Calculation of r

One usually calculates r_{AB} from a pedigree. One must know the number of separate common ancestors, a, and the number of generational links, L, along each pathway connecting A and B through each common ancestor. For autosomal genomes in diploid species,

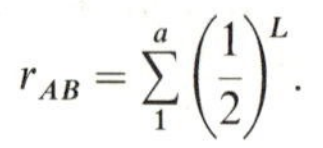

$$r_{AB} = \sum_{1}^{a} \left(\frac{1}{2}\right)^{L}.$$

For example, in the following pedigree A and B have two common ancestors, their father, F, and mother, M.

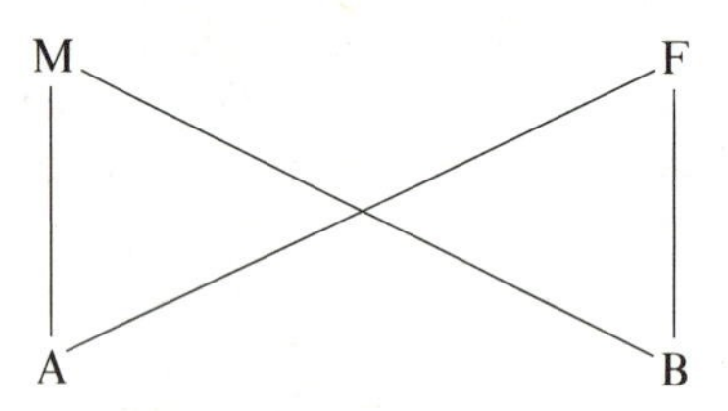

The pathway from A to B through the father has two links, so $L = 2$. Similarly for the mother. Therefore,

$$r_{AB} = (1/2)^2 + (1/2)^2 = 1/4 + 1/4 = 1/2.$$

Thus, $r_{AB} = 1/2$ is the relatedness of full sibs. If there were only one common ancestor A and B would be half-sibs and $r_{AB} = 1/4$.

It is possible in an inbred population for r to be greater than 1/2; but since we assume absence of inbreeding, we shall use $r = 1/2$ for full sibs and parent-offspring relatedness.

estimates as gene-equivalents, following the method introduced by Brown (1975a:735); however, this product can be converted to offspring-equivalents by multiplying by two, as done by West Eberhard (1975), who developed independently the same approach. These expressions provide a framework and method that is useful for comparing any pair of alternative strategies.

Offspring Rule Compared to Hamilton's Rule. We refer to the criterion used above as the offspring rule to distinguish it from Hamilton's rule. In general form the offspring rule may be stated as:

$$Gr_H < Lr_B,$$

where G is the "gain" through indirect fitness if helping is chosen, and L is the "loss" through direct fitness by not choosing to attempt breeding. If we now compare the offspring rule, rearranged as

$$G/L > r_B/r_H,$$

to Hamilton's rule,

$$G/L > 1/r_{HP},$$

the essential difference can be seen to lie in r_B, which is the relatedness of parent to offspring and is normally 1/2. The difference arises from the fact that in Hamilton's rule gains and losses are measured in "fitness" of individuals in the same generation, while in the offspring rule they are measured in offspring, hence the name. "Fitness" refers to the generation of the decision-maker, but offspring are removed by one generation, hence the r_B or 1/2 in place of the 1. The denominator is also affected; r_{HP} is the relatedness of the helper to the parent, rather than to the parent's offspring. In the case of helpers rearing full sibs, no problem arises because relatedness to parents and to sibs is equal ($r_{HP} = r_H = 1/2$). The significance of this difference was not apparently realized by field workers until 1975. The offspring rule can be derived mathematically from Hamilton's rule and may be considered to be a restatement of the same principle for a different set of circumstances.

The difference between the two rules in practice can be illustrated with a simple example. Consider a female Stripe-backed Wren nearly one year old with a choice between helping her father and stepmother as a nonbreeder or leaving to find a mate and territory to breed in a pair or trio. As a breeder in a pair or trio she could expect $L = 0.40$ offspring per year. Her expected survival would be equal under each option (data from Rabenold, 1984). A quartet (pair and two nonbreeders) could expect 1.86 offspring per year; the difference attributable to the second helper would be $G = 1.86 - 0.40 = 1.46$ offspring per year. The helper is a half-sib of the offspring ($r_H = 1/4$), and her average relatedness to the breeders is $r_{HP} = (0 + .5)/2 = 1/4$. Therefore,

$$G/L = 1.46/0.40 = 3.65,$$

$$r_B/r_H = 0.50/0.25 = 2, \text{ and}$$

$$1/r_{HP} = 1/0.25 = 4.$$

According to the offspring rule she should help and not breed, since $3.65 > 2.00$. According to Hamilton's rule (applied incorrectly for comparison) she should breed and not help, since $3.65 < 4.00$. In fact, most such females choose to help and not breed. In this example, simply plugging the numbers into Hamilton's rule taken from his original papers yields a criterion that differs by a factor of two from that of the offspring rule. Hamilton's rule is, of course, not wrong; it simply requires intelligent use. The offspring rule is a version of Hamilton's basic relationship that is suitable for problems of this sort.

Grafen (1984) has adopted the offspring rule but made it look like the original Hamilton's rule, even though it really is not the same. Substituting our G for his b and our L for his c, his version is $GR - L > 0$. In preserving the appearance of Hamilton's rule while applying it to field data he has redefined the symbol for relatedness, R, so that $R = r_H/r_B$. Mathematically, however, his version yields identical results to mine, provided one remembers that Grafen's R is not Hamilton's (1964) r_{HP}.

The Present-Future Tradeoff. The preceding conceptual framework is useful (1) for animals that breed only once (semelparous) or (2) for considering the genetic consequences on present reproduction without regard for the future, or (3) if G and L are based on lifetime reproductive success. All species with helpers, however, may breed repeatedly in successive years (iteroparous). Furthermore, most communally breeding birds exhibit delayed reproduction. These facts indicate that in evaluating how natural selection operates on helping we must consider the effects of helping on future survival and reproduction.

To do this we divide direct and indirect fitness into present and future components as shown in Table 4.2. Using lower-case for future and capitals for present, our criterion for helping to be selected over nonhelping is:

$$D_H + d_H + I_H + i_H > D_B + d_B + I_B + i_B.$$

This expanded version of the offspring rule is a general formulation that is useful for comparing any two alternative strategies. Any strategy (H, B, or any other) has the four components (D, d, I, i), although, as in the present case, some of them may be zeros.

Table 4.2

Components of inclusive fitness for the breeding and nonbreeding (helping) options of a potential helper at age X in a singular-breeding species. Assume (1) a steady-state population, (2) if breeding is chosen, bird does not help again, (3) if helping is chosen, bird chooses breeding the following years, and (4) one breeding pair per unit.

B = breeder. H = helper. R = recipient of help. y = future ages. m_x = age-specific fecundity of breeder. m_y = future fecundities if breeding is chosen. m'_y = future fecundities if helping is chosen. r_B = relatedness of breeder to its own offspring, usually 0.5. l_y = survivorship at future ages if breeding is chosen, when $l_x = 1.0$. l'_y = survivorship at future ages if helping is chosen when $l_x = 1.0$. r_{HR} = relatedness of helper to recipient. h_x = age-specific increment to recipient's m_x.

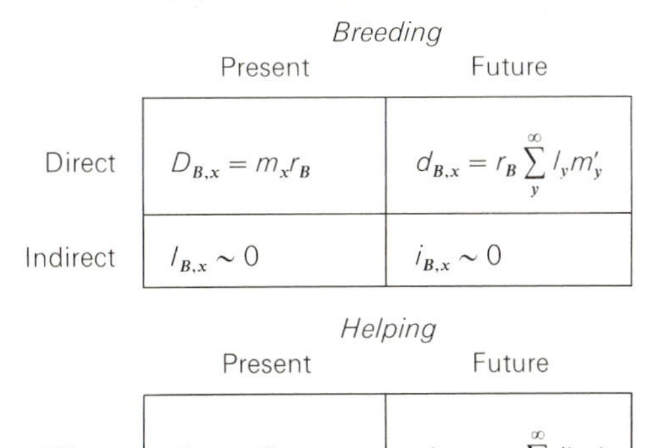

Breeding

	Present	Future
Direct	$D_{B,x} = m_x r_B$	$d_{B,x} = r_B \sum_y^{\infty} l_y m'_y$
Indirect	$I_{B,x} \sim 0$	$i_{B,x} \sim 0$

Helping

	Present	Future
Direct	$D_{H,x} \sim 0$	$d_{H,x} = r_B \sum_y^{\infty} l'_y m'_y$
Indirect	$I_{H,x} = \sum_R h_{x,R} r_{HR}$	$i_{H,x}$: see text

Some explanation of these terms may be useful.

$D_H - D_B$. This is the expected difference in fecundity between the H and B strategies, weighted by relatedness (1/2). It should be negative if we assume that helpers do not breed ($D_H = 0$). These quantities may be expressed as nestlings, independent offspring, or whatever is convenient. Of course, the closer the unit of measurement is to mature offspring, the better, unless the quality of mature offspring matters (see Chapters 7 and 15).

$d_H - d_B$. This is the expected difference in fecundity for future components of lifetime reproductive success weighted by $r = 1/2$. A positive result might be due to higher survivorship as a helper than breeder, or to higher future fecundity after being a helper. A higher probability of getting a territory as a result of helping (as opposed to surviving) would be reflected in a higher fecundity at future ages. A negative result can occur in altruism.

$I_H - I_B$. This is the expected difference in marginal gain in fecundity of nondescendent relatives in the present reproductive season weighted by the appropriate relatedness, as discussed above. When breeders do not help, then I_B approximates zero.

$i_H - i_B$. This is the expected difference in future effects on indirect fitness. A few examples may be useful. When the young counted as marginal gain above become helpers, they typically help relatives, such as parents or sibs. The weighted marginal gain for which they are responsible in the future is counted here. Such aid may be crucial for parents in some cases. Another example may arise when older helpers are more effective than younger helpers because of their experience when younger. This category also includes enhanced survival of breeders due to reduction of their cost of reproduction by helpers. For estimates of its magnitude see Reyer (1984) and Rabenold (1985).

A shorter version of the expanded offspring rule is available for singular-breeders. It relies on the facts that in these species breeders do not help and helpers do not breed. Therefore, $D_H = I_B = i_B = 0$, and the criterion for choosing to help becomes:

$$d_H + I_H + i_H > D_B + d_B.$$

This form is useful in considering field data and will be used in Chapter 13. Strictly, i_B need not be zero if a parent's offspring help each other. Nevertheless, it is a useful simplification when such effects are negligible.

Life-history Formulation. Each component of expected inclusive fitness can be expressed in familiar life-history terms weighted by relatedness. This allows these components to be measured in natural populations and then evaluated in the unified currency of inclusive fitness. Some illustrative examples are shown in Table 4.2. Of course, such terms as m_x and l_x can themselves be subdivided into their components. This formulation will be employed later in chapters dealing with tests of inclusive fitness theory (e.g., Chapters 13 and 14).

Double Accounting

The offspring rule may be used in two ways: (1) as a guide to the conditions under which each strategy should be *selected*, and (2) as a model for generating *a priori predictions*. The following discussion compares these two usages to each other with respect to accounting procedures and to the conventional population genetics or life-history approach, which we consider first.

Life-history Approach. Suppose we wish to know when a helping genotype is selected over a nonhelping genotype. Assume that breeding status is denied to all individuals by forces beyond their control for a year, after which they breed every year until death. Under the nonbreeding circumstances, individuals of the helping genotype help their parents 90% of the time, but those of the nonhelping genotype simply coexist with their parents without ever helping. Thus, most breeders of the helping genotype would have offpring as helpers; breeders of the nonhelping genotype would have no helpers.

For the helper genotype to increase in frequency, individuals of the helper genotype must produce mature offspring at a faster rate than nonhelpers. In other words, the lifetime reproductive success of individuals of the helper genotype must exceed that of the nonhelping genotype. With this approach the benefits of helping appear only as enhanced values of fecundity, m, and survivorship, l, during the breeding ages. For example, of all the young produced by Grey-crowned Babblers in one study, roughly 40% were attributable to helpers, a conspicuous enhancement. Since the helper genotype outproduces the nonhelping genotype, the former would be selected. With this approach indirect benefits need not be calculated, and the benefits of helping are allocated to the breeders rather than to the helpers.

Inclusive-fitness Approach. Now, if we use instead Hamilton's principle in the form of the offsping rule to analyze the same problem (whether or not helping is selected), the benefits of helping are allocated to the helpers and *not to the breeders.* The average breeder gets credit only for offspring that it would produce unaided, not for those additional offspring credited to the helpers. To credit these additional offspring to both the helpers and the breeders would be double accounting. To avoid this error, the values of the future-direct components in Table 4.2 should be adjusted *downward* in the manner just described. This approach also leads to the conclusion that the helping genotype will be selected.

Predictions about Individuals. The offspring rule may also be used to make *a priori* predictions about the behavior of an animal. Consider the example of the Stripe-backed Wren, which is discussed at some length in Chapter 11 (where references are given). These wrens typically remain as nonbreeding helpers in their natal territories for a few years before beginning to breed. The number of years in nonbreeding status varies with local conditions for each sex. Does a wren take indirect fitness into account in making its decision to initiate breeding? Assume a wren has a choice between breeding and helping to rear full sibs. Since the values of r are equal, they cancel out; we need consider only the number of young for which the wren is responsible. Future effects will be disregarded for simplicity.

A female wren in a quartet is responsible as a helper for 1.46 young. As a breeder it could produce 0.40 young without help but up to 3.30 with help. If it took indirect fitness into account, the wren should terminate helping to begin breeding only to breed with at least two helpers (yielding 1.86 young) so long as both parents remain. If it does not consider indirect fitness in its decision, it should accept any breeding opportunity with a nonzero chance of success. If we did not count the effects of helpers on reproduction as part of a breeder's fitness, we would have to credit it with only 0.40 young; and we could not predict its behavior with respect to choice of group size, since all sizes would be treated equally at 0.40 young. An individual wren can adopt only one strategy at a time. It is not credited twice (as a helper and then as a breeder) for the same help, so double accounting is not a problem.

Each of these three approaches is useful in its own way. The way in which the effect of helpers is counted differs with each approach. In the life-history approach only breeders receive credit. In the inclusive-fitness approach only helpers receive credit for their help, not the breeders. When making predictions about individuals, the effects of helping may be credited either to helpers or breeders, depending on the options.

In the first two cases the goal is to determine the conditions under which helping is selected. Such analyses require consistent accounting across the entire population, and the conclusions are statements about a population. The third case differs from the first two in that its goal is to make predicitions about individuals, not populations.

The bottom line is: The above considerations must be taken into account when using the offspring rule and the expressions in Table 4.2. Useful discussion of the perils of double accounting has been provided by Grafen (1982, 1984).

Altruism and Mutualism

Since there is considerable confusion about altruism and mutualism, it is useful to provide precise definitions in terms of the above components of inclusive fitness.

The concept of altruism was not accepted into evolutionary biology until Hamilton (1963) drew attention to its interest as a theoretical problem. Altruism is defined in terms of net absolute gains and losses to the parties involved. Therefore, the *currency* in which these gains and losses are defined must be specified. There are at least three definitions of altruism in use in evolutionary biology. They differ in the currency with which gains and losses are specified. Only Hamilton's original concept will be used in this book, but mention must be made of some of the others.

Hamiltonian Altruism. Altruism was defined by Hamilton (1964) in terms of *net* effects on the *direct fitness* of the participants. An act is altruistic if its net effects on the performer or donor lower its direct fitness ($D + d$) relative to the population while its effects on the recipient raise its direct fitness. This concept is attractive theoretically and empirically. It allows a deficit in direct fitness to be compensated by a gain in indirect fitness. These effects can be measured empirically. An act is altruistic if:

$$\text{for the performer } D + d < 0,$$

$$\text{while for the recipient } D + d > 0.$$

Note too that indirect fitness is not mentioned in the definition. The definition is given purely in terms of the effects of the behavior on direct fitness. By this definition parental care is normally not altruistic because its net effects on direct fitness are positive: its direct benefits outweigh its direct costs. Alexander (1974) referred to Hamiltonian altruism as phenotypic altruism, although the phenotype of the behavior is not involved in the definition. Brown (1975a) called it individual-fitness altruism, referring to the currency.

Triversian Altruism. Altruism was defined by Trivers (1971) in terms of net effects on *inclusive fitness.* Trivers adopted this usage in his treatment of "reciprocal altruism" (see Chapter 14). An act is altruistic if its net effects on the performer or donor lower its inclusive fitness ($D + d + I + i$) relative to the population while its effects on the recipient raise its inclusive fitness. Losses in inclusive fitness can only be compensated by gains in inclusive fitness at a later time. An act is altruistic if:

$$\text{for the donor } D + d + I + i < 0,$$

$$\text{while for the recipient } D + d + I + i > 0.$$

Again the phenotype of the behavior or trait is not part of the definition. Again parental care is not altruistic because the parent receives a net gain in inclusive fitness (entirely through the direct component). Alexander (1974) referred to Triversian altruism as genotypic altruism, although the genotype of the behavior is not part of the definition. Brown (1975a) called it inclusive-fitness altruism, referring to its distinguishing feature.

Alexandrian Altruism. Some authors prefer to regard parental care as altruism. Following Alexander's (1974) usage, Krebs and Davies (1981) defined altruism so as to include parental care; the donor suffers a cost (not necessarily a net loss) and the recipient enjoys a benefit. The fact that the parent typically receives a net gain for its cost is disregarded. The disadvantage of this definition is that any social behavior that benefits a recipient at some cost to the donor must be altruistic (Brown, 1982b). Since virtually all cooperative or mutualistic behaviors have some cost (e.g., parental care), they must all be called altruistic if one is to be consistent with the definition.

Our definitions of altruism (following Hamilton) and mutualism may be portrayed graphically (Fig. 4.3A). This figure employs the direct-fitness-space approach that was introduced by D. S. Wilson (1975). Note that the axes are net effects on direct fitness ($D + d$) compared to a population average. The word reciprocity has been used by some authors for relationships that might better be described as mutualism (Ligon and Ligon, 1978), in my opinion. Caution in the use of reciprocity is advisable for two reasons. In the dictionary sense, (1) spite can be reciprocal (for example, if both players defect in a game of Prisoners' Dilemma—see Chapter 14); (2) reciprocity has two senses and this may cause confusion. It is commonly

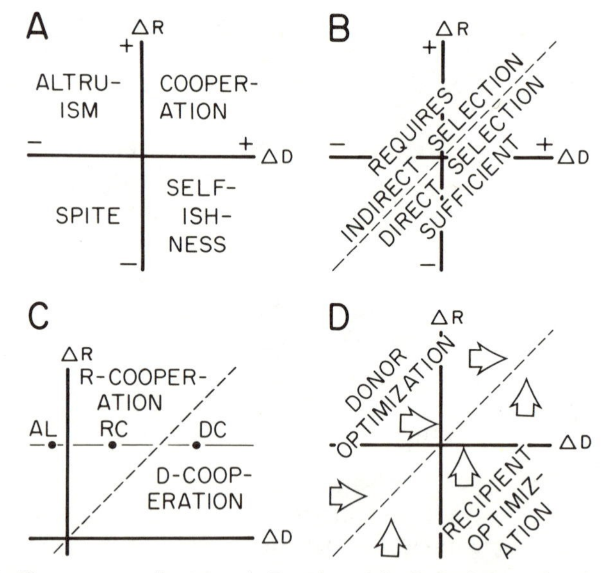

Figure 4.3 Fitness space diagrams. Δ-R = change in direct fitness of recipient. Δ-D = change in direct fitness of donor. A. Categories of social relationship based on net effects on direct fitness. B. Cost-benefit conditions for direct selection to favor a donor trait when there is no between-group genetic variance. C. Three hypothetical locations of helping behavior. AL = altruism; RC = R-cooperation; DC = D-cooperation. For further explanation of B, C, and D and of fitness space diagrams generally see Chapter 12, Figure 12.7, and D. S. Wilson (1975). D. Directions of donor optimization and recipient optimization. (From Brown and Brown, 1981b.)

used in a restricted or narrow sense (Waltz 1981) to mean that some kind of score is kept by each player (e.g., the Tit-for-Tat strategy of Axelrod and Hamilton, 1981), but even Trivers used it for situations that had previously been termed mutualism, i.e., the broad sense, in which score-keeping is not required.

The terms spite and selfishness are from Hamilton (1964, 1970). The quadrant marked mutualism was not named by Hamilton but has been called cooperation (Brown, 1975a; Emlen, 1978) and mutualism (West Eberhard, 1975). The terms, mutualism and cooperation, require the assumption that recipients behave as donors. This is reasonable since both recipients and donors benefit. Strictly, however, the fitness-space diagrams in Figure 4.2 refer to consequences of unilateral actions of the donor. Therefore, the terms cooperation and mutualism are less than ideal.

Conclusions

This chapter has provided a simple comparison between inclusive fitness theory and classical fitness theory, and a means to frame testable hypotheses with which the new contributions to fitness theory can be evaluated empirically. We have also defined some basic terms in the evolutionary biology of social behavior and pointed out sources of confusion in the literature.

5 Delayed Breeding Sets the Stage for Helping

Three Basic Questions

In order to understand the ecological factors that predispose a bird to perform alloparenting, it is necessary first to re-emphasize that helping occurs in a variety of situations and may, therefore, be favored by different combinations of factors in different cases. For example, even in one species, the Acorn Woodpecker, helping occurs by breeding males who share a female, by breeding females who lay eggs in a joint nest, and by nonbreeders. In all three cases the woodpeckers may care for young not their own, but in each case the sets of ecological factors predisposing individuals to help are different, though perhaps overlapping. Note that in Acorn Woodpeckers helpers may be breeders or nonbreeders, illustrating a general point made repeatedly by Skutch (see quotation introducing Chapter 1) and most other reviewers. Clearly, in this example—as in others—breeding status varies independently of alloparenting behavior.

In some species helping is typically restricted to birds of nonbreeding status, as in the singular-breeding species of Table 2.2. Even in these species, however, it does not follow that a nonbreeder must be a helper. In the Australian Mapgie, for example, some nonbreeders live outside of communal units and have no opportunity to help, while others live in communal units but do not help (Carrick, 1963, 1972). In the Northwestern Crow (not strictly a communal breeder) nonbreeders may associate with their parents but rarely help (Verbeek and Butler, 1981). Even in a typical singular-breeding species, such as the Grey-crowned Babbler, some nonbreeders help at the nest while others probably do not (Brown et al., 1978).

It has often been stated that nonbreeding helpers are typically offspring from a previous brood who have not dispersed. Before color-banding was employed extensively to study communal breeding the characterization of nonbreeding helpers as nondispersers might have seemed safe; now it is not. In plural-breeding species, such as the Mexican Jay, helping by nonbreeders may occur after dispersal. Even in singular-breeding species, helping after dispersal is not rare, especially when dispersal tends to involve same-sex sibs (Chapter 6).

The general point here is that to understand the ecological determinants of helping it is necessary to treat breeding, dispersal, and helping as three separate and independent variables. Each of these may be represented as a two-state variable. A bird may breed (B^+) or not breed (B^-); disperse (D^+) or not disperse (D^-); and help (H^+) or not help (H^-). Using this symbolism, we may frame questions that involve linkages between these variables or questions that do not assume linkage. Most importantly, we may treat each of these phenomena independently of the others. It is essential for clear thinking that linkage not be assumed in all cases.

For singular breeders a common approach is to contrast helping while not breeding and not dispersing, with dispersing, breeding, and not helping. This approach was used by Woolfenden and Fitzpatrick (1984) for the Scrub Jay. It may be represented as:

$$B^-D^-H^+ \text{ vs. } B^+D^+H^-.$$

This formulation assumes that all three variables vary in concordance, e.g., nonbreeders are invariably helpers. It is, therefore, difficult to tease out the separate components of either option.

A more heuristic approach is to consider each variable separately. For example, the options with regard to helping or not among nondispersing nonbreeders may be expressed as follows:

$$B^-D^-H^+ \text{ vs. } B^-D^-H^-.$$

With this approach nonbreeding and nondispersal are controlled, and attention is focused on helping *per se*. An approach much like this was used successfully in studies of the Pied Kingfisher (Reyer, 1980a, 1984; discussed in Chapter 13).

Linkage between B^-, D^-, and H^+ is especially clear in singular-breeding species, such as many jays, wrens, and babblers. In these species non-breeding and nondispersal may be viewed as probable *preconditions* for the evolution of helping. To understand the evolution of helping in these species we first need to examine the ecological factors that favor delayed breeding and delayed dispersal. In doing so it is important to decouple the question of helping from the questions about delayed breeding and delayed dispersal. Conditions that favor delayed breeding may also be conducive to helping by breeders. In this chapter we examine delayed breeding as a subject that is both interesting for its own sake and as a precondition for helping by nonbreeders. In the following chapter dispersal receives similar treatment. The general approach of this book is to consider three basic questions: (1) Why delay breeding? (2) Why delay dispersal? (3) Why help? These questions are, of course, interrelated; but unless each is considered separately the relationships between them cannot be fully appreciated.

As an example of an interaction consider the following. It is usually considered that delayed breeding and nondispersal "set the stage" for helping, but helping may feed back positively in this system to strengthen the causes of nonbreeding and nondispersal. By increasing the production of young in a population, helping increases the excess of births over deaths of breeders (*a* in Fig. 3.2), thus increasing the equilibrium surplus of nonbreeders (N_F) in a population with limited breeding territories.

Why Delay Breeding?

The presence in a population of animals that appear to be quite capable of breeding but do not do so is a longstanding problem in population biology (Cole, 1954; Stearns, 1976). The phenomenon is rather common in seabirds but can be found even in small passerines. It is pronounced in many communally breeding birds. Some early writers viewed delayed breeding as a kind of voluntary restraint by individuals to prevent overpopulation (Skutch, 1967; Wynne-Edwards, 1962). The nonbreeders were interpreted as a reserve to be used in case the breeding part of the population was depleted. These arguments, which depended upon the deme or species as a unit of selection, were effectively rejected by Lack (1954a, b, 1968) and others.

In their place a variety of factors has been suggested. Most of these are listed for easy comparison in Table 5.1. Although it is convenient to list these factors one by one, I would like to emphasize that two or three of them may be important simultaneously, and that interactions between them may be critical.

Unlike the earlier theory for delayed breeding in seabirds (IA), which was based on the detrimental effects of early breeding on lifetime reproductive success, theories advanced for communally breeding birds have been based upon *shortage of a "resource,"* such as suitable territories (II), females (III), or a "labor force" (V). Therefore, it is the *exclusionary* effects of competition for limited resources that have been considered to be important in "setting the stage" for helping behavior. This is not to deny a role for environmental modulation; all of these exclusionary effects plus other factors in Table 5.1 may be *modulated by temporal or spatial variations in the environment. All of the factors in Table* 5.1 *are in some respect ecological.* The factors and some of the interactions are discussed below.

Table 5.1

Factors suspected to delay breeding by diminishing potential success (I–VI) or by benefiting nondescendent kin (VII).

I. Lack of skill in:
  A. foraging and associated behavior
  B. territorial and other agonistic interactions
  C. predator avoidance
  D. care of young
  E. construction of nests and sleeping structures
II. Territorial behavior of rivals
III. Shortage of opposite sex, usually of females
IV. Costs of dispersal due to increased risk of:
  A. energy shortage in unfamiliar environments
  B. predation in unfamiliar environments
V. Deficient labor force
VI. Environmental depression:
  A. of food levels and foraging success, interacting with IA–D, II, IV
  B. by increased predation, interacting with IC and IVB.
  C. causing increased mortality of females, exacerbating a skewed sex ratio, III
VII. Helping parents or sibs
VIII. Variance enhancement (parental manipulation)

The Skill Hypothesis

History and Natural History. The explanation of delayed breeding in birds that was most favored by Lack's group depends on learning (Ashmole, 1963). The idea is simple: young birds have not acquired sufficient skill at foraging to enable adequate provisioning of nestlings. Since the effort to breed when young would consequently be unlikely to be successful and might conceivably reduce survival of the breeder (a point deserving more study), it would be more productive over a lifetime to refrain from breeding until an older age. Reproductive restraint in the immature years was thus thought to maximize lifetime reproduction, not curtail it (Curio, 1983). The need for time to learn foraging skills has been invoked by Rowley (1977) to explain delayed breeding in the White-winged Chough. The skill hypothesis has not been formally modeled or applied previously to communally breeding birds, although all workers seem to be aware of it. It is especially useful for nonterritorial species, such as colonial seabirds, to which habitat saturation hardly applies. The possibility that skill supplements other factors delaying breeding in group-territorial birds deserves consideration. It may not be of widespread importance, but at least it should be subjected to tests instead of simply ignored.

Convincing evidence exists for at least the first part of the skill hypothesis. In some birds with a recognizable immature plumage, adults have been shown to exhibit a higher efficiency at foraging than do immatures. This has been shown for pelicans (Orians, 1969), herons (Recher and Recher, 1969), terns (Dunn, 1972; Buckley and Buckley, 1974), cormorants (Morrison et al., 1978), and various gulls (Schreiber and Young, 1974; Searcy, 1978; Burger, 1981).

Even more convincing is the observation in three species of gulls that the number of years for which a typical member of the species wears a plumage recognizable as immature (1, 2, or 3) equals the number of years in which foraging efficiency is below that of adult-plumaged birds (MacLean, 1982, 1986). Furthermore, improvement occurs with each year of immaturity (Fig. 5.1), and the degree of improvement needed to reach the adult state is greater for the species that take longer to mature. For further discussion of deferred breeding in seabirds see Ainley et al. (1983). A test of this sort is desirable for communally breeding birds, and Figure 5.1 provides an example of what could be done.

Many species of group-territorial birds have distinctive coloration preceding the normal reproductive ages, as do the seabirds mentioned above. A typical example is the Grey-crowned Babbler (Counsilman and

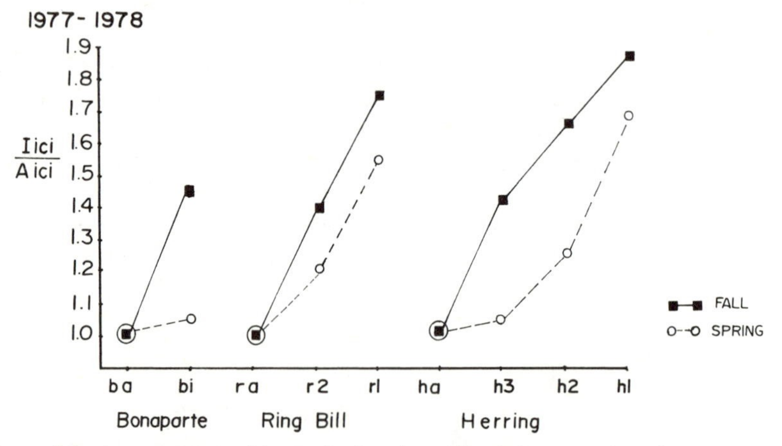

Figure 5.1 Improvement with age in foraging skill of three species of gulls on the Niagara River. Foraging success is estimated from the average interval between successful dives, known as the intercatch interval (ici). It is expressed here as the ratio of average immature ici to average adult (Iici/Aici). For the Bonaparte's Gull ba = adult, b1 = first year. For the Ringbilled Gull ra = adult, r2 = second year of life, r1 = first year of life. For the Herring Gull ha = adult, h3 = third year, h2 = second year, h1 = first year. Note improvement from fall to spring and from young to old. (From MacLean, 1982).

King, 1977). As shown in Figure 5.2, the iris changes from dark brown in the one-year-olds through intermediates to pure yellow in birds four or more years old. As the iris lightens and becomes yellower, the probability of breeding in the age group increases. Similar correlates of nonbreeding status involving color of plumage, iris color, or bill color are found in a variety of other group-territorial, communally breeding birds, for example, Mexican Jays (Hardy, 1961), *Cyanocorax* jays (Fig. 6.2; Hardy, 1973), and Brown Jays (Skutch, 1935). These observations are consistent with a variety of hypotheses for delayed breeding and are not particularly good evidence for the skill model.

Another observation suggesting that the skill hypothesis may apply to communal birds is the smaller body size in the nonbreeding early ages. This is apparent in Grey-crowned Babblers (Fig. 5.3), Mexican Jays, Tufted Jays (Crossin, 1967), and White-winged Choughs (Rowley, 1977). First-winter individuals of the latter species average 88% of adult weight, with an increase each year for at least two more years. A smaller body size when young might ease occasional energy shortages of young animals,

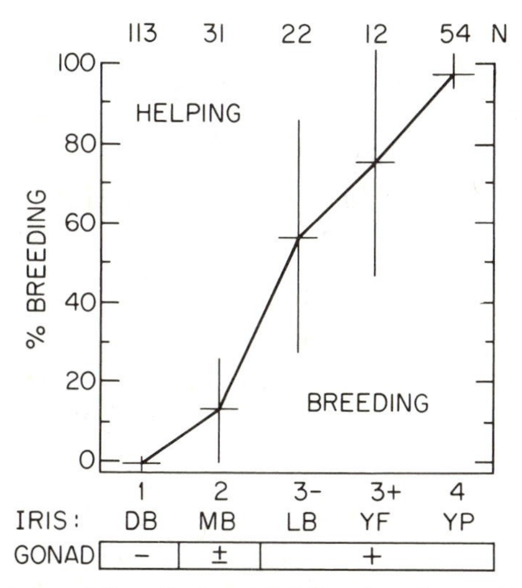

Figure 5.2 Percentages of Grey-Crowned Babblers breeding and helping at progressively older ages. Data are for the 1976 breeding season, Meandarra, Australia. Mean and 95% confidence intervals are shown for the sample sizes (*N*) above. Iris colors are DB (dark brown), MB (medium brown), LB (light brown), YF (yellow with dark flecks), and YP (pure yellow without flecks, the adult condition). Approximate ages for each iris age-class are shown in years. Gonads are small and nonfunctional (−) for DBs; functional and adult size (+) for LB, YF, and YP; and intermediate (±) for MB. (From Brown, 1985.)

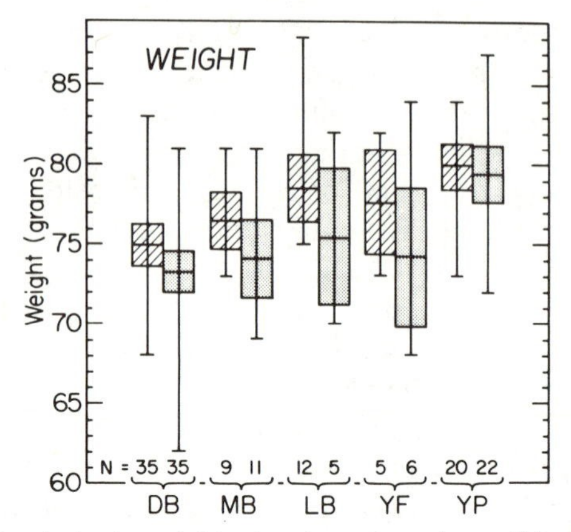

Figure 5.3 Variation in body weight related to sex and age (iris color) in the Grey-crowned Babbler. For each iris-sex category the mean is shown by the horizontal line; the range, by the vertical line; and two standard errors on each side of the mean, by the rectangle. Males = hatched area. Females = stippled area. Iris colors and approximate ages: DB = dark brown; 3–18 months old. MB = medium brown; 1.5–2.5 yrs. LB = light brown; 2.5–3.5 yrs. YF = yellow with dark flecks; 3–3.5 yrs. YP = pure yellow; 3.5 yrs or older. N = sample size. (After Brown et al. 1982a.)

or it might simply reflect poor conditions. These postulated relationships of body size to immaturity remain to be tested.

There may be skills to be learned in deciding what foods to eat, where and when to find them, and how to catch or handle them, especially in omnivorous or insectivorous species such as jays, wrens, and babblers, and in the first two months of life, when these birds are inefficient foragers. Species that store food in autumn to be used in winter, such as jays, might improve their cognitive map of storage sites with experience. Learning of such skills should be especially important in the first year of life, but Figure 5.1 shows that skills may also continue to improve subsequently in gulls.

A Model of Skill-Environment Interaction. The dependence of age at first breeding (α) on the interaction between food availability and learned skills in foraging can be illustrated by the simple model shown in Figure 5.4. Assume that the potential rate at which a bird can provision a nest with young, R, is determined by two factors, the abundance of food, Q, which varies locally and yearly with environmental conditions, and the animal's skill, S, at obtaining it. Thus,

$$\text{provisioning rate} = R = SQ = \text{skill} \times \text{abundance}.$$

Assume too that skill improves at a rate dependent on age, x, and a coeffi-

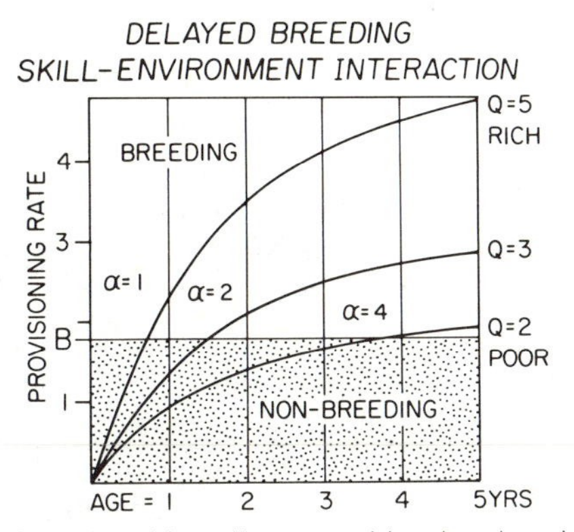

Figure 5.4 A model for delayed breeding caused by slow learning of foraging skills. The potential rate at which nestlings could be provisioned, R, is determined by the age, x, and the speed of learning. A threshold value of R at which breeding is energetically feasible is shown at B. In richer environments (higher values of Q) breeding is feasible at earlier ages.

cient of learning, l, as follows:

$$S_x = 1 - e^{-xl}.$$

Therefore,

$$R_x = Q(1 - e^{-xl}).$$

Now if there is a threshold level of R below which breeding is not attempted, B, we have a situation in which age at first breeding, α, is determined jointly by the environment and the animal's skill as shown in Figure 5.4. In a rich environment or good year, breeding can begin earlier than in a middle-quality environment or year; and in poor environments, breeding becomes feasible even later.

If breeding status is dependent on foraging skills that develop with age, as above, then in some environments we should expect to find that the percentage of individuals breeding in a population increases with age. Figure 5.2 shows such a case. If resource abundance varies temporally, we should expect a larger percentage of younger individuals to breed under better conditions.

The learning coefficient, l, influences the speed with which provisioning rate, R, approaches the asymptote, Q, set by the richness of the environment. If l is high, learning is rapid. In this case, most learning occurs early, perhaps before age one year, and environmental richness is less likely to affect the various age classes differentially. Either all ages breed, or they do not.

I chose the particular mathematical function that is illustrated because it embodies the following important characteristics: (1) increasing foraging ability with age, (2) rapid rate of improvement early, and progressively slower rates with age, (3) allowance for different rates of improvement, (4) limitation of success rates by environmental factors, and (5) simplicity. No doubt other mathematical representations could express these qualities. Similarly, other models might be used to incorporate different or additional properties. The present model seems intuitively to fit available knowledge of the ontogeny of foraging relative to age at first breeding. Empirical tests might lead to a different and more accurate model.

The same model could, of course, be employed to describe improvement in social skills or any other skills important for breeding success.

We can also use the concept of potential provisioning rate to model length of breeding season. Several factors may act to lengthen the breeding season, as defined by the period during which R is above B. This is shown in Figure 8.5 as the period in which $E > K$. These include group size (assuming it affects the energy budget), skill, and resource abundance, as influenced by factors intrinsic to the territory such as vegetation, and extrinsic factors such as rain. A correlation between group size and length of breeding season was found by Brown and Brown (1981b) in the Grey-crowned Babbler. Of course, some of these factors may be correlated with each other. In this case, group size was also correlated with age of the breeders (Fig. 11.2) and weakly with some measures of vegetation thought to reflect territory quality (Brown et al., 1983).

In concluding this section I would like to remind the reader that the skill model is *not* a theory for the evolution of helping behavior or communal breeding. It is presented as only one of several concepts that may be useful in thinking about *delayed breeding.* Since helping has sometimes been confounded with delayed breeding, this point deserves emphasis.

I would also like to underscore the point that my presentation of this model does not represent a judgment of its importance in communal birds. Although this point remains to be determined, it has long been my suspicion that habitat saturation is likely to be more important for communal birds generally (Brown, 1969a). Nevertheless, we cannot reject the skill model until it is tested.

Territorial Behavior

A Model of Habitat Saturation. Assuming now that an individual possesses sufficient foraging skills for breeding, its next requirement in most communally breeding birds is a territory that is suitable for breeding. "Suitable for breeding" is a statement implying that food resources, nest sites, shelter from predators, and other requisites for *successful* breeding are present. When such suitable habitat is fully occupied, thereby preventing additional individuals from breeding, these deprived individuals may be referred to as surplus, and the breeding habitat may be said to be saturated. This model of habitat saturation is presented in Figure 5.5.

If we allow the number of suitable territories to be represented by T_S and the number of competitors for them by N_T, then the number of surplus competitors, or nonbreeders, N_F, will be determined in a straightforward manner:

$$\begin{aligned}\text{number of nonbreeders} &= N_F = N_T - T_S \\ &= \text{number of competitors} - \text{number of territories.}\end{aligned}$$

Habitat saturation may be said to occur for specified habitats (e.g., all suitable habitats) when the following condition is met:

$$\text{number of competitors} = N_T > T_S = \text{number of suitable territories.}$$

This may occur through variation in N_T, T_S, or both. In variable environments both may vary widely. The degree of excess (not degree of saturation) may be expressed as N_F/T_S.

In a review of the effects of territorial behavior on population density in birds generally I suggested a role for territoriality in delaying breeding (Brown, 1969a). I then interpreted group territoriality coupled with helping

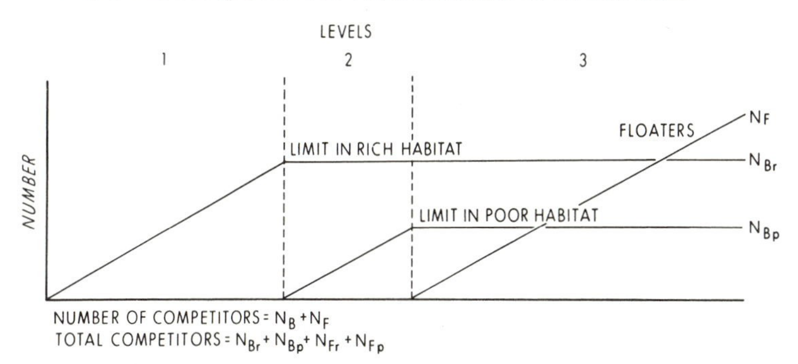

Figure 5.5 Habitat saturation: hypothetical effects of territoriality on breeding densities in two habitats, assuming fixed minimum territory sizes. N_B = number of breeders; N_F = number of surplus individuals or floaters; N_{Br} and N_{Bp} = number of breeders in rich and poor habitats, respectively; N_{Fr} and N_{Fp} = number of floaters in rich and poor habitats, respectively. (From Brown, 1969b.)

by nonbreeders as an adaptive response by individuals to conditions that cause a surplus of potential breeders. In presenting evidence supporting this theory Stacey (1979a) termed these conditions habitat saturation. The importance of habitat saturation as a precondition for the origin of nonbreeding helpers in wrens had previously been independently suggested by Selander (1964) in a brief passage in a taxonomic monograph. I later quoted this passage in order to bring it to the attention of persons in this field (Brown, 1978b).

Territoriality may prevent some individuals from breeding by limiting the number of breeders, but territorial behavior does not necessarily create a surplus. If a population is recovering from a period of low density, suitable habitat may not be limiting (Level 1 in Fig. 5.5). If there is a range of territory qualities, the best should be occupied first, with overflow into poorer habitats, a condition identified by Level 2 in Figure 5.5. Only if all suitable habitats are normally monopolized by breeders does the surplus fraction (N_F) of the population become important (Level 3 in Fig. 5.5).

The model in Figure 5.5 portrays a simple situation in which territories are incompressible, causing dividing lines between levels to be sharp. In nature we expect that such sharp boundaries would be unusual. The creation of a surplus, however, is not dependent on incompressible territories. It is sufficient that there be a density limit for breeders that is reached in nature.

A high survival rate of the nonbreeders facilitates a high equilibrium density of nonbreeders, hence potential helpers (as explained in Chapter 3; Fig. 3.2). High survival also mitigates the cost of delaying reproduction. "The loss in reproductive potential due to evolution of delayed maturity, although large in species with low survival rates, is much less severe in species with high survival rates. . . . Adaptation to the existence of a surplus by means of delayed maturation is, consequently, more likely in a species having a high rate of survival" (Brown, 1969a). In a population at equilibrium the number of young that must be produced yearly to just replace annual losses to the breeding population depends upon the rate of survival of the young and adults and the age at first breeding (see equation in Brown, 1969a).

The fate of the surplus varies considerably among species. In some species, such as the Red Grouse (*Lagopus scoticus*), a species without helpers, the mortality rate of floaters is higher than that of breeders, which tends to keep the number of floaters low (Watson, 1985). In Mexican Jays the survival rate of one- and two-year olds (mostly nonbreeders) is as high as that of jays of breeding age. In the Stripe-backed Wren the annual survival rate of male nonbreeders ($s = .78$) is even higher than that of breeders ($s = .64$) (Wiley and Rabenold, 1984; Rabenold, 1985), resulting in a large fraction of the population having the status of nonbreeding helpers.

Nonbreeding birds in noncommunal species tend to lurk as floaters in the interstices of territories or live inconspicuously in subordinate positions in a large home range that overlaps more than one territory (*Zonotrichia capensis*, Smith, 1978). I argued in 1969 that group territoriality results when two conditions are met: (1) rising defense costs due to nonbreeders and habitat saturation make it economical for owners to share their territories with individuals that would share the costs of defense; (2) for the nonbreeder the chance of obtaining breeding status in the future is enhanced by remaining in its natal group under such conditions. The first theme is elaborated in Chapter 8, and the second in Chapter 6.

Evidence for Habitat Saturation. The effects of territorial behavior on a population may be divided into two types, spacing out and exclusion. At Level 2 in Figure 5.5 territoriality causes spacing out within a habitat and spreading of the population into less desirable habitats. These spacing mechanisms tend to cause an ideal free distribution (Fretwell and Lucas, 1970) in which individuals have equal reproductive success, as demonstrated for the Great Tit (*Parus major*) (Brown, 1969b). We are concerned here primarily with the exclusionary effects that characterize Level 3.

The exclusionary effects of territoriality were a focus of attention in the 1960s, when the effects of territorial behavior on population densities came under scrutiny (reviewed in Brown, 1969a). The existence of a surplus of nonbreeders in all-purpose-territorial species was demonstrated most cogently by various removal experiments (reviewed in Brown, 1975a; Davis, 1978). Similar experiments have not seemed important for communally breeding birds because the existence and age-sex composition of a surplus of nonbreeders is evident from studies of color-banded birds (e.g., Brown, 1963a; Carrick, 1963; Rowley, 1965a; and many later studies).

Not obvious, however, are the *ecological causes* of nonbreeding. Several kinds of evidence have been used to justify the conclusion that territorial behavior is an important factor causing delayed breeding.

A *comparison between spatially separated populations* was used by Stacey (1979a) to justify his conclusion that territorial behavior was important in the Acorn Woodpecker. He employed data of his own for populations in Arizona and New Mexico (mainly Water Canyon, N.M.) and data of MacRoberts and MacRoberts (1976) for a California population (Hastings Reservation). Group size was larger in California, as shown in Table 10.1. Stacey wrote about the unsaturated population as follows:

> The results of this study strongly support the hypothesis that habitat saturation plays an important role in the occurrence of communal breeding in the acorn woodpecker. Several independent lines of evidence, including the presence of unoccupied territories, the relatively high frequency with which reproductive positions became available in established groups, and the occurrence of unisexual

groups during the breeding season, all indicate that birds in Water Canyon would have a higher probability of finding space in which to breed than individuals that dispersed from groups at the Hastings Reservation.

Thus the data indicate a *higher level of competitor pressure* for breeding positions in California than New Mexico. The responses to saturation in California are of two kinds. Some individuals share a territory and a mate (see Chapter 10); and some delay breeding altogether, even though some Acorn Woodpeckers can breed successfully in their first year. The fact that some first-year individuals of each sex may breed either in a pair or group suggests that the skill hypothesis does not apply for Acorn Woodpeckers. Chapter 10 provides more recent data on the California population that support Stacey's conclusion. Emlen (1984) has examined a somewhat larger data set on the Acorn Woodpecker and reached a similar conclusion.

In the Hall's and Grey-crowned Babblers, which inhabit rather arid parts of Australia, suitability varied greatly among occupied territories, and was correlated with group size and age of parents (Brown and Balda, 1977; Brown et al., 1983). In both of these species unoccupied areas were deficient in preferred vegetation types for nesting and foraging. Potential breeders had a choice of attempting to breed in unoccupied poor habitat or remaining with their parents as nonbreeders. Consequently, particularly in the Grey-crowned Babbler, the habitats suitable for breeding could be described as saturated. The better occupied areas were conspicuously and continuously defended. Marginal territories were sometimes abandoned. In the better habitats of the Grey-crowned Babbler, the threshold for breeding (Fig. 8.5) was reached earlier and breeding continued longer than in marginal habitats, although this effect may have been caused by the larger numbers of helpers in the better territories (Brown and Brown, 1981b). Rainfall varied dramatically on the study area within and between years. In the breeding season of 1976, a good year for rain, all groups bred just after the rains; but in 1977, a drought year, few groups bred (Dow, pers. comm.). Nonbreeding groups typically waited in their territories for conditions to improve. These studies provided correlative evidence that the number of territories actually suitable for breeding varied with yearly rainfall in an area of unpredictable rain.

A positive correlation between territory suitability and group size has also been found for the Superb Blue Wren, a species found often in areas of unpredictable rainfall. Nias (1984) interpreted this correlation as evidence for saturation of the preferred habitat, namely, blackberry brambles. A lack of unoccupied dense-understory vegetation prevented some wrens from breeding as pairs and probably caused many to remain as nonbreeders.

In some species replacement of mates provides further evidence for habitat saturation. Competition for vacant breeding positions is intense in several communal birds that delay breeding (e.g., Scrub Jay, Woolfenden and Fitzpatrick, 1984; Acorn Woodpecker, Koenig, 1981a; Hannon et al., 1985).

The Role of the Environment

Two Types of Environmental Effects. The number of nonbreeders (hence potential nonbreeding helpers) in a population is influenced by environmental variability in two ways, which can be portrayed with a simple equation.

Recall the following relationship:

$$\text{number of nonbreeders} = N_F = N_T - T_S$$
$$= \text{number of competitors} - \text{number of territories.}$$

Variation in N_F caused by variation in T_S stems from events in the same year, such as effects of rain on food abundance, which affect the *conditions* for breeding by determining the number of suitable territories. Therefore, I refer to it as same-year determination. Variation in N_F caused by variation in N_T arises from events in previous years that have affected the production of *competitors*. This is called here previous-year determination. Of course, good conditions may attract more competitors; but I ignore this interaction for simplicity.

Same-year determination of the number of nonbreeding helpers was apparently first explicitly discussed by Orians et al. (1977b). They observed that the levels of abundance of insects suitable as prey for icterid blackbirds were an order of magnitude lower in Argentine marshes than in marshes of western North America. Their parallel observation that helping is common in the Argentine Brown-and-yellow Marshbird but not seen in North American marsh-nesting blackbirds suggested to them that low food levels might be conducive to nonbreeding helper status in two ways. (1) The helpers might be prevented from breeding by their low rate of energy intake but not prevented from allofeeding, and (2) the breeders could easily profit from such help. Since this icterid is colonial, it might be possible to exclude territoriality and habitat saturation as a cause of nonbreeding in the helpers. If helpers were of both sexes, then a shortage of females could be rejected and the skill hypothesis would be a prime contender.

Previous-year determination of the number of nonbreeding helpers in a population operates by affecting the number of competitors, N_T. If no young were produced in the previous year, no young would be available to help, regardless of how poor the rain and resources were in the present season. If many young were produced, then many potential helpers should be expected in any year, but especially in a low-resource year. The effect should be most pronounced when good years precede bad years.

An example of previous-year determination of number of yearlings is shown in Figure 5.6. The number of yearlings in a given year is the product of the number produced as nestlings in the preceding year and their survival rate. The number of yearlings on the study area varied from 23 in 1973 to almost none in 1981. In Mexican Jays yearlings are almost invariably nonbreeders, and over 90% of them contribute somewhat to brood care. The effect of the current year's resources in predisposing them to become helpers would be undetectable since 90% of them would help anyway.

Similar determination of the number of helpers is to be expected in other species with virtually obligate nonbreeding for the first year, as in Grey-crowned Babblers (Fig. 5.2). Yearly variation in the proportion of surplus nonbreeders in Mexican Jays does occur, but the effect is determined mainly in the preceding year, not by climatic effects in the current year. The proportion of surplus birds varies substantially from year to year (Fig. 5.7), and is highly dependent on breeding success in previous years.

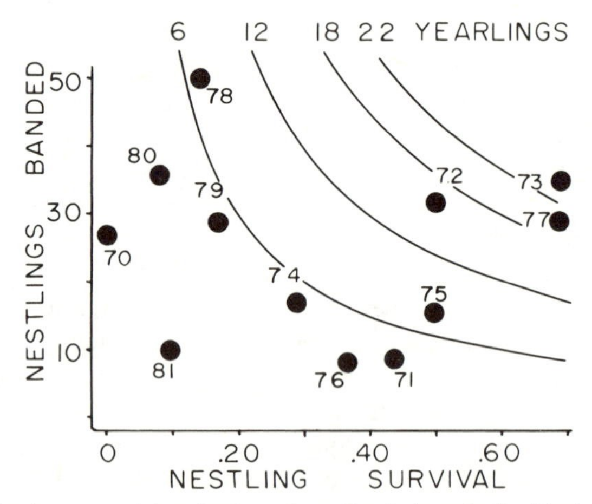

Figure 5.6 Previous-year determination of yearling helpers in the Mexican Jay. Yearly variation in number of nestlings produced (Y-axis) combines with variation in survival of those nestlings (X-axis) to determine the number of yearlings in the following year. The number of yearlings is indicated by the contour lines, which connect points having the same number of yearlings designated above. The actual number of yearlings in the population is larger because of immigration. Each year class (black dots) is designated by a number. (From Brown, 1986.)

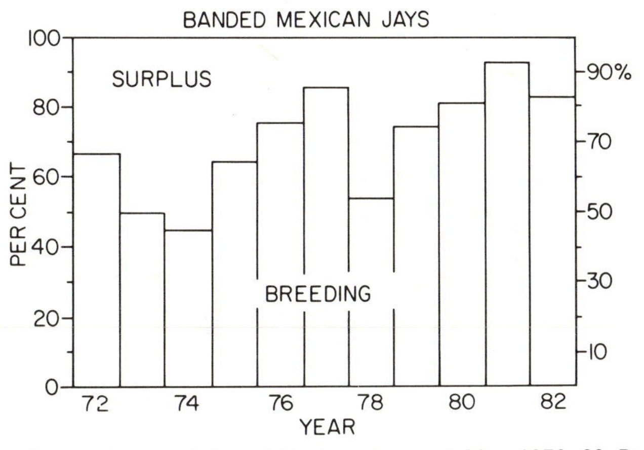

Figure 5.7 The surplus population of Mexican Jays on 1 May, 1972–82. Data are for birds from the preceding year class or older. Birds of the year are not included. Percentage breeding was estimated for the banded birds only. The percentage of birds not banded ranged from 17% at the beginning to zero in 1982.

Thus, conditions that affect breeding success can influence the proportion of nonbreeding helpers in the population in two ways: (1) by same-year effects on rate of food intake causing some *territories or foraging areas* (for colonial species) to be subthreshold for breeding, as described by Orians et al. (1977b) and Emlen (1982a, 1984; Emlen and Vehrencamp, 1983, 1985); (2) by previous-year effects on *production* and survival of potential helpers, as described here. The two mechanisms are not mutually exclusive. Both may act together. We expect type 1 mainly in quick-breeding species in which nonbreeding among yearlings is facultative and potential breeders easily revert to nonbreeding helpers. We expect type 2 in species in which nonbreeding among yearlings is virtually obligate.

Arid Climates. In hot, arid climates, where many communally breeding birds live, rainfall may be highly erratic. Is a separate theory needed for these regions? Emlen (1982a) argued that no previous theory applied to communal species in such climates and that one was needed. Later, however, he generously admitted that my 1974 model based on territorial exclusion definitely included such climates (Emlen, 1983). Territoriality is, in fact, typical of many birds breeding communally in such climates, including the Acorn Woodpecker and the species most important in formulating my model, the Mexican Jay.

To explain delayed breeding in hot, arid climates, Emlen (1982a) employed a simple model of mine in which breeding success of a bird that survived until the inception of the breeding season was dependent on two factors, F, the probability of obtaining a mate and being able to breed, and L_{max}, the number of offspring that could be reared if a suitable territory were available. Thus,

$$\text{expected reproductive success} = FL_{max}.$$

I use the notation of the original version (Brown, 1978a:139) rather than Emlen's. Noting that models of habitat saturation were concerned with F, Emlen argued that variation in L_{max} provided a different explanation of delayed breeding (as had Orians in effect).

Are the data consistent with this hypothesis, and do they allow rejection of alternative hypotheses? As evidence for an influence of L_{max}, Emlen described a correlation between rainfall and proportion of nonbreeding helpers in a population of bee-eaters. The interpretation of this correlation is complicated by the fact that both F and L_{max} should vary in parallel as a function of rainfall in a species with feeding territories. Therefore, simple correlations of this sort cannot be used as evidence for the relative importance of L_{max} in species subject to yearly variation in number of territories suitable for breeding. Since the White-fronted Bee-eater studied by Emlen does have feeding territories (Hegner et al., 1982) and since Emlen (1982a) did not discuss this potential complication, the results were subject to more than one interpretation; however, whether through F or through L_{max} there seems to be little doubt about the importance of rainfall as a determinant of breeding status. I suggest that a more complicated mathematical model is needed to represent the interaction of these variables.

It seems to me that in Acorn Woodpeckers and White-fronted Bee-eaters we may be dealing with the *same phenomenon*, namely, spatio-temporal variation in territory suitability, which determines simultaneously (1) the number of *suitable* territories, T_S, and, consequently, the probability of getting one, and (2) the chances of success on a given territory in a given year. Both Acorn Woodpeckers and White-fronted Bee-eaters live in variable environments which affect F through temporal variation in L_{max}, as well as through spatial variation in L_{max}. Emlen's study included both spatial and temporal variation but emphasized the latter. Stacey's (1979a) study emphasized spatial (i.e., geographic) variation, but temporal variation in suitability of territories for Acorn Woodpeckers is well known and documented by Trail (1980). Emlen (1984) himself treated them similarly in a later review.

The role of habitat saturation in stable habitats (if there are any) is widely accepted, but there seems to be some uncertainty about its role in arid regions of variable rainfall. In respect to species for which "environmental unpredictability . . . precludes the attainment of any steady-state population whose density remains at, or accurately tracks, the carrying capacity," Emlen (1982a:34) wrote, "We cannot speak in terms of habitat saturation, or cite a shortage of nest sites or territory openings as a driving factor in the evolution of helping." Territorial behavior, however, is characteristic of many birds that breed in such environments, such as the Mexican Jay, Acorn Woodpecker, Grey-crowned Babbler, White-fronted Bee-eater, Stripe-backed Wren, and White-browed Sparrow-weaver. In some cases the number of *suitable* territories has been shown to vary with the weather. In other words, *habitat saturation occurs at different levels in different years*, depending on conditions, thus giving rise to the observed correlations between numbers of helpers and environmental conditions (as explained further below). Emlen's assertion that habitat saturation plays no role in delaying breeding in such species is inconsistent with this evidence.

Colonial versus Territorial Social Systems. The above distinction between stable and unstable environments has proven not to be heuristic since the exclusionary effects of territoriality (habitat saturation) affect F in unstable environments as well as in stable ones. I propose instead that a more useful distinction is between *territorial* and *nonterritorial* species, where territory is understood to include a defended foraging area. Most colonial species are nonterritorial, but the White-fronted Bee-eater is an exception and holds feeding territories away from the colony (Hegner et al., 1982).

The advantage of this distinction is that *habitat saturation resulting from territoriality can be reliably excluded in the nonterritorial species.* This means that F is unimportant (provided nest sites are not limiting and other density-dependent factors are not acting) and L_{max} becomes important, as envisioned by Orians et al. (1977b) and Emlen (1982a). Thus I agree with Orians and Emlen that L_{max} represents a neglected cause of delayed breeding, but I propose that its importance is most clearly seen in the contrast between nonterritorial, colonial species and territorial, noncolonial ones, rather than in the contrast between stable and unstable environments. An advantage of the first dichotomy is that it is qualitative and clear cut. A disadvantage of the second view is that stability varies on a gradient, and does not provide a qualitative dichotomy free from semantic problems.

Stability of the Breeding Population. In the older literature on territoriality it was argued that stability in the number of breeders in a habitat was caused by territoriality (e.g., Kluyver and Tinbergen, 1953; Glas, 1960; reviewed in Brown, 1969a); and Lack (1966), unlike most authors, used stability as his principal criterion for acceptance of territorial behavior as a factor limiting breeding population density. A degree of stability in communally breeding birds is commonly observed. In the Scrub Jay the density of breeders was not constant, but it was not correlated with the total number of jays (Woolfenden and Fitzpatrick, 1984:47). Territory size was in fact negatively correlated with breeder density; so territories were compressible. Nevertheless, the number of breeders on the study area was impressively stable *relative* to the number of Scrub Jays there.

This stability probably depends on the absorption of nonbreeders into pre-existing groups, resulting in large fluctuations in mean unit size in contrast to a *relatively stable number of units or territories*. This is illustrated in Figure 17.1 for the Mexican Jay. Figures 12.2 and 12.3 show the age structure and year-class structure of this population, revealing considerable yearly variation in recruitment of yearlings into the population. This variation resulted in considerable variability in the numbers and proportions of breeders and nonbreeders (Figure 5.7). Thus in Scrub Jays and Mexican Jays stability is relative rather than absolute. Stability in number of breeders is probably greater in the Scrub than Mexican Jay because the number of breeders per territory is a constant in Scrub Jays but varies in Mexican Jays.

Insufficient "Labor Force"

Life-history theorists are accustomed to the concept of the cost of reproduction, the idea that reproductive effort now may reduce future fecundity (Williams, 1966b). In communal birds, however, the net effects of reproduction may, under the right circumstances, be positive for the future rather than negative, at least for established breeders (Brown and Brown, 1981b). This is because the young produced by the parents typically serve a year or more as helpers for their parents, thereby reducing the cost of reproduction for the parents and raising their reproductive success (details in Chapter 11). The value of a territory then is determined not only by its resources but also by its "labor force," the value of which can be estimated from the number of helpers. By the same token, it may be more profitable for an individual to delay breeding until it is possible to obtain a territory together with helpers than to breed sooner without helpers. This would be especially likely if survival as a helper were higher than as a novice breeder without helpers.

Such a case has been described for the Stripe-backed Wren by Wiley and Rabenold (1984). In this species, territories with suitable habitat appear to be available (Rabenold, 1984, 1985); but territories with a suitable "labor force" are in short supply. The choice for a one- or two-year old male is to breed early with too few helpers, low success ($m_x = .40$ young per year), and higher mortality ($s = .64$); or to delay breeding to a later age when a suitable "labor force" becomes available and to act as a helper in the meantime, in which case survival would be higher during the delay ($s = .78$). Breeding success after the delay could be higher ($m = 2.40$). In terms of expected lifetime reproductive success in this example the benefits of waiting outweigh the costs at least to age 6 years for males but only to age 4 for females, who have lower survival as helpers than males. Males in this species typically delay breeding until age 2–4, but females only until age 1–3. The data are consistent with the above interpretation.

Sex Ratio

It is often assumed for the sake of generating an hypothesis that delayed breeding is adaptive and that its benefits outweigh its costs. When breeding is delayed in males because all females of breeding age are taken, delayed breeding is not necessarily adaptive. Such males may be simply the losers in a competition for a limited number of females. Their remaining options in species with only one breeding female per unit are (1) to roam in search of a mate indefinitely, (2) to act as a helper in a subordinate, nonbreeding capacity in a stable social unit, and (3) to share a female with another male (polyandry; see Chapter 9).

How often does a shortage of females arise? Many species of passerine birds show an apparent excess of males among adults (reviewed in Brown, 1969a), and this is likely to be true of many communally breeding species. In some of these, such as the Mexican Jay, an excess of males occurs among birds of breeding age; but this is unlikely to be the cause of delayed breeding, since both sexes delay.

Only when the helpers are almost exclusively male does this explanation of delayed breeding become attractive, and even then other causes may be more important (as in the European Bee-eater; see Chapter 13). An excess of males occurs in certain bee-eaters (Fry, 1972), the Superb Blue Wren (Rowley, 1965a), the Splendid Wren (Rowley, 1981a), the Pied Kingfisher (Douthwaite, 1973, 1978; Reyer, 1980a), two species of nuthatch (Norris, 1958), Red-cockaded Woodpecker (Gowaty and Lennartz, 1985), and other species listed in Table 2.2. In the species listed above there is no good evidence of mate sharing yet. Although a definitive answer awaits genetic studies, there is reason to believe that mate sharing is uncommon to rare in these species. Therefore, a shortage of females remains a plausible hypothesis for nonadaptive delayed breeding in these species.

Why are females scarce relative to males as adults? Studies on this question in avian communal breeding systems are rare, but present evidence suggests a tantalizing variety of causes. The evidence available points toward a higher mortality rate in females than males long after fledging.

Events before fledging may be important in the Red-cockaded Woodpecker. In Scrub Jays females experience higher mortality than males during the ages when dispersal occurs, but not as yearlings or as breeding adults (Woolfenden and Fitzpatrick, 1984). Pied Kingfishers had an even sex ratio at fledging but 1.8 males per female among adults (Reyer, 1980a). In Mexican Jays the sex ratio begins to diverge from 1:1 when breeding begins and continues with advancing age; the sex ratio among old breeders (10+ yr) approaches 100% male. Only females incubate and brood the young in Mexican Jays, a role that exposes them to greater risk than males. In the Splendid Wren too the female does all the incubation and suffers a higher mortality than the male (female: $s = 0.43$; male: $s = 0.71$; Rowley, 1981a). Female Florida Scrub Jays, however, have the same annual survival as males (0.82).

In Stripe-backed Wrens the survival rate of males is higher than females only when in the helper stage and especially in the second year—the modal dispersal age (Wiley and Rabenold, 1984). The higher mortality of nonbreeding females (Table 3.2) was attributed to the risks of dispersal, which are avoided by males typically by not leaving their natal unit. Breeding adults of both sexes had relatively high annual rates of survival (Table 3.2).

In the Red-cockaded Woodpecker events *before* fledging may affect the sex ratio. Gowaty and Lennartz (1985) found that 65% of nestlings were male in pairs without helpers, as opposed to 54% in pairs with helpers. Although this difference was not quite significant, the fact that helpers in this species are mainly males suggests the possibility that pairs lacking helpers would favor males among their nestlings in order to have a helper the next year. Helpers were associated with increased fledging success in this species, but only in females not known to have bred before on the study area. Presumably most of these were first-time breeders although this is questionable. Therefore, not all pairs without helpers need helpers. Interestingly, broods of new females contained significantly more males than those of females with a longer history on the study area. One way to interpret this is that females needing helpers were producing sex ratios most likely to result in helpers the following year (Emlen et al., 1986); however, other factors must be considered. (1) If male nestlings are less likely to starve in times of food shortage than females, it is only among broods needing help (those of relatively incompetent females) that such differential starvation should occur. (2) A female who reared young successfully when naive might not need helpers the following year because she would then be experienced and competent. Thus the value to her of favoring males in order to have a helper would be diminished, if not abolished. (3) With a population sex ratio of 99 males to 69 females among nestlings, females would be more valuable with respect to their chances of reproduction. (4) By the reasoning explained in Chapter 15 it is cheaper for parents to obtain helpers by variance utilization than by variance enhancement. Furthermore, a helper would have to provide a large benefit to a parent to justify to the parent the sacrifice of a year's reproduction by one of its young (male or female).

Dangerous Dispersal

When dispersal entails a higher risk of mortality than staying home, it might be better to stay home awaiting a chance to breed nearby rather than to undergo a greater risk of death for a territory farther away. With this mechanism the *habitat need not be saturated.* Vacancies might exist simply because birds were reluctant to look for them beyond the borders of contiguous territories. Little attention has been given to this hypothesis, and no examples of it have been described.

Altruism and Parental Manipulation

In at least three singular-breeding species the male breeder has been observed to be dominant to all others in his social unit (Scrub Jay, Woolfenden and Fitzpatrick, 1977; Grey-crowned Babbler, King, 1980; Striped-backed Wren, Rabenold, 1985). Dominance hierarchies have also been demonstrated for Mexican Jays in which old males often dominate their groups (Barkan et al., 1986). It has been generally assumed that this dominance prevents the nonbreeding helpers from breeding without going so far as to drive them out. The dominant gains helpers, hence greater reproductive success, in these examples, by tolerating the presence of the nonbreeders. This situation can be represented by the model of Vehrencamp (1979), which is shown in Figure 5.8.

The model assumes that individuals are better off on average in a middle-sized group than in a larger or smaller group. Vehrencamp envisioned the breeders as dominant and able to increase their own fitness, W_a, by lowering the fitness of the subordinates, W_w. The subordinates were assumed to accept this reduction in their fitness so long as the result was not lower than what they could achieve by leaving, W_1, as shown in Figure 5.8. If the subordinates were related to the dominants, the gain by the dominants would be an indirect gain for the subordinates, in which case the subordinates could make up for their net loss in direct fitness below W_1 by their gain in indirect fitness. This is a case of altruism in theory, although

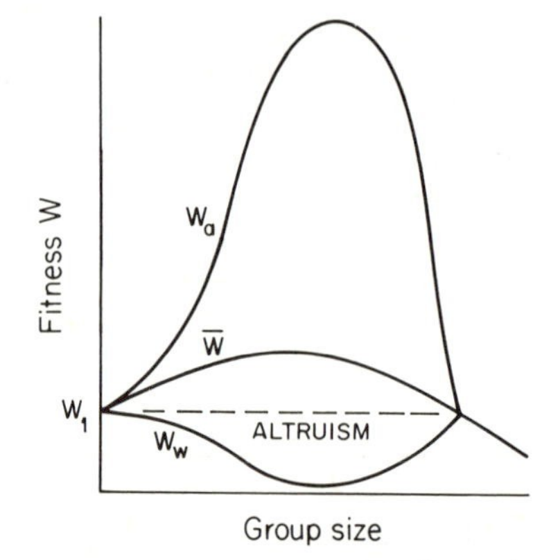

Figure 5.8 Direct fitness of dominants and subordinates in groups containing individuals related by $r = 1/4$. $\bar{W}$ is the average direct fitness of a group member, W_a is the direct fitness of the dominant, and W_w is the fitness of subordinates. Individuals whose fitness is below W_1 are Hamiltonian altruists made possible by indirect kin selection. (After Vehrencamp, 1980. The word "altruism" has been added.)

this was not emphasized by Vehrencamp. If the subordinate helpers were offspring of the dominants, this case would also be parental manipulation (Alexander, 1974) by suppression, which is the view that Vehrencamp prefers. There are alternatives to the parental-suppression model which fit the alternatives to the parental-suppression model which fit the facts better. These are discussed within the conceptual framework of variance utilization and variance enhancement in Chapter 15.

Altruism, whether caused by parental manipulation or not, is one hypothesis for delayed breeding. Personally, I am reluctant to invoke altruism as a cause for delayed breeding when so many plausible alternatives remain, and I have never done so. "The existence of delayed breeding is . . . easily explained without kin selection. Presumably, breeding is delayed in avian communal breeding systems for the same reasons as in other birds . . ." (Brown 1978a:135). Although some researchers have implied that models invoking kin selection for helping necessarily imply that breeding is delayed through indirect selection (Woolfenden and Fitzpatrick, 1978, 1984; Koenig and Pitelka, 1981), this is not true. In fact, my models invoking indirect selection (then called kin selection) to explain helping did *not* explain delayed breeding through indirect kin selection (see Appendix). Delayed breeding was explained in these papers by habitat saturation and shortage of females (Brown, 1969a, 1974, 1978a, etc.). This is not to say, however, that either Woolfenden and Fitzpatrick or I have eliminated altruism from consideration as an explanation of delayed breeding. It might still prove to be important in some cases.

Some authors in rejecting altruism have treated kin selection and ecological factors as a dichotomy in explanations of helping (e.g., Koenig and Pitelka, 1981:273). This argument for rejecting altruism is not acceptable. While there may be a dichotomy between genetic and nongenetic factors, in practice these factors are so intricately interrelated that they cannot easily be separated. For example, a correlation between frequency of helpers in a population and rainfall or some other environmental factor does not suffice to reject relatedness as a factor. To see why this is so, consider Hamilton's rule. The decision to be altruistic depends in part on relatedness, but it also depends on ecological factors, namely, the direct costs and indirect benefits to the potential altruist. In years of poor rainfall, we should expect that the criterion for altruism would be satisfied for more individuals because they give up less and gain more in inclusive fitness by choosing altruism. Consequently, on the basis of this hypothesis we would predict more altruism in years of low rainfall. Emlen (1982a) reported more bee-eater helpers in years of low rainfall, a result that is consistent with an important role for both genetic and nongenetic factors. In general, although correlations of helping with environmental factors do show the relevance of ecological factors, they in no way weaken the case for altruism. To test the importance of indirect selection other approaches are needed (see Chapters 11–13).

I would like to suggest that it is debatable whether Vehrencamp's model of parent-offspring relationships is better viewed as altruism achieved through indirect selection or as parental manipulation, which she prefers. An offspring may be said to act altruistically if it gives up an available higher direct fitness for a lower direct fitness when helping. This is true for all points below W_1 in her model. Therefore, this is altruism; but is it also parental manipulation? When applied to social insects, parental manipulation implies that the offspring cannot breed sexually anywhere and that its ability to breed has been depressed by the parent, compared to the unmanipulated condition. In Vehrencamp's model, however, offspring can breed anywhere except in their natal territory, provided one assumes that W_1 represents a threshold for breeding (at or above W_1 breeding is possible outside the territory). This assumption represents the situation in territorial birds generally, whether they have nonbreeding helpers or not. For example, it describes the Scrub Jay in California, which lacks helpers, just as well as the Scrub Jay in Florida, which has them. The ability of the offspring to breed outside their natal territory may be reduced in communal birds, but it has not been depressed by the parents compared to the unmanipulated condition. In this perspective there is actually a lessening of parental intolerance in making the step toward helping, not an increase. Since in this model the young could breed simply by leaving the parents, their decision to stay is not imposed by the parents but is a straightforward case of altruism and not a typical case of parental manipulation.

Falsifiability

The astute reader may now be somewhat exasperated at the lack of rigorous proof for even the most popular hypotheses listed in Table 5.1. This situation is partly a consequence of the failure of previous authors (myself included) to enumerate fully and to explore the possible alternative working hypotheses that might explain delayed breeding ecologically. Until a hypothesis is clearly stated it is likely to be overlooked and unlikely to be rejected. Indeed, the history of the skill hypothesis is a good example; in none of the more recent studies on communal breeding has the skill hypothesis been even considered, let alone rejected.

It is worth considering how each of the hypotheses listed in Table 5.1 could be falsified (rejected); however, I hope that a few examples will suffice. For brevity I consider now only the first two listed, namely (1) the skill model and (2) the habitat saturation model. Of course, we may immediately reject habitat saturation for the small number of nonterritorial species; but for group-territorial species, which are the vast majority, we must ask how the two hypotheses can be either falsified or evaluated relative to each other. The correlational evidence cited above in support of habitat saturation (correlations between percentage of nonbreeders and an environmental factor) is also consistent with the skill model. Worsening of the environment lowers both Q in the skill model and T_s (through effects on L_{max}) in the habitat saturation model, thus allowing both models to predict more nonbreeders as conditions worsen.

Fortunately, the two hypotheses may be distinguished by experiment. If breeders are removed from a saturated territory, the saturation hypothesis allows us to predict that some nonbreeders would breed as a result of opened vacancies. The skill hypothesis, in contrast, does not allow this prediction; removal of other individuals would not make a bird a better forager. Such removal experiments have been attempted on a small scale (Stacey, 1979a), but a fully satisfactory field test remains to be done.

A second type of test can be easily performed. Foraging success should be a function of age if the skill hypothesis is correct. This can be tested by observations of foraging success of breeding and nonbreeding birds of known age. Such tests have been carried out for noncommunal birds (Fig. 5.1) and might be feasible for some communal species.

Failed Breeders

Although failure of breeding cannot be regarded as a cause of delayed breeding in the usual sense, a breeding failure can force a breeder to delay its next attempt until the next breeding season. Renesting may occur if the failure occurs early enough; but if the failure comes too late for renesting, the failed breeder may find itself confronted with young of close kin and no possibility of having young of its own. Being physiologically prepared to feed young, it may then direct its feeding efforts toward young not its own. This scenario is known in Mexican Jays, where it is common (unpublished data), and in some colonial species, such as White-fronted and European Bee-eaters (Emlen, Avery, pers. comm.), as well as in other species. The bee-eater species are discussed further in Chapter 13.

Age of Role Shift from Nonbreeder to Breeder

Variation in α among Species. In the comparative perspective, the age of role shift from nonbreeding to breeding, α, is a critical variable. Because nonbreeders in communal birds tend to become helpers, the number of helpers in many species is determined primarily by the number of nonbreeders. The latter is determined, at least in a comparative perspective, by the age of transition to breeding age, α. For example, in the genus *Aphelocoma*, α ranges from one year in California populations of *A. coerulescens* to three or four in Arizona populations of *A. ultramarina*, with intermediate values in other populations (Table 5.2). Unit sizes vary accordingly, being larger with higher alphas. Similar variation in α also occurs in other genera of jays (Fig. 6.2) and wrens (Selander, 1964).

Delays of one or two years are now known to be common among neotropical jays (Hardy, 1976; Raitt and Hardy, 1976, 1979; Lawton and Guindon, 1981), Australian and Asiatic babblers (King, 1980; Brown and Brown, 1981b; Gaston, 1978a,b; Zahavi, 1974), neotropical wrens (Selander, 1964; Wiley and Rabenold, 1984; Rabenold, 1985), African shrikes (Grimes, 1980), weaver finches (Lewis, 1981; Collias and Collias, 1978a), Australian mud-nest builders (Baldwin, 1974, 1975; Rowley, 1977), ground hornbills (Kemp and Kemp, 1980), and Galapagos mockingbirds (Grant and Grant, 1979; Kinnaird and Grant, 1982).

Annual survival rates in species with long-delayed breeding tend to be rather high, as in the Yellow-billed Shrike, Stripe-backed Wren, Grey-crowned Babbler, Jungle Babbler, Scrub Jay, and Mexican Jay (survival

Table 5.2

Comparison of four populations of *Aphelocoma* jays in two species, the Mexican Jay and the Scrub Jay. α = age at first breeding. (From Brown, 1985).

Population	Approximate α	Typical Unit Sizes	% Breeding	Pairs per Territory	Youth Signals	Dispersal Mechanism
A.c. Calif.*	1–2 yr	2	90	1	–	Long and short movements
A.c. Fla.**	2–3 yr	2–5	70	1	–	Budding; short movements
A.u. Texas***	2–? yr	2–4+	50	1	–	?
A.u. Az.*	3–4 yr	5–20	18–59	2–4	+	Inheritance; short movements

* Brown, 1963a, 1985
** Woolfenden, 1975
*** Ligon and Husar, 1974; Strahl and Brown, unpublished data.

rates and references in Table 3.2), all of which normally delay breeding until age two or three or more years. In these species survival rates run from 0.68 to 0.90. Species in which females commonly breed at age one include the Pied Kingfisher, New Mexico Acorn Woodpecker, Superb Blue Wren, and Splendid Wren. In these species annual survival is lower, running from 0.43 to 0.62 (Table 3.2). As a result of the combination of high survival and nondispersal of nonbreeders, social units of such species tend to accumulate a corps of potential helpers that are usually offspring of one or both parents. Group size tends to be lower in those species listed as having survival rates in the lower range.

Individual Variation in α. In addition to variation among species in α and in consequent behavior of individuals, there is conspicuous variation within species, as already suggested for babblers in Figure 5.2. Table 5.3 reveals some of the variation among individual Mexican Jays. A few rare individuals have bred successfully at age two (with an older mate), but the fraction of birds breeding at a given age does not reach an asymptote until age four, at about 66% (A). Breeding success (at least one young has reached banding age) does not reach an asymptote until age five (B). Even for the best breeders, breeding success varies greatly from year to year. The fraction of older birds for whom we have no evidence of nesting (perennial helpers) is quite large in Mexican Jays. It decreases with age (C) but does not have an asymptote of zero. Instead, even by age three, over half have never been associated with a nest of their own; and at older ages around 10% seem to be perennial nonbreeding helpers. In terms of successful nesting, even at age five over 50% of the individual jays have never fledged

Table 5.3

Developmental changes in breeding status in an Arizona population of the Mexican Jay. Data are for all birds of precisely known age (banded as nestlings or yearlings), 1969–81. As the differences between the sexes are small, the sexes are combined here. (From Brown, 1985.)

	Age in Years								
	1	2	3	4	5	6	7	8	9–12
A.	0	23	42	66	60	62	70	71	100
B.	0	.07	.23	.55	1.18	.72	.65	1.07	1.43
N_1.	141	102	79	64	40	29	20	14	7
C.	100	79	51	22	10	12	13	11	
D.	100	97	88	72	54	44	43	39	
N_2.	145	108	82	64	41	32	23	18	

A. Percentage of the number of birds alive at the stated age (N_1) for whom a nesting attempt was observed at that age. Data are for successful and unsuccessful nests. A small fraction of nest attempts went undetected; the actual percentage should, therefore, be slightly larger.

B. Age-specific reproductive success. Nestlings of banding age (about 14 days) produced per bird.

N_1 Sample sizes for *A* and *B*.

C. Percentage of the birds that lived at least to the stated age who had no nesting record by that age, i.e., before ever breeding.

D. Percentage of the birds that lived at least to the stated age who had never fledged young.

N_2 Samples sizes for *C* and *D*.

young. Only by age six or seven is a bird likely to have fledged at least one young.

Reliance on dominance as an exclusive key to future breeding by an individual appears tentatively to be unreliable in Mexican Jays. One individual that bred at age two was a male who was not the dominant male in his unit. The potential causes of such variation in α, including dominance, are currently being studied.

Conclusions

Three major phenomena of avian communal breeding systems are interrelated but require separate analyses: (1) delayed breeding, (2) reduced or delayed dispersal, and (3) helping or alloparenting. In the case of nonbreeding helpers, the first two factors are permissive and conducive to the third (Brown, 1969a, 1974).

In this chapter we have considered several hypotheses to explain delayed breeding, before going on to consider dispersal and alloparenting. In animals that provision their young, foraging skill can in theory be a limiting factor. This skill depends upon age, learning, food abundance, and their interaction. These factors are formulated into an alternative hypothesis for the explanation of delayed breeding in animals generally. The potential importance of such a hypothesis has been largely overlooked.

The most popular hypothesis is based on the exclusionary action of territorial behavior, which prevents some individuals from gaining access to space (and its contained resources) needed for breeding, the so-called habitat saturation hypothesis. Various other hypotheses offer insights in terms of sex ratio, dangerous dispersal, labor force, altruism, and parental prevention.

In this chapter, the treatment of the role of the environment in delaying breeding differs somewhat from that of previous authors. Orians et al. (1977b) and Emlen (1982a) pointed out that low food levels can create a situation in which some individuals may find it profitable to breed while others do not, the latter becoming potential helpers. Thus they recognized the environment as a factor affecting the individual's decision whether or not to breed. I take this idea a little further by pointing out that fluctuations in environmental suitability for breeding cause a situation conducive to helping exclusively through *interaction with a behavioral variable that sorts individuals by age*, so that the younger individuals choose not to breed while the older ones choose to breed. Variability in the environment may modulate the porportion of the population that breeds, by interacting with other variables such as foraging skill, the exclusionary effects of territorial behavior, the risks of dispersal, and other factors.

6 Reduced Dispersal Sets the Stage for Helping

Many communally breeding species are characterized by reduced dispersal, as well as delayed breeding and alloparental care. Dispersal is a critical process in the life of a bird because it often occurs in the transition from immaturity to breeding status. In order to place these events in a broader perspective, we consider briefly not only the movements of dispersal, but also the ecological processes that select for different types of strategies used to obtain breeding status. This sociobiological view supplements the classical approach, which emphasizes the movements themselves and the genetic population structures that result from them.

Population Consequences of Dispersal Strategies

In the spectrum of social systems, from asocial to highly social breeding, the disposition of the nonbreeders is important. The nonbreeders may be divided for simplicity into two categories, those that leave and lack territories, termed *floaters*, and those that stay in their natal unit, here assumed to be helpers. The fraction of the population existing in each of these categories, breeders, floaters, and helpers, is shown for a range of species in Figure 6.1.

In colonizing species (thought to be selected for high fecundity and much dispersal) a surplus of nonbreeders is rare (Fig. 6.1A). In noncolonizers (low fecundity, less dispersal) a surplus is more likely to accumulate. Estimates of the proportion of surplus birds in noncommunal birds are difficult and speculative, but Smith (1978:579) estimated that 50% of her marked population of Rufous-collared Sparrows (*Zonotrichia capensis*) were floaters; and removal experiments suggest that the proportion of floaters may be substantial in many species (Brown, 1969a; Klomp, 1972; see also Rappole et al., 1977), as in Figure 6.1B. Such conditions in a sedentary population create a situation in which it may be more favorable for the parents of a young bird to let it remain with them rather than to cause it to leave "before its time."

In most colonial birds there is no feeding territory, and the young may or may not associate with their parents. These are not necessarily colonizing species in the sense used above. Here the word colony refers to the dispersion pattern created by the clumping of nests. In the Pinyon Jay most young remain in the colony home range but do not aid their parents (Balda and Bateman, 1971), a situation not far from B in Figure 6.1. In the Pied Kingfisher 12% of one population and 31% of another was composed of helpers, and probably even more nonbreeding birds existed in the population (Reyer, 1980a). A larger percentage of the population may occur as helpers in territorial species (Fig. 6.1D). For the African Yellow-billed Shrike (Grimes, 1980), about 80% of the studied population was composed of nonbreeding helpers. In the Mexican Jay the percentage of nonbreeding helpers in the population over a nine-year period varied from 7 to 55 (Figure 5.7; Brown, 1986). The extreme shown in F of Figure 6.1 is not found in birds, but it corresponds roughly to the Naked Mole-rat (*Heterocephalus glaber*, Jarvis, 1981) and to many eusocial insects.

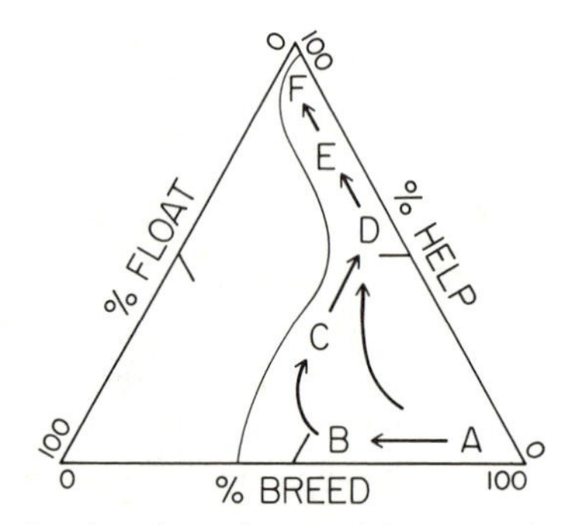

Figure 6.1 Population fractions in various social systems involving helpers, floaters, and breeders. Axes are percentages. At each point in the triangle the percentages sum to 100. A. High-dispersal species with high fecundity and no floaters or helpers. B. Territorial species with floaters but no helpers. C. Communal-colonial birds and other social systems with mixtures of breeders, helpers, and floaters. D. and E. Territorial communal birds with variable numbers of helpers but few if any floaters. F. Eusocial species (Naked Mole Rat): many helpers, few breeders or floaters. Arrows suggest trends. (From Brown, 1985.)

The population genetic consequences of the shifts outlined in Figure 6.1 are examined in Chapter 12.

Waiting for Alpha: General Advantages of Staying Home

The prerequisite for reproductive status is survival in the nonreproductive period. Given the delayed onset of breeding in many communal breeders, *where would a nonbreeder find a better place to survive until breeding age than with its parents in a home range already familiar* to it and possibly improved by it through storage of food or dormitory construction? If the home range is temporarily unsuitable, nonbreeders may simply accompany the breeders in short wanderings to places probably known to the older unit members, as in White-winged Choughs (Rowley, 1977), Acorn Woodpeckers (MacRoberts and MacRoberts, 1976), and Superb Blue Wrens (Rowley, 1965a), and return when conditions improve. Even in some species in which immature yearlings do not help to feed nestlings, as in the migratory Blue-snow Geese (*Chen coerulescens*) (Cooke et al., 1975), and the nonmigratory Australian Magpie (Robinson, 1956), the young commonly share the same home range as their parents through their first year. Obviously, it is in the parents' interest for their young to survive, and if any risks must be taken to ensure the survival of their young, the parents and siblings are the individuals that should be most likely to take them. Living in a group when not breeding is a common strategy for survival among vertebrates (Pulliam and Caraco, 1984); both breeders and nonbreeders should benefit from flocking during the nonbreeding season and perhaps in some circumstances also in the breeding season. In brief, it is not difficult to suggest advantages to both parent and offspring of continued association between them until the offspring can breed.

Another advantage to the individual nonbreeder of remaining in its natal unit may lie in *easier access to breeding status.* Since the first extensive color-banding study on a communal species by Rowley (1965a), it has been evident that individuals can use nonbreeding helper status as a stepping stone to breeding status either in their own or a neighboring territory. This transition has been observed in nearly every territorial species that has been studied long enough to detect it (Superb Blue Wren, Rowley, 1965a; Scrub Jay, Woolfenden, 1975, 1976; Mexican Jay, Brown and Brown, 1984; Green Woodhoopoe, Ligon and Ligon, 1979; Yellow-billed Shrike, Grimes, 1980; Jungle Babbler, Gaston, 1978b; White-winged Chough, Rowley, 1977; Acorn Woodpecker, MacRoberts and MacRoberts, 1976; Stripe-backed Wren, Rabenold, 1985; and also in one colonial species, the Red-throated Bee-eater, Dyer and Fry, 1980).

The idea that remaining in the natal territory is an individual strategy for attaining breeding status under severely competitive conditions has been suggested several times with slight variations (Selander, 1964; Brown, 1969a, 1974; Woolfenden, 1975, 1976; Koenig and Pitelka, 1981).

In addition to these advantages based on future direct fitness, individuals that maintain a mutualistic (cooperative) or altruistic relationship with close kin during their prereproductive period would benefit their indirect components of inclusive fitness.

The Predispersal Period: Prolonged Immaturity

Skutch (1935) wrote:

> The Brown Jays possessed one peculiarity which convinced me would make them particularly interesting to watch. Their bills, feet and bare rings surrounding the eyes were not all of the same color, as is the case with most birds, but were so variously marked with yellow and black that it seemed that I should be able to recognize individuals—and the difficulty of recognizing individuals, among other species, is one of the chief handicaps which face the serious bird-watcher. I devoted most attention to the bills, which proved to afford the best recognition marks. Some of the Jays, at the beginning of the breeding season, had bills which were entirely bright yellow; the bills of others, perhaps the majority, were tipped or streaked with black, but hardly two were pied in exactly the same manner. Still other bills were uniformly black. Later I learned that the nestlings' bills, feet and orbital rings are uniformly yellow, and that they turn black with age in an irregular fashion, apparently taking two years or more to become entirely black.

The retention of traits that are characteristic of a brief and early developmental stage into older ages has received many names and much discussion (Gould, 1977; Pierce and Smith, 1978). Both neoteny (Brown, 1963a) and paedogenesis (Geist, 1977) have been used to describe this phenomenon in communal birds, but since the use of such terms seems to cause more confusion than clarity, I will not assign them here. This trend can be identified by comparison among species. It is exemplified in the genus *Aphelocoma*, which retains the nestling bill coloration into the second year of life in the most social population (Arizona Mexican Jay) while losing it at the time of fledging in less social forms (Table 5.2). Prolonged immaturity in conjunction with characteristic visual signals is strikingly illustrated in the *Cissilopha* jays (Fig. 6.2) and babblers (Fig. 5.3). The same concordance of plumage or soft parts and breeding status is also common in noncommunal breeders, such as the species of gulls in Figure 5.1.

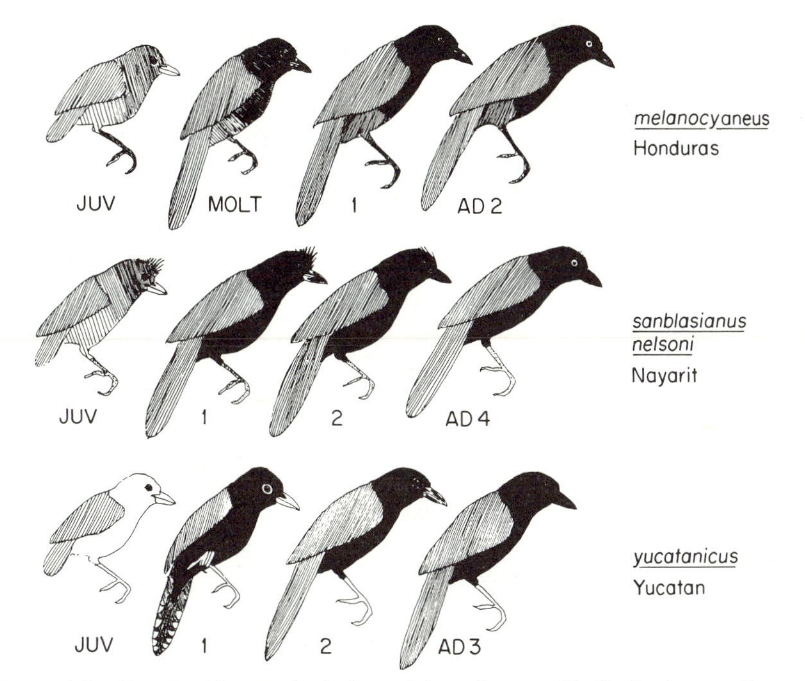

Figure 6.2 Signals of immaturity in jays of the subgenus *Cissilopha* (genus *Cyanocorax*). Each row shows the ontogenetic sequence of appearances (plumage and soft parts) in one species, beginning with the juvenile plumage (JUV) at left and ending with the full adult condition (AD) at right. The numbers show representative ages for the intermediate appearances and, for the adult, the age at which the full adult appearance is normally first acquired. A light bill and, in some species, a dark iris characterize the pre-adult appearance. (After Hardy, 1973; from Brown, 1985.)

In communal breeders this prolongation of juvenile characteristics probably involves also a prolonged dependence of the young on their natal social unit. Mexican Jays can sometimes be seen begging for food and being fed up to 12 months after hatching, although this is unusual. They are probably dependent in other ways too for a much longer period. The young of communal species profit from the vigilance of the unit, from the resources of the territory that is defended by the unit, from the nests and body heat of the group if they roost together, and in other ways. This association may be interpreted as extended parental care merging into parental facilitation (Chapter 7), which presumably benefits both parents and offspring.

A conspicuous aspect of this prolonged dependency is an extended period of immaturity. Among communal species that defer breeding we can distinguish both obligate and facultative immaturity. For example, in the Grey-crowned Babbler (Fig. 5.3) the gonads are tiny and probably nonfunctional in the first year in both sexes, intermediate in size in the second year, and reach full size and function by the third year. Obligate immaturity corresponds to the period of nonfunctional gonads. Facultative immaturity corresponds to the later period during which it is thought that failure to breed is caused by the social or energetic environment, rather than by a physiological limitation of gonadal function, although the two are not entirely separable.

Dispersal Strategies

Why do some surplus birds stay with their parents and others leave? The answer to this question for a given species involves its dispersal strategy. I refer to the set of tactics followed by an immature or maturing individual of a species in gaining and keeping breeding status as its dispersal strategy. The range of options for an individual is illustrated in Figure 6.3. The history of an individual may be described as a sequence of social states that begins at "birth" and tends to move toward "breeder." The optimal path followed by an individual should maximize its inclusive fitness. Since its own reproduction (direct fitness) is potentially the largest component of inclusive fitness for individuals in avian and most mammalian helper systems, an individual should move through the states in such a way as to minimize the risks involved in achieving breeding status and to maximize the payoff on arrival there.

The optimal state sequence involves a tradeoff between the danger of inbreeding depression when breeding in or near the natal unit and the

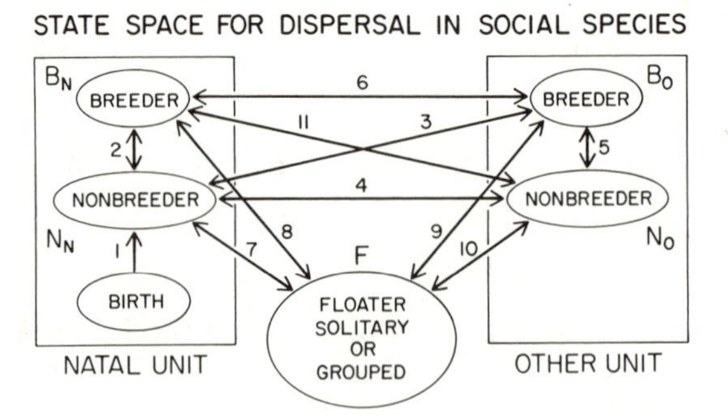

Figure 6.3 State space for dispersal strategies in communal species. In its movements between birth and breeding an individual may move among five principal states. In risk-aversive strategies individuals minimize the number of moves to "breeder." In risk-prone strategies, individuals make more moves, may live as floaters for a while, have greater flexibility to find high breeding payoffs, and suffer more risks. (From Brown, 1983a.)

dangers of dispersal (increased risk of starvation, predation, and environmental hazards). Five possible states are designated in Figure 6.3: nonbreeding in natal unit, N_N; nonbreeding in other unit, N_O; floater outside of any breeding unit, F; breeder in natal unit, B_N; breeder in another unit, B_O. The risks of staying in a state or of moving between states may be expressed in a matrix of transitional survival rates $s_{i,j}$, where i and j designate successive states. The overall survival after a given state sequence would then be the cumulative product of these rates, $l_x = \Pi s_{i,j}$. Additional subscripts for age, sex, group size, etc. could be added.

Reproductive success (m) by definition is achievable only in the breeding states. Inbreeding depression is more likely in B_N than in B_O, but other factors may reduce or outweigh this consideration, causing $m_N > m_O$. This is a strong possibility for one sex if the other sex usually leaves the unit, a common but not universal condition in social animals (Greenwood, 1980). Of course, other factors also influence reproductive success, such as age, experience, environment, unit size, and sharing of a mate.

The combination of risks and reproduction in a sequence of states determines the optimal state sequence for an individual. By analogy with life-history theory we may write an expression for average lifetime reproductive success, LRS, as

$$\text{LRS} = \sum_{y=0} l_x m_x,$$

where y designates the latest in a sequence of states, as designated in Figure 6.3. This simple expression hides a host of complicating factors, but it emphasizes the importance of social states in an individual's life history.

The number of possible sequences is almost infinitely variable, but several types of sequence are prominent among communal birds. Helping behavior is often associated with *reduced dispersal*, usually by avoiding paths through F (floaters in Fig. 6.3). Paths through F are typical of nomadic species and north-temperate-zone species. Nomadism is favored in theory by cyclical food supplies, large clutches, high juvenile survival, low adult survival (Andersson, 1980), and good locomotor capacity. These factors collectively increase an individual's likelihood of being in a position in which the chances are good of being able to find a better situation than its present one. I suggest that a critical variable is the *dispersal benefit* ratio Z_x for a given social state or age x. It is the expectation at status or age x of lifetime reproductive success in other locations, $\text{LRS}_{O,X}$, divided by the expectation associated with staying in the natal unit, $\text{LRS}_{N,x}$, taking into account the risks of survival as well as the payoffs in reproduction, and the penalty of inbreeding depression (present and future):

$$Z_x = \text{LRS}_{O,x}/\text{LRS}_{N,x}.$$

Communal birds as a group are noted for having low rates and distances of dispersal at all ages; so it would seem that in them Z_x is low at all ages *compared to noncommunal birds*, but especially when young. Finding the risks of staying at home to be less than leaving home, they minimize their stay in F, using their natal unit as a base for finding a new unit, and making use of other unit members to aid them in surviving and ultimately achieving breeding status (Brown, 1969a; Rowley, 1981a).

Most communal birds defend group territories. In them dispersal tends to be highly conservative. Some, however, are colonial or even nomadic (e.g., wood swallows, Rowley, 1976). In these species it appears that many or most young of helping age at least try to stay with their parents or sibs (i.e., their natal unit), though they may sometimes fail in this (Reyer, 1980a). In short, it seems that pathways 2, 3, and 4–5 are favored over others in Figure 6.3. Among communal birds there has been a reduction in the number of state changes involved in dispersal and an increased reliance on other unit members, thus reducing overall risk.

Delayed Breeding as a Dispersal Strategy. Many dispersal strategies may be regarded as having originated as devices to obtain breeding status and mates with the aid of other unit members, a thesis partially anticipated in Brown (1969a, 1974) and Woolfenden and Fitzpatrick (1978). The goal of breeding status is achieved differently depending heavily on the demographic characteristics of the species. A spectrum of strategies may be recognized that is arrayed in part according to adult survival rate, namely, of birds one year old or more. Species with high survival tend to have "patient strategies," These are characterized by delayed maturity and prolonged association of young with parents and other relatives. This strategy was first described in comparative perspective for the New World jays (Brown, 1974). In these species the young stay with their parents on their natal territories for one year in some species, two or more years in others; and in some species the offspring may even breed on the natal territory in the company of their parents (Mexican Jay, Brown and Brown, 1981a). Further examples of species with delayed breeding are described in Chapter 5.

Mate Sharing as a Dispersal Strategy. An alternative to delaying breeding is to breed earlier but make some sort of compromise with the classic pair-on-territory situation. One such compromise that is fairly common, especially for males, is the sharing of a mate. A common feature of such species is a relatively *low survival rate.* This penalizes long delays in breeding and puts a premium on "*impatient strategies*," extreme measures to obtain breeding status at a relatively young age. The result is that breeders in mate-sharing species can breed earlier but are denied the relatively large amounts of help that they might receive from their own offspring under delayed breeding. Instead the potential nonbreeding helpers attempt breeding at an earlier age than in longer-lived species.

In *polyandry*, a female may be shared by two or more males, who also share in parental care, with the female at the same nest (Table 2.2 lists examples, which are further discussed in Chapter 9.) In *polygynandry* (see Chapter 9) two or more females may be shared by two or more males in a socially bonded group, as in lions (*Felis leo*; Bygott et al., 1979), Acorn Woodpeckers (Koenig et al., 1984), and Pukeko (Craig, 1980a,b). Polyandry and polygynandry are thought to arise when males that are potential breeders face habitat saturation. Their options then are to wait for a chance at monogamy in the future or to accept a female-sharing status in a cooperative unit in the present. Since it is easier for cooperative males to gain and keep a territory or harem in lions and birds (Brown, 1969a; Woolfenden and Fitzpatrick, 1978; Craig, 1979; Birkhead, 1981), even males that are not totally excluded from breeding may find sharing the most profitable option. As the above examples show, mate sharing is not restricted to early breeders. It does, however, allow breeding at an earlier age than would be possible under strict monogamy. This subject is developed further in Chapter 9.

Joint Nesting as a Dispersal Strategy. Sharing of a nest by two or more females is another alternative to nesting separately that increases the options for a dispersing bird and may reduce or eliminate dispersal. Joint nesting probably makes it impossible for a parent to discriminate reliably its own offspring from those of other unit members. Egg recognition has been shown in the Ostrich (Bertram, 1979), and female Groove-billed Anis discriminate against eggs of other females until their own are laid (Vehrencamp, 1977); but evidence of parental discrimination after hatching is lacking for all joint-nesting species. Apparently the uncertainty of genetic relatedness is compensated for in such cases by the reliability of cooperative parental care.

Joint nesting may be associated with polygynous, polyandrous, polygynandrous, or monogamous mating systems. Polygynous joint nesting with cooperative care of young is found in the Ostrich, Magpie Goose (Frith and Davies, 1961), and occasionally other waterfowl (*Cygnus*, Malcolm, 1971; *Anas*, Duebert, 1968). Joint-nesting Pukeko may be polygynandrous or polygynous (Craig, 1980a,b; Wettin, 1984). Joint nesting with cooperative care is found in the Crotophaginae (Davis, 1942; Vehrencamp, 1978), which are monogamous (Vehrencamp, 1977); Australian White-winged Choughs (Rowley, 1977), which are apparently polygynandrous; Apostle Bird (Rowley, 1976); and occasionally babblers (*Pomatostomus*, Counsilman, 1979; *Turdoides*, Gaston, 1978a,b), which are mainly monogamous. The mating system of the occasional joint-nesting units of Brown Jays is uncertain (Lawton and Lawton, 1985). In addition to these species, the polygynandrous species listed under mate-sharing (above) may also be regarded as joint-nesting. Joint nesting may be adaptive when nest sites are limited, and it obviously reduces the risks of incubation (if any). Risks of nutritional penalties to the young arising from long intervals between feedings are also reduced when feedings are shared (see game theory model in Chapter 14). Attempts of individuals to nest separately within their unit would sacrifice these benefits and risk parasitism by rival females. Joint-nesting groups, therefore, seem to have some built-in safeguards against "cheating" and do not seem to require kinship between breeders to promote cooperation. Nevertheless, in some of these species kinship is close and seems likely to be important. In the White-winged Chough, the offspring remain in the unit a few years before breeding, and in *Crotophaga*, sons tend to remain and breed in their natal unit using the same nest as their parents (B. S. Bowen, pers. comm.) See Chapter 10 for further development of this theme.

Social Dispersal. In some communal birds and mammals dispersal is often in groups of two or more. The individuals in a social propagule (dispersing group) may belong mainly to the same sex (Native Hen, Ridpath, 1972; lions, Bygott et al., 1979; White-browed Sparrow-weaver, Lewis, 1982a; Yellow-billed Shrike, Grimes, 1980; Green Woodhoopoe, Ligon and Ligon, 1978, 1983), but may include both sexes in some species (Acorn Woodpeckers, Koenig and Pitelka, 1981). In these social propagules individuals are often sibs or half-sibs, and this indirect-fitness benefit augments the direct-fitness advantage for some breeders and nonbreeders. Social dispersal may be associated with mate sharing or delayed breeding. For discussion of social dispersal in the Green Woodhoopoe see Chapter 14.

Dispersal Options in Colonial Systems. Most colonial species with helpers also depart from the system of group territoriality with delayed breeding (Reyer, 1980a; Douthwaite, 1973; Fry, 1972, 1977; Emlen, 1978; Balda and Bateman, 1971; Maclean, 1973). Generally, colonial species lack an all-purpose or feeding territory and so are not prevented from breeding for territorial reasons. Delays in breeding are more likely to be due to a shortage of females or to inadequate foraging skills relative to availability of food (Brown, 1985; Orians et al., 1977b). The surplus of nonbreeders is usually small, and it is often confined to males (except in *Merops*). The result is that helping behavior is usually not as conspicuous in colonial species as in group-territorial species. In colonial species with helpers, many or most pairs lack helpers. Even when helpers are present the number of helpers per pair is usually small compared to many group-territorial species.

The above sketch has important exceptions. In bank-nesting species (some bee-eaters and kingfishers) or in lodge nesters, even though breeding is not prevented by habitat saturation, nest digging or building may be energetically expensive. In the White-fronted Bee-eater, Hegner et al. (1982) have discovered that feeding territories are held cooperatively away from the nesting area. In brief, several factors contribute to making helping a viable option for some nonbreeders in colonial species.

Conclusions

In many species of birds individuals delay breeding without remaining with their parents. These species include many colonial seabirds as well as small, migratory landbirds with all-purpose territories. Nondispersal, therefore, is not an inevitable consequence of nonbreeding. In communal birds reduced dispersal is a common strategy. Short-distance dispersal is almost a rule in such species. On the other hand, delayed dispersal is characteristic mainly of singular-breeding, nuclear-family species. Many mate-sharing and/or joint-nesting species do not delay breeding and dispersal conspicuously. Communal birds in general depart from expectations based on north-temperate-zone birds: that is, they usually do not combine early and extensive dispersal as individuals with monogamy. Instead they compromise this "ideal" north-temperate-zone pattern in one or more of several ways which appear to be conducive to helping behavior. These compromises include delayed breeding, sharing of a mate, joint nesting, and social dispersal.

7 Territorial Inheritance as Parental Facilitation

Dispersal typically precedes breeding in vertebrates and is often crucial if the young animal is to obtain a favorable breeding position. In group-territorial species, however, the chances that a single young and naive individual, striking off on its own, can find good territories that are unoccupied are indeed slim. Therefore, as we saw in the preceding chapter, other dispersal strategies are often employed; and these typically involve other individuals of the group from which the disperser originates.

The major way in which the group is involved is the *parent's tolerance* of the young within the parental territory and group after nutritional independence. This behavior by the parents probably facilitates survival of these offspring during their pre-breeding period, and facilitates the eventual acquisition of full breeding status by providing time and a secure base from which the dispersers can gather information about nearby opportunities. This *post-independence involvement of the parents in the dispersal and acquisition of breeding status by their offspring* has been termed *parental facilitation* (Brown and Brown, 1984). In this chapter we first describe what has been learned about various types of parentally facilitated dispersal and then present a simple model that encapsulates the essence of these observations. Finally, we consider the ecological settings that favor such relationships.

Delayed Dispersal in Group-territorial Birds

In 1974 I proposed a theory for the evolution of delayed breeding, reduced dispersal, and helping in jays. With modification the theory is applicable to many other group-territorial species (but not to nest-sharing or non-territorial species). The theory was based upon delayed dispersal as it had been observed in the Superb Blue Wren (Rowley, 1965a) and in Mexican Jays (Brown 1963a, 1970, 1972), and on logical assumptions about habitat quality and habitat saturation. The theory generated fourteen testable predictions, some of which were related to dispersal. Three of them, to be discussed here, were as follows:

Territories differ significantly in productivity. The rank order of flocks in respect to production of young will tend to remain the same from year to year.

Adaptations for retaining good territories in the same genetic lineage will be favored. The territory occupied by a flock will show little change in location, areas, and boundaries over intervals of many years. Larger flocks will show less change than smaller ones.

Genetically the future occupants of a territory are descended from the present occupants, especially for larger flocks. Genetic differences between flocks (and between demes created by local topography) will tend to remain the same from year to year. (Brown, 1974).

Let us now examine the results of more than ten years of subsequent research on dispersal in group-territorial birds with these predictions in mind. We begin by describing several studies on dispersal and end this section by identifying a variety of philopatric dispersal strategies.

Australian Fairy Wrens. Rowley has studied two Fairy Wrens (*Malurus*), the Superb Blue Wren in eastern Australia and the Splendid Wren in western Australia (1965a, 1981a). In these species the males are very brightly colored and the females are drab. Surprisingly perhaps, it is the cryptically colored sex that suffers the highest adult mortality, at least in Splendid Wrens (annual survival: 0.71 for males, 0.43 for females). Consequently, males outnumber females among birds of breeding age.

Males and females can breed when one year old, and females of any age that lack a mate are unusual. Therefore, it appears that females are usually not prevented from breeding. So the factors hypothesized in Chapter 5 to delay breeding do not affect females strongly in these species. Dispersal in females consists of departure from the natal area and searching over a wide area. This depart-and-search strategy is common among noncommunal birds. It may lead to a floating population if territories are limited, but in Superb Blue Wrens a population fraction of floating females is too small and transient to be conspicuous. In terms of the diagram in Figure 6.3 the pathways used in the *depart-and-search* strategy are 1–7–9 mainly and sometimes simply 1–3.

Males face a different set of options in *Malurus*. These are: (1) search for a vacant territory far or near; (2) breed on this territory if a female joins; (3) if not, occupy it as a bachelor-territory; (4) or stay home, with or without alloparenting. Most males in *Malurus* do not move far from home (where hatched), and most successful breeding territories are only one or two territories distant from home. Breeding in a neighboring territory is common for males. In this behavior males differ strikingly from females—a finding that has proved to be common in communal birds (Greenwood, 1980).

These data suggest that males in *Malurus* are adopting mainly two strategies, stay-and-foray and depart-and-search. In the *stay-and-foray* strategy the male stays home but uses it as a relatively safe base from which to explore nearby territories for breeding opportunities. If it finds one, the male takes it; if not, he alloparents. In the *depart-and-search* strategy the male leaves home, searches widely, takes a territory, and advertises for a female. This leads in the first breeding season to either a bachelor situation or monogamy. The stay-and-foray pattern follows the route 1–3 in Figure 6.3, thus avoiding the floater status.

Scrub Jay. Dispersal in this species has been studied by Woolfenden and Fitzpatrick (1978, 1984) in the scrub oak areas of central Florida. Both sexes exhibit a stay-and-foray pattern. The depart-and-search pattern appears to be rare. Both sexes stay home at least a year, but males tend to stay longer before leaving and to go shorter distances. Figure 7.1 shows that many males breed as close as the neighboring territory, while few females do.

A distinctive feature of dispersal in this population is the method of acquiring a territory by *budding* off the home territory. This is done only by some males. If the home territory is large enough, a young male may defend a part of it as a separate territory from that of his parents and attract a female of his own. This strategy requires wresting some space away from a neighboring group to make room for the new territory. If

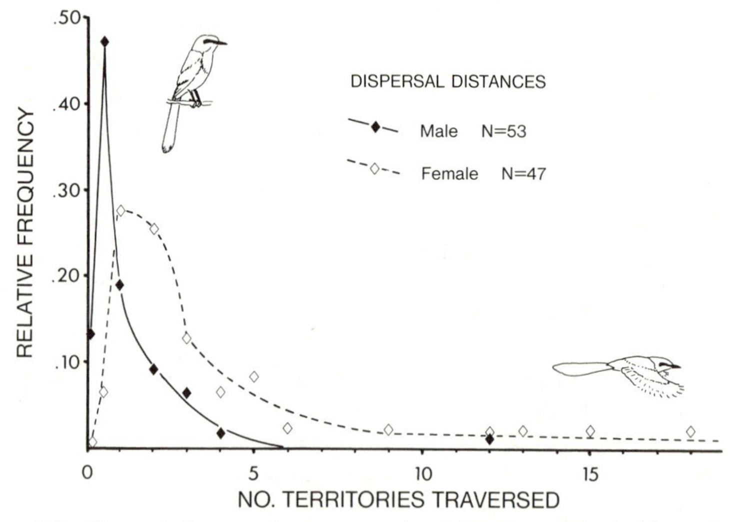

Figure 7.1 Dispersal distances between natal and breeding territories, for male and female Scrub Jays in Florida. Distance is measured in territories (average width of one territory is 339 m). Dispersal distance of one-half indicates partial inheritance of natal territory (budding). (From Glen E. Woolfenden and John W. Fitzpatrick, *The Florida Scrub Jay: Demography of a Cooperative-breeding Bird.* Copyright © 1984 by Princeton University Press. Figures reprinted with permission of Princeton University Press.)

aid from the parents is involved in this process, as it probably is, it may be interpreted as parental facilitation.

Stripe-backed Wren. This species has been studied by Wiley and Rabenold (1984; Rabenold, 1984, 1985) in the savannas of central Venezuela. Both sexes delay breeding at least a year and remain at home on their natal territory. The females then may travel a moderate distance (two or three territories) to find a breeding site. Apparently most females obtain their breeding territories by the stay-and-foray method.

Males follow an even more conservative strategy than in those species considered so far. Most males never leave home to breed; they simply await the demise of their father or other male breeder. This *never-leave* strategy is not so bad as it might seem because survival while waiting is higher than when breeding (see Table 3.2). Inheritance of breeding status and territory goes in order of age and is said to be relatively peaceful. Males then follow the pathway marked 1–2 in Figure 6.2.

Mexican Jay. Dispersal in the Mexican Jay has been studied in the Chiricahua Mountains of Arizona in pine-oak woodland (Brown and Brown, 1984). This jay lives in larger groups than any of the preceding species; and unlike them, it commonly has more than one breeding pair in a territorial group. Because of limited immigration, pairs may form between birds already in the group without inbreeding. The difference between the sexes in dispersal is slight. Both males and females may follow the never-leave pattern or the stay-and-foray pattern (Fig. 7.2). Neither sex

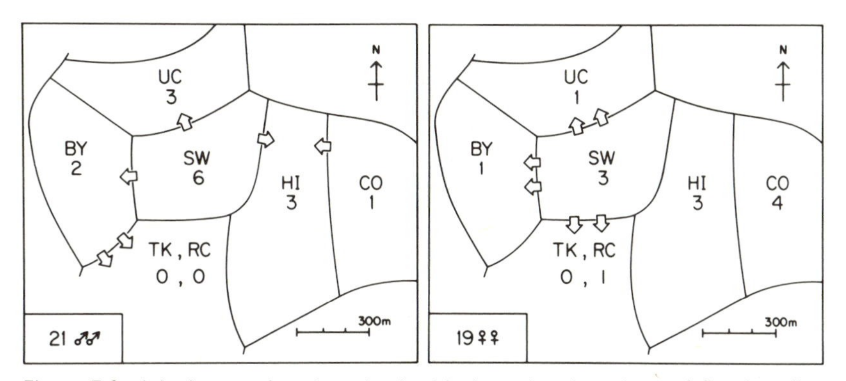

Figure 7.2 Inheritance of territory in the Mexican Jay. Locations of first breeding records are shown for jays banded as nestlings 1969–80 in the mapped territories. Letters designate group territories. Aside from the insertion of a new territory, TK, between SW and RC in 1979, boundaries were approximately as shown 1972–81. Numbers show individuals whose first breeding record was in their natal unit in the indicated territory (*stayers*). Each arrow represents an individual and connects its natal territory to its first breeding territory (*leavers*). Note the lack of a strong sex bias in dispersal. (From Brown and Brown, 1984.)

has moved more than one territory away from home to breed, although a longer movement by a nonbreeding bird has been observed. About 15% of birds that survived their first winter changed groups even though they rarely breed before age three years. These individuals may reach breeding status in their new territory, following pathways 1–4–5 in Figure 6.3.

The Trend to Greater Parental Involvement

The preceding examples were selected to illustrate a correlation in group-territorial birds between group size or sociality and involvement of the young with their natal territory prior to, and sometimes including, their age at first breeding. Perhaps not surprisingly, in a comparative perspective, as the young remain longer with their parents the groups become larger and more complex, leading at times to *plural* breeding and strife within groups (Chapters 15, 16). All of these cases may be interpreted as involving parental facilitation.

The trend can be seen clearly within a single genus of jays, *Aphelocoma* (Tables 7.1, 5.2). Scrub Jays on the mainland of California at the Hastings Reservation (pers. observ.) leave their natal territories as soon as they can forage for themselves in mid-summer. They may form flocks of up to 30 individuals, which roam freely in food-rich areas, as the territorial pairs are no match for a large flock. These young jays are essentially floaters (Fig. 6.3). They wander quite far and may or may not find a territory and

Table 7.1

Comparison of dispersal strategies in some *Aphelocoma* jays. See also Table 5.2.

	PIONT		*Dispersal Patterns*			
Populations	male	female	DS	SF	B	NL
1. Scrub Jay, California	0 yr	0 yr	m, f	—	—	—
2. Scrub Jay, Florida	1–4 yr	1–2 yr	—	m, f	m	—
3. Mexican Jay, Arizona	1–d	1–d	—	m, f	—	m, f

PIONT = Post-independence occupancy of natal territory.
d = death. NL = Never leave.
DS = Depart-and-search. m = male.
SF = Stay-and-foray. f = female.
B = Budding.
1. Brown, J.L., pers. obs. Hastings Reservation, 1957; Atwood 1980a,b.
2. Woolfenden and Fitzpatrick, 1978, 1984.
3. Brown and Brown, 1984.

mate in their first year. The Scrub Jay on Santa Cruz Island, California (Atwood, 1980a,b) and the Steller's Jay in the San Francisco Bay area (Brown 1963b, 1964) are rather similar with respect to dispersal. These species exemplify the depart-and-search pattern of dispersal, which is common in temperate-zone birds and noncommunal birds.

As described above, the Florida Scrub Jay and Mexican Jay exemplify the trend of increasing involvement of young with their parents from the depart-and-search pattern of dispersal in California Scrub Jays to the never-leave pattern, with increasingly larger groups and plural breeding (Fig. 2.3).

The ability of some birds in their first year of life to take priority at food sources ("dominance") over established breeders has so far been found only in Mexican and Pinyon Jays (Balda and Balda, 1978; Barkan et al., 1981, 1986) and might be related to the inheritance of territory—a point currently under study.

Environmental Effects. That social changes of this sort may require little or no genetic change is suggested by the occurrence of two of these stages within the same species (Table 7.1). In the Green Jay the South American subspecies have helpers while the Central American ones do not (Alvarez, 1975). In the San Blas Jay of Mexico the southern subspecies are plural-breeding while the northern ones are singular-breeding (Hardy et al., 1981). Similarly, Skutch (1960) described Brown Jays as singular-breeding in Guatemala, while Lawton and Lawton (1985) found two or more breeding females incubating in Costa Rica. In Hall's Babbler (Brown and Balda, 1977), Jungle Babbler (Gaston, 1978b) and Grey-crowned Babbler (Counsilman, 1980; Brown et al., 1983), group sizes vary with habitat differences within and between populations. In the Mexican Jay the Texas population differs from the Arizona one in having smaller group sizes and probably being singular-breeding (Strahl and Brown, unpubl. data).

That environmental change alone may be responsible in some cases is shown by the Galapagos Mockingbird. In a year of extremely heavy rain, the 1983 El Niño, this normally singular-breeding species exhibited plural breeding, with some groups showing joint nesting and others, separate nesting (P. R. Grant, paper at Meeting of A.O.U., 1983). In this case the same population living on the same territories as in earlier years exhibited a change in social organization concordant with the environmental change.

Another case of a population difference in social structure that seems unlikely to have a genetic cause was described for the Bicolored Wren in the Venezuelan llanos (Austad and Rabenold, 1985). Two subpopulations 1–2 km apart and only partially separated by pastures were contrasted. In the denser subpopulation, inhabiting open palm savanna, groups were larger, usually with helpers; and the helpers were of both sexes. In the sparser subpopulation, inhabiting mixed shrub-woodland and palm savanna, groups were smaller, usually without helpers; and nearly all helpers were male. In this case the denser population was more productive probably because of greater protection from predators caused by safer nest sites. Nests in the denser area were on trees with longer trunks, affording more isolation and visibility, and in areas that were more often flooded during breeding.

Colonial and Wife-sharing Systems

The preceding discussion of involvement of parents sharing their territory with their offspring does not apply to all communally breeding birds. Most colonial birds do not have feeding territories. This is readily understandable for species that obtain their food over large bodies of water. For example, the Pied Kingfisher nests in colonies in Africa, where it has been studied in Kenya by Reyer (1980a, 1984). Pied Kingfishers fish over the open waters of Lake Victoria and Lake Naivasha. The females breed at age one year and typically leave their natal colony to do so, thus employing the depart-and-search dispersal strategy. A male at this age may adopt one of four behaviors: (1) breed, (2) help his own parents to feed their young (primary helpers), (3) help nonparents (secondary helpers), (4) not breed and not help. The order of male preference for these options is apparently as stated. For all of these options the males tend to return to their natal colonies. The alloparents are nearly all first-year males.

In these kingfishers some parents do become socially involved with their yearling offspring, but in this case the parents appear to profit in terms of breeding success more than the young. It is understandable that the parents should utilize aid proffered by their offspring, as long as it does not reduce parental long-term reproduction as represented in their offspring. So long as the long-term interests of the parent in its own yearling offspring are not jeopardized by the parents' solicitation of aid from them, then the parents should accept such help. The young are not forced to help their parents, however; they could just as easily help nonparents or refuse to help. We shall return to the Pied Kingfisher in our discussion of altruism (Chapter 13).

Another example of dispersal in communal birds in which the parental territory appears unlikely to be involved is in the Lonnberg Skua of the South Pacific near New Zealand, which was studied by Young (1978). This skua breeds when there is a plentiful supply of nesting petrels flying to their nests through the breeding territory of the skua. During the nonbreeding season these territories contain relatively little food and the young spend their time foraging at sea. They continue this for several years until the age of first breeding. There appears to be a large surplus of breeding-age birds but only a few breeding territories. In some cases a territory may be defended by two males who share copulations with a single female. This may be a case in which a single male would find the cost of territorial defense to be sufficiently great as to justify sharing it and the female with a second male (see Chapters 8 and 9).

In both of these examples, a natal territory that is used by the young, presumably as an aid to survival in the nonbreeding season, is absent. Either there is no territory used for feeding, as in the Pied Kingfisher, or the territory is abandoned by the young at the end of the breeding season, as in the Lonnberg Skua.

A Model of Parental Facilitation

The trend toward greater involvement of parent and offspring while the latter seek breeding status may be regarded as alternatively (1) selfish parasitism by the young of their altruistic parents, (2) exploitation of the altruistic young by the selfish parents (Vehrencamp, 1979, 1980), or (3) as mutualism (Brown and Brown, 1984). In Chapter 15 we consider the possibility of parent-offspring conflict (alternatives 1 and 2). Here we simply assume that parents who act to maximize the number of their grandoffspring will be selected over those who do not. Basically, this arrangement is mutualistic since, up to a point at least, aid to one's offspring tends to increase one's grandoffspring. Some conflict, however, is not precluded. We also assume that *not all offspring are treated in the same way.* Some may be favored more than others.

In a comparative perspective the trend to greater parent-offspring involvement may be interpreted as a series of increases in the tolerance of parents for their maturing offspring. Since these changes parallel increasing delays in breeding, it is not unreasonable to deduce that our series exemplified by *Aphelocoma* parallels increasing difficulty of the offspring in attaining breeding status. Under such circumstances, efforts by the parents to facilitate this crucial process in their offspring may be expected to increase. In short, in a series of progressively more difficult conditions for dispersal parents become increasingly tolerant of their young. Tolerance is manifested first as increases in the period of association and of group size, and then as tolerance for plural breeding.

This then is the basic idea of parental facilitation in cross-species perspective. In general terms, *parental facilitation* refers to the "facilitation by the parents of the process of achieving breeding status by their young"

(Brown and Brown, 1984). Included in this concept are (1) increased *survival* of offspring as a consequence of not being evicted, (2) better chances of eventually achieving reproductive status associated with the use of the *home range as a base* from which to gather information about nearby birds and territories, and (3) in some species, the possibility of *breeding on the home territory* either after the demise of the same-sex parent or (4) while this parent breeds on the same territory.

Breeding on the home territory has several possible advantages. Not only are the risks of dispersal eliminated, but the chances of eventually getting a territory may be improved. Furthermore, an inherited territory is likely to be a better than average territory simply because its potential for at least some reproductive success has already been confirmed and because more potential breeders are produced on good territories than bad ones. In addition, parents are likely to treat their own offspring more considerately than other birds with whom they might share a territory.

We now consider how parental facilitation might work within a plural-breeding species, the Mexican Jay, with respect to breeding on the natal territory. Assume that the dominant breeders have some control over the number of offspring that stay in the group to breed, *stayers*, and the number that leave, *leavers*. The number of retained offspring is probably constrained by a variety of mechanisms that might reduce both their fitness and their parents' fitnesses. These mechanisms need not be specified here, but their result must be explicit in the model. In the case shown in Figure 7.3 the maximum number of young tolerated by the parents is two. All offspring beyond two must leave.

A specified constant number of independent offspring, F, divisible into all possible combinations of stayers, S, and leavers, L, may be described by a curve,

$$F = S + L,$$

in which all points have the same number of nutritionally independent offspring, i.e., the same parental fitness except that quality is not taken into account. In Figure 7.3 these curves are marked F. We now postulate that leavers have a fitness of 1 and and that stayers have a fitness that is a multiple, A, of the fitness of leavers. The number of grandprogeny, G, of a parent may then be expressed as

$$G = L + SA.$$

A set of points having the same G but different combinations of L and S may be described by a parental isograndfitness curve. These are marked G in Figure 7.3.

If a parent can produce up to F independent offspring (3 in Fig. 7.3), how should it divide them among S and L? The optimal solution to this

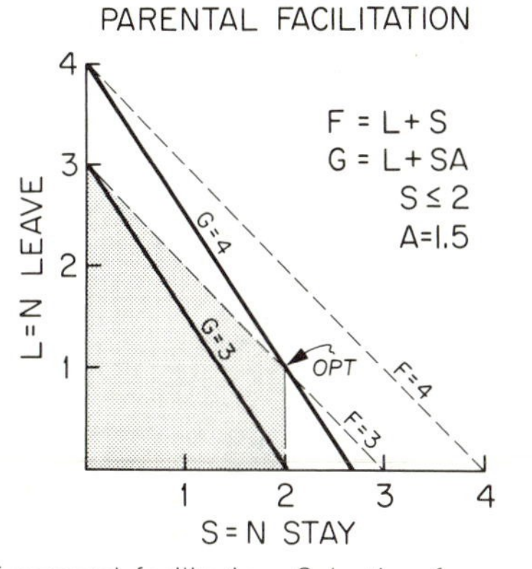

Figure 7.3 A model of parental facilitation. Selection for parental grandfitness may prevail over selection for simple reproductive success under certain conditions when the parent can induce a higher fitness in its mature offspring by retaining them than by shedding them. The *thin dashed lines* show sets of points with equal numbers of mature offspring (parental isofitness curves). The *heavy lines* show sets of points with equal numbers of mature grandoffspring (isograndfitness curves). The *vertical line* shows the constraint that no more than two mature offspring can breed in their parents' territory. Fitness and grandfitness increase as the distance from the origin increases on the quadrant shown in this figure. Consequently, the highest achievable grandfitness is on that isograndfitness curve farthest from the origin that touches the feasible region. The feasible region (stippled) is bounded by its isofitness curve ($F = 3$) and the following constraints: $0 \leq L$, $0 \leq S \leq 2$. For parental fitnesses less than 4 in which all offspring leave ($S = 0$), parental grandfitness is lower than for the parental fitness of 3 in which two offspring stay. In this example, staying offspring are 1.5 times as fit as leaving offspring. The fitness values in this example were chosen purely for mathematical convenience. No implications about population size are intended. (From Brown and Brown, 1984.)

(linear programming) problem may be obtained in two stages. First, we define the *feasible region* for the parent as all possible combinations of non-negative *S* and *L*, given an upper limit to *S* (which is 2 in Fig. 7.3). Second, we find the curve with the *largest isograndfitness* that intersects the feasible region. The solution for the feasible region shown in Figure 7.3 is that the parents should produce three offspring and allow two to stay, forcing one to leave. This provides a total of four grandprogeny, which is more than the parents could obtain by any other combination of *S* and *L* in this example.

Key assumptions of this model are that some offspring be allowed to stay in the natal territory and breed in the presence of the parents and other adults, while other offspring leave and breed elsewhere. These assumptions have been verified. Of 40 Mexican Jays that were first banded as nestlings and later observed breeding between 1969 and 1981, 15 of the 21 males and 13 of the 19 females bred first in their natal territory

(Fig. 7.2). In most cases they had lived their entire lives on their natal territory and continued to breed there. Of the males, 8 ultimately fledged young in their natal territory; of the females, also 8. These individuals were not restricted in their movements to a fraction of the territory; they shared all of it with their other unit members, as is normal in the Mexican Jay (Brown, 1963a). Dispersal in this species is extremely reduced; of these 40 birds, not one moved farther than a neighboring territory to breed, and most did not move at all. Some longer moves could have been detected, as the study area's neighboring territories were watched each year for the presence of banded birds. Immigration from outside the study area was common. A more detailed analysis of dispersal is in progress. The present data, however, are sufficient to establish one of the most extreme cases of *territory inheritance* and philopatry yet described for birds (Greenwood, 1980; Greenwood and Harvey, 1982).

Do these stay-at-home breeders only breed in the absence of their parents? Briefly, offspring of either sex may nest successfully in their natal unit's territory, but as of 1982 only for fathers were there records of breeding successfully in the same unit as successful nests of sons and daughters. There is a suspicious lack of such records for mothers, but sample sizes are small.

Finally, this particular model (but not all parental facilitation models) assumes that stayers achieve higher reproductive success than leavers. We found that reproductive success per year averaged higher for adults (banded as nestlings) breeding in their natal territories than for those breeding outside their natal territories. To avoid complications caused by differences in reproductive success correlated with age, the comparison was restricted to birds of age five years or older, since only then is reproductive success independent of age (Fig. 11.3). The advantage of staying over leaving was evident for both sexes but was significant only when both sexes were combined.

The concept of parental facilitation provides a useful supplement to currently popular views of parent-offspring relations. I use *parental facilitation* to apply to the facilitation by parents of the process of achieving breeding status by their young. Therefore, it is *not simply extended parental care* in the usual sense, since the normal types of parental care, such as feeding young, are not involved. It does, however, have features in common with earlier models concerning offspring quality. Both Smith and Fretwell (1974) and Brockelman (1975) employed a similar tradeoff between offspring quality and number in their treatments of egg size and clutch size. In their models, as in ours, higher fitness can be achieved by a parent under some circumstances by increasing offspring quality at the expense of quantity.

Most previous treatments of philopatry (Greenwood, 1980; Greenwood and Harvey, 1982) and inheritance of territories in birds have treated these phenomena as strategies of individuals to acquire breeding status. Parents tended to be depicted as *reducing* the direct fitness of their offspring, rather than increasing it (Vehrencamp, 1979, 1980). Emphasis has been on the potential for parent-offspring conflict, not on the potential for mutualistic relationships (Emlen, 1982b).

An alternative interpretation was suggested by Brown (1974:76). In this view it is to the advantage of both parents and offspring to facilitate the handing down of the most productive territories from one generation to the next within the same genetic lineage—roughly analagous to the inheritance of capital in human societies. This is not to say that parent-offspring conflict never occurs (see Chapter 15).

Evidence is mounting that *within* a population territories of communal birds vary significantly in correlates of the production of young (Brown and Balda, 1977; Gaston, 1978b; Vehrencamp, 1978; Trail, 1980; Brown et al., 1983). In the Mexican Jay, as well as in some other communal birds, social units differ greatly in their production of mature young (Fig. 7.2). Some units are highly productive and are *genetic exporters.* They export genes through the production of young, some of whom leave and breed elsewhere (as in the SW unit in Fig. 7.2). By the same token, some units are *genetic importers.* Because of low reproductive success they are unable to produce enough offspring to control the resources, and these units are populated mainly by individuals hatched in other units (e.g., RC unit in Fig. 7.2). Many units might fall between these two extremes. A similar situation, also fulfilling the requirements of the 1974 model, was described for the White-browed Sparrow-weaver (Lewis, 1982a). Of course, relatively more young and more future breeders are produced by the genetic-exporting units. It is the behavior of individuals in exporter units and the composition of these units that determine the direction of evolution in the population. Therefore, it is useful in understanding the evolution of communal social systems to pay special attention to the most productive units—as opposed to considering average units.

The crucial factor in productive units is control of the most productive space. Individuals controlling such space would leave more grandoffspring by facilitating the transfer of resources to their own offspring than by forcing all offspring to leave or by forcing them to be merely helpers.

In attempting to explain why stayers have more reproductive success than leavers in this particular case we need postulate nothing more than a difference in productivity of the areas where they breed. However, the situation is probably not that simple. Following the logic of optimal group sizes (Chapter 8) we may expect the advantage of staying to diminish as more individuals choose to stay. The difference might diminish to zero, as in the "ideal-free" situation (Fretwell and Lucas, 1970) as depicted in the optimality model shown in Brown (1969b). We have, therefore, conditions in which the best strategy for an individual is a function of what others in the population are doing. At equilibrium we expect some individuals to stay and others to leave, with the reproductive success of each to be equal. It seems unlikely, however, that such an equilibrium would be more than transient. Our data do not suggest any kind of population equilibrium (Figs. 12.2, 12.3).

If it is truly beneficial to parents to tolerate the persistence of their offspring even to breeding age, then we might expect parents to treat offspring more tolerantly in the Mexican Jay than in the congeneric Scrub Jay. Consistent with this prediction it has been found that young Mexican Jays often have priority over adults in access to food in winter (Barkan et al., 1981, 1986). This has not been reported for the Scrub Jay.

In the model in Figure 7.3 the number of grandoffspring can be increased without reducing F simply by allowing some to stay. For example, if $F = 3$, then G can be raised from 3 to 4 by this means. Thus in the model parents are *not required to give up anything*, and the cost to them is regarded as being negligible. It is possible under certain circumstances, however, *to have more grandoffspring while having fewer nutritionally independent offspring*. For example, in Figure 7.3 in moving from singular breeding with $F = 3.5$, $L = 3.5$, $S = 0$ and $G = 3.5$ to plural breeding at the optimum ($S = 2$, $L = 1$), the number of grandoffspring increases to $G = 4$ while the number of offspring is reduced from 3.5 to 3. Although this might seem to be a sacrifice by the parent for its offspring and would be counted as such in conventional models, it is not. The *parent is actually profiting by substituting a smaller quantity of higher quality offspring for a larger quantity of lower quality offspring*, as in the models of Smith and Fretwell (1974) and Brockelman (1975). Here quality is determined in part by the parent's increased tolerance. This is not, therefore, a case of parent-offspring conflict. Although the parent gives up some offspring, this reduction does not constitute altruism in the sense of Hamilton (1964), since there is no net loss to the direct component of fitness as expressed in the grandoffspring. *If quality of offspring is considered, there is also no net loss in fitness as expressed in offspring.*

The present models and data demonstrate the inadequacy of models that are phrased strictly in terms of numbers of offspring or lifetime reproductive success. Even the number of mature offspring is not a reliable index of fitness in communal birds. Many mature birds in communal species do not achieve breeding status at all or they may experience a detrimental delay. Reproductive success of mature birds is often compromised, reduced, or curtailed by factors which the parents might influence. In cases involving quality of offspring, such as in parental facilitation, variance utilization, or variance enhancement (see Chapter 15), it is heuristic to consider also the number of grandoffspring, as in models of parent-offspring conflict (Parker and Macnair, 1978).

Conclusions

In this chapter I have explored some of the more interesting and controversial aspects of dispersal. As dispersal of the offspring becomes progressively more difficult, the value of involvement of their parents and their resources increases. It is interesting to speculate on how this affects parent and offspring in various types of communal breeding, but this will be deferred until Chapter 15 after I have discussed the costs and benefits of adding members to a group in the next chapter. Here I have illustrated the value of using grandoffspring as a measure of direct fitness in situations that involve variance in reproductive potential by the offspring.

The concept of parental facilitation is introduced. Parental facilitation refers to the post-independence involvement of the parents in the achievement of breeding status by their offspring. This may occur in various ways. In the Florida Scrub Jay, parents facilitate acquisition of a territory by some sons by aiding them in taking over parts of adjacent territories and by giving up part of their own territory. In the Mexican Jay, offspring are allowed to breed on the natal territory. Another example of retained offspring that might be interpreted as parental facilitation is described for Groove-billed Anis in Chapter 10. The cost to the parents of parental facilitation is poorly known but is expected to be negligible.

8 Mutualism, Cost-sharing and Group Size

The addition of a breeding or nonbreeding helper to a group of two may effect a variety of costs and benefits beyond care of young. This is especially true in group-territorial species. Much has been learned about energy budgets of birds in recent years (Mugaas and King, 1981). Field studies of flocking in the nonbreeding season have yielded important insights into the energy costs and benefits of group living (Pulliam and Caraco, 1984). Yet questions of energy costs and benefits in communally breeding birds are almost untouched. The optimality model presented in this chapter is intended to stimulate further work in this area by describing some likely relationships between the relevant variables in a way that leads to the formation of testable hypotheses.

The models described here maximize fitness for *individuals, not groups.* Conflict between individuals is described in terms of the model.

Group Size May Depend upon Individual Energy Budgets

This chapter begins by treating the case of a single breeding pair living in a socially bonded unit with a variable number of nonbreeders, some or all of which may function as helpers, participating in parental or alloparental care as well as other group functions. We consider first only effects on fitness mediated by energy, which may be crucial for reproduction and survival. Assume territories to be incompressible areas of constant size but variable quality. The effects of relaxing this assumption are discussed later. Territories are ranked in quality from best, Q_{max}, to worst, Q_{min}, from right to left in Figure 8.1.

The basic fitness of a single hypothetical male, $F''_{i,1}$, on territory i may be considered to be a function of territory quality, Q_i, disregarding for the moment the effects of resource depletion, of defense costs, and other complications (Fig. 8.1). We employ the first subscript, i, to designate territory rank, and the second, j, to designate group size ($j = G_j$). The fitness of our hypothetical male alone on territory i ($F''_{i,1}$) will then be (1) diminished by the cost of defending his territory by himself, D, (2) augmented

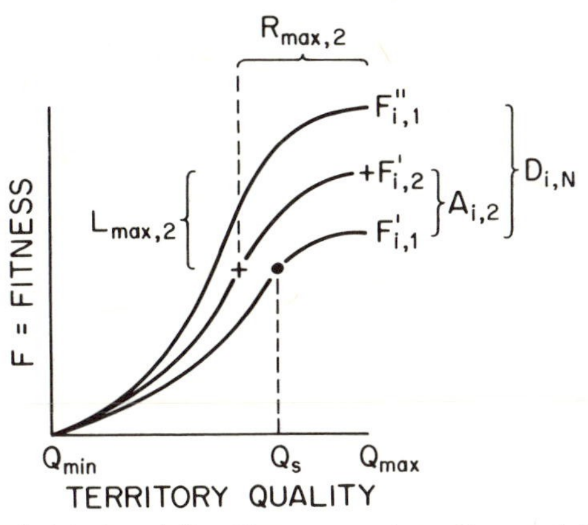

Figure 8.1 A threshold of sociality. The curves show fitness of the defenders of a territory as a function of territory quality and number of defenders. The basic fitness of a hypothetical male on a specified territory, $F''_{i,1}$, depends on the quality of the territory, Q_i. From $F''_{i,1}$ defense costs, $D_{i,N}$, are subtracted, and the advantage gained by sharing these costs among members of a group of size two is added ($A_{i,2}$) to give the function, $F'_{i,2}$. The reduction in quality (horizontal axis) caused by adding one member to the best territory, Q_{max}, is shown by $R_{max,2}$. The corresponding loss to fitness (vertical axis) is shown by $L_{max,2}$, which may be found by following $F'_{i,2}$ between the + signs. The lower + point, which represents the fitness of a male sharing the best territory in a unit of two, yields a fitness equal to the fitness of a lone male on a territory of quality Q_s. This point on $F'_{i,2}$ is referred to as the threshold of sociality for this territory at the specified population density, N. (From Brown, 1982a.)

by sharing defense costs with added group members, A, and (3) diminished by resource depletion, $L_{i,j}$, caused by adding members to his territorial social unit. Baseline existence costs, such as for body care and thermoregulation, are subsumed under $F''_{i,1}$.

The potential effect of defense costs on the fitness of our hypothetical male, $F''_{i,1}$, is shown in Figure 8.1. Fitnesses before defense costs are shown as $F''_{i,1}$; after defense costs but without other costs as $F'_{i,1}$.

Specifically, defense costs, $D_{i,N}$, for territory i will be a function of attractiveness to intruders of that territory (approximated by its quality, Q_i) and of the population density, N (Fig. 8.2). This function will increase monotonically with population density and will slope more steeply upward as more territories are occupied, becoming steepest when all suitable territories are occupied and a surplus of competitors exists. At this point the available habitat may be said to be saturated. Therefore, $D_{i,N} = f(Q_i,N)$. Above saturation densities, D should be a more linear function of the density of nonbreeders. The density of nonbreeders approximates an equilibrium that depends on and fluctuates with their rates of production and mortality (Fig. 3.2; Caughley, 1977).

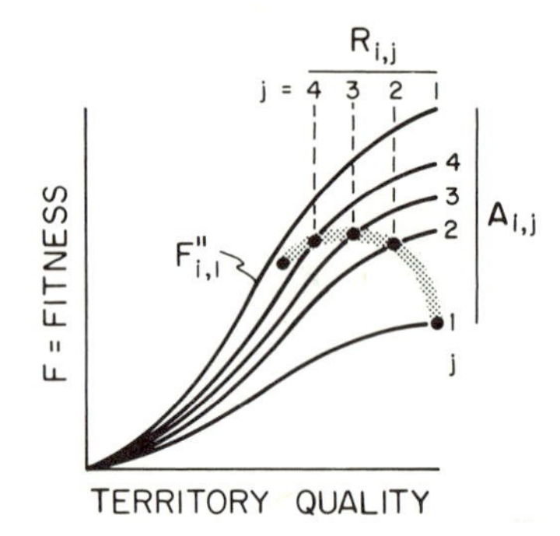

Figure 8.2 Optimal unit size in a group territorial animal. The top curve ($F''_{i,1}$) shows the basic fitness of a hypothetical single male on a territory of quality Q_i without defense and depletion costs. Curves 1–4 show fitness of a male with defense costs shared among all unit members (not necessarily equally) and with resource depletion. Unit sizes are $j = 1$ to 4 as indicated. Reduction in territory quality caused by resource depletion at progressively larger group sizes is indicated by the vertical dashed lines. (From Brown, 1982a.)

As members are added to the group we assume that they enter as subordinates to the original owner, who remains dominant. In nature this is a frequent case (e.g., in Scrub Jays, Woolfenden and Fitzpatrick, 1977; Pukeko, Craig, pers. comm.). Assume that the dominant admits additional members only when it is to the dominant's advantage to have a larger group, and that subordinates enter or stay only when it is to their advantage (as in Brown, 1969a; Gaston, 1978c). It may occur, therefore, that the optimal group size for the dominant may differ from that for a subordinate. Furthermore, if rank is important to subordinates, each of them may have its own optimal group size.

A Threshold of Sociality

Consider now the decision of an individual whether to take a vacant, low quality territory alone or join a social unit in the best territory as a subordinate (Fig. 8.1). Augmenting the size of a social unit is assumed to deplete the energy resources of its territory. The amount of resource depletion, R, in terms of its effect on territory quality, is determined by the particular territory, i, and the size of the group, G_j; $R_{i,j} = f(Q_i, G_j)$. The reduction in quality for a given territory is the horizontal distance, R, along the Q axis. The corresponding loss to fitness, $L_{i,j}$, may be read from the vertical axis. It is not unlikely that $L_{i,j}$ is inversely related to Q_i, with richer territories being less vulnerable to depletion than poorer ones. The fitness reduction, $L_{i,j}$, also depends upon the slope of the relevant portion of $F'_{i,j}$.

In this model, adding a member is assumed to reduce defense costs. In nature adding a member may instead or in addition increase territory size (Craig, 1979; Woolfenden and Fitzpatrick, 1978); but here for simplicity we have assumed territory size to be constant, thus avoiding the complications of spiteful superterritories modeled by Parker and Knowlton (1980). In either case there is thought to be an increment of fitness; so our method provides an equivalent to the effects on fitness of enlarging the territory.

If defense costs are shared equally by all members of the unit, then fitness will be augmented above $F'_{i,1}$ by

$$A_{i,j} = D_{i,N}(G_j - 1)/G_j, \tag{1}$$

as in Figure 8.1. We may now express the expected fitness of an individual on territory i in a unit of size G_j as follows:

$$F_{i,j} = F''_{i,1} - D_{i,N} + A_{i,j} - L_{i,j}. \tag{2}$$

The combined effects of $D_{i,N}$ and $R_{i,j}$ on fitness in a group of two in the best territory may now be compared with the fitness of a lone individual on the best unoccupied territory. In Figure 8.1 the fitness of an individual in a group of two on the best territory is identical to the fitness of a single individual on a territory of Q_s. For this set of conditions Q_s designates a *threshold of sociality*. More precisely, Q_s identifies a pair of points on $F'_{i,2}$ and $F'_{i,1}$ at which a unit of two becomes a feasible alternative to a unit of one.

By analogous procedures we can identify for any specified territory a threshold for units of three compared to two or of any size, $n + 1$, compared to n.

Optimal Unit Size for an Individual in a Cooperative Social Unit

The optimal unit size for an individual on territory i existing at population density N is the one in which its fitness, $F_{i,j}$, is maximized. An example is shown in Figure 8.3 for the best territory. Similar maxima may be found for territories of lower quality. Of the four variables important in this system, population density probably has the most pervasive influence (Brown, 1969a) since it influences defense costs, D (Fig. 8.2), and, therefore, the benefit derived from sharing defense, A. Density alone, however, is insufficient to explain the existence of group territories. This holds true even when variations in territory quality are considered, since the optimum can in theory be a unit size of one at any combination of density and territory quality (though not in the particular example of Fig. 8.3). Indeed, certain birds, such as nectar specialists, never (or rarely?) show group

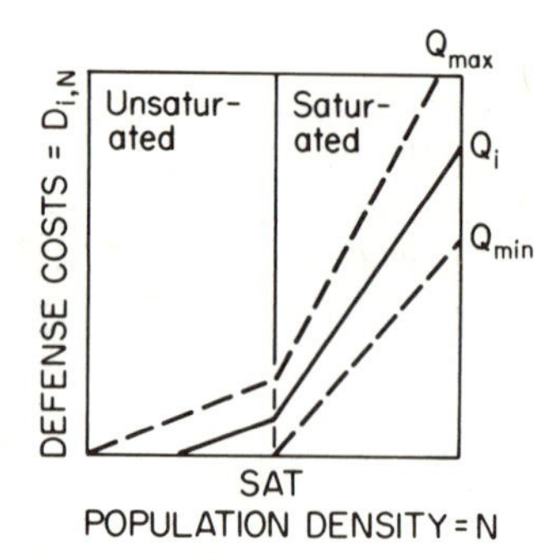

Figure 8.3 Defense costs of a hypothetical male alone on a specified territory are a function of population density, N, which influences intruder pressure, and territory quality, Q_i, which influences relative attractiveness of a particular territory. Above saturation (SAT) all suitable territories are occupied; the function should, therefore, rise more steeply than below SAT. (From Brown, 1982a.)

territoriality despite large variations in density and territory quality (Brown, 1978a).

Critical to group defense in the model are the ways in which resource depletion, R, and defense sharing, A, are influenced by unit size. Optimal unit size at a given population density occurs when net profit, $A_{i,j} - L_{i,j}$, is maximal (Fig. 8.4), where $L_{i,j}$ is the reduction in fitness due to the effect of a particular unit size on the quality of a particular territory. Data with which to estimate $L_{i,j}$ are unavailable, but it is clear from equation (1) that the gain from adding a member diminishes toward an asymptote as unit size, G_j, increases, and that the gain is relatively small at unit sizes above three or four. Consequently, equation (1) places a strong constraint on increases in group size under the present assumptions. The effect of $A_{i,j}$ is to favor increases for small unit sizes but to make larger unit sizes largely disadvantageous and dependent chiefly upon very low values of $R_{i,j}$. This prediction is not evident from previous verbal models.

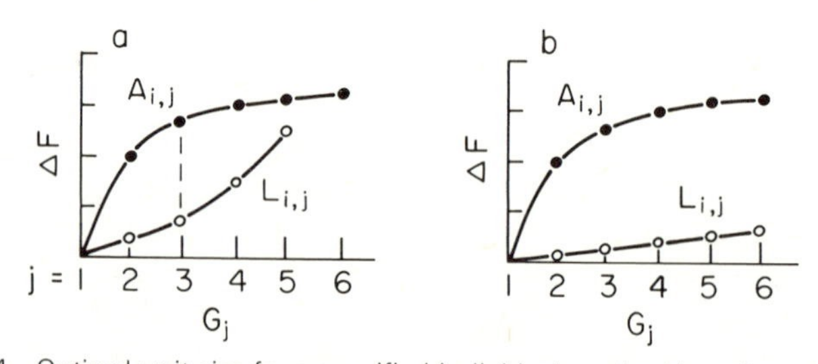

Figure 8.4 Optimal unit size for a specified individual on the ith territory occurs when benefit minus cost $(A_{i,j} - L_{i,j})$ is maximal. In (a) the optimum is at $G_j = 3$. In (b) the optimal unit size is not clearcut and is at or above six. (From Brown, 1982a.)

If adding unit members not only reduces defense costs but also increases area in a useful way, then the constraint, $A_{i,j} \leq D_{i,N}$, inherent in equation (1), is removed. The benefit curve, $A_{i,j}$, would then tend toward a higher asymptote, making larger groups more profitable.

The effects of resource depletion on fitness in group-territorial animals have received virtually no attention in the literature on group territoriality. They could, however, have significant effects on optimal unit size. If we assume—in the absence of data—a linear function for $L_{i,j}$ such that $L_{i,j} = a_i G_j$, where a_i indicates the *rate of change* in fitness due to resource depletion as a function of unit size, then by combining equations (1) and (2) and substitution:

$$F_{i,j} = F''_{i,1} - D_{i,N} + \frac{D_{i,N}(G_j - 1)}{G_j} - a_i G_j$$

and

$$\frac{dF}{dG} = D_{i,N} G_j^{-2} - a_i.$$

Setting $dF/dG = 0$ and solving for G_j, we obtain $\hat{G}_j$ the optimal unit size for a particular territory, that is, the unit size that maximizes F_i.

$$\hat{G} = \sqrt{[D_{i,N}/a_i]}. \qquad (3)$$

In this formulation optimal unit size depends upon the relationship between defense costs and resource depletion. Larger groups are favored by large defense costs (which may be caused by higher population densities) and/or by lower rates of increase of resource depletion with unit size.

For simplicity, a has been treated as a constant for a given territory; however, a is probably a variable that is a function of unit size. In addition, it probably is more severe in poorer territories because food might more quickly reach critically low levels there. These considerations would give a more importance in determining optimal unit size.

Dominants and Subordinates May Differ in Optimal Unit Size

Optimal unit size for a dominant may differ from that for its subordinates. Dominance is expected to influence many aspects of behavior within social groups. Since a dominant can often take food discovered by subordinates, but rarely vice versa (Rohwer and Ewald, 1981; Baker et al., 1981), effects of resource depletion may be less for a dominant than for a subordinate (Pulliam, 1976). In this case, $L_{i,j}$ might be less for a dominant than a subordinate. The effect would be similar to changing this function from the condition in Figure 8.4a to that in Figure 8.4b or to reducing the value of a in equation (3); it would yield a larger optimal unit size for a dominant than for a subordinate. Conflicts of interest are further discussed below.

On the other hand, since a dominant may bear a disproportionately large share of defense costs (as in some wagtails, Davies and Houston, 1981), addition of a subordinate might not increase $A_{i,j}$ for a dominant as much as for another subordinate. Alternatively, during the breeding season a dominant might have more to gain through reduction in defense costs than would a subordinate. The curve for $A_{i,j}$ in Figure 8.4 would then be shifted down or up accordingly, with concomitant effects on optimal unit size.

Vigilance and Risk Dilution

The cost of vigilance (and other antipredator behaviors) can be treated similarly to the cost of territorial defense. There is an initial vigilance cost for a single individual, V, which is considered to be a monotonically increasing function of predator pressure, u. The addition of members to the group who serve as "watchdogs" (Ricklefs, 1980; avoiding the anthropomorphic term, "sentinel") benefits the individual by an amount, S_j, which is a monotonically increasing function of group size potentially reaching an asymptote as $S_j = V$ (Pulliam, 1973). There may be an additional benefit against predation through the selfish herd effect (Hamilton, 1971) and an additional cost due to the greater attractiveness of larger groups to predators (Vine, 1971). Since these tend to cancel each other out, their combined effect will for simplicity be subsumed under S_j; that is, the net reduction of V due to being in a group is assumed to rise from a minimum when solitary to a maximum no greater than V when in a large group. The net effect then is to augment $F_{i,j}$ by $-V_u + S_j$. We may now extend equation (2) as follows:

$$F_{i,j} = F''_{i,1} - D_{i,N} + A_{i,j} - V_u + S_j - L_{i,j}. \qquad (4)$$

A Key Factor for Group Territoriality is Associated with Reproduction

Group territoriality appears to be known only in animals that *breed* in units of three or more while holding the group territory. Animals that defend territories during the nonbreeding season but do not breed in them, such as some warblers (Rappole and Warner, 1976), sandpipers (Hamilton,

1959; Myers et al., 1979), many nectar-feeding birds (Gill and Wolf, 1975) and a wagtail (Davies and Houston, 1981), do not defend them in units larger than two birds. The Galapagos Mockingbird (*Nesomimus parvulus*) was considered to be an exception (Hatch, 1966) to this generalization, but Grant and Grant (1979) showed that it too may breed in units of three or more. These facts suggest that a key factor for group territories is associated with reproduction.

Two general factors may be involved; both concern the allocation of increased energy profits that typically become available at the season of reproduction. First, the increased availability of energy for aggression in the population during the reproductive season increases intruder pressure, causing increased defense costs. We assume here that the function, $D_{i,N} = f(Q_i,N)$, is for the breeding season. A correction factor or a different $f(Q_i,N)$ for $D_{i,N}$ would be needed for the nonbreeding season.

Second, a new category of behavior appears in the breeding season, namely, production and care of young. Group territoriality is unknown in highly fecund species and appears to be confined mainly to taxa with significant male parental care and/or protection of young (being unknown in lek birds and mammals and in other species with unrestricted promiscuity). Therefore, it seems that care of young is more critical than fecundity in group-territorial species. This is consistent with the observation that feeding of young by helpers occurs in virtually all nonprimate species with group territories.

For simplicity in dealing with reproduction, I now limit the model to males of avian species in which males neither incubate eggs nor brood young, and to the period of parental care. A male will then have an additional energy expense, Y. This can be reduced by the activities of helpers in an amount, H, that is a function of territory quality and unit size. We assume here a unit consisting of a mated pair and zero or more helpers. Consequently, $H_{i,j} = f(Q_{i,j},G_j)$, where $H \leq Y$.

We now have the following expression for the fitness of a breeding, group-territorial male as a function of population density, territory quality, and unit size:

$$F_{i,j} = F''_{i,1} - D_{i,N} + A_{i,j} - V_u + S_j - Y + H_{i,j} - L_{i,j}. \quad (5)$$

In this model, the recouping of losses through the aid of nonbreeding unit members is similar in territorial defense $(-D_{i,N} + A_{i,j})$, vigilance $(-V_u + S_j)$, and parental care $(-Y + H_{i,j})$. Removal of helpers experimentally has established that they have a positive effect on fitness, at least in one typical group-territorial species (Brown et al., 1982b), but no field study has succeeded in measuring the relative importance of territorial defense, vigilance, and care of young for this benefit. This model allows any or all of these effects of unit size to be important.

Inclusion of the effects of vigilance and care of young complicates the modeling of optimal unit size only slightly, provided similar assumptions are made. If vigilance and care of young are also shared equally among all unit members, then following the methods used above (see Brown, 1982a), one can obtain an expression for the optimal unit size:

$$\hat{G} = \sqrt{\frac{D_{i,N} + V_u + Y}{a_i}}. \tag{6}$$

Thus optimal unit size is determined jointly by the shared tasks and by the depletion function, much as in equation (3) except that more tasks are involved. Consequently, the general relationships shown in the figures remain valid provided $(D_{i,N} + V_u + Y)$ is substituted for $D_{i,N}$ and the summed benefits $(A_{i,j} + S_{i,j} + H_j)$, for $A_{i,j}$.

Returning to the question of why group territoriality occurs only in species that breed in territorial groups, expression (6) provides some insight. In breeding, more tasks are shared and, therefore, greater benefit accrues, thus raising $\hat{G}$ by increasing the numerator. Furthermore, since food is more abundant when breeding, it is possible that a is smaller, also raising $\hat{G}$.

Equation (3) is basically an energy balance expression that describes fitness as a function of energy at a given time. A deficiency of equation (3) and of most energy budget approaches to the economics of territoriality is that no account is taken of the period of time during which a condition of high energy profit obtains. If we assume available energy before reproduction, $E_{i,j,N}$, to be given by:

$$E_{i,j,N} = E''_{i,1,N} - D_{i,N} - A_{i,j} - V_u + S_j - L_{i,j}, \tag{7}$$

where all terms are defined in terms of energy rather than fitness, then we can imagine that a threshold, K, exists above which breeding is attempted (a common idea, e.g., Lack, 1968). The period, $P_{i,j,N}$, during which $E_{i,j,N} > K$ is the time in which breeding may occur. The longer is $P_{i,j,N}$, the more broods may be attempted and young reared by that group (Fig. 8.5). This model is not intended to mimic the environmental control of reproduction; its goal is to illustrate the relationship between group size and length of breeding season.

The time needed per brood, $B_{i,j}$, may also be a function of group size; $B_{i,j} = f(G_j)$. The number of broods may now be expressed as $P_{i,j,N}/B_{i,j}$. For cases in which number of clutches or litters is of prime importance, as in the Grey-crowned Babbler (Brown and Brown, 1981b), "annual

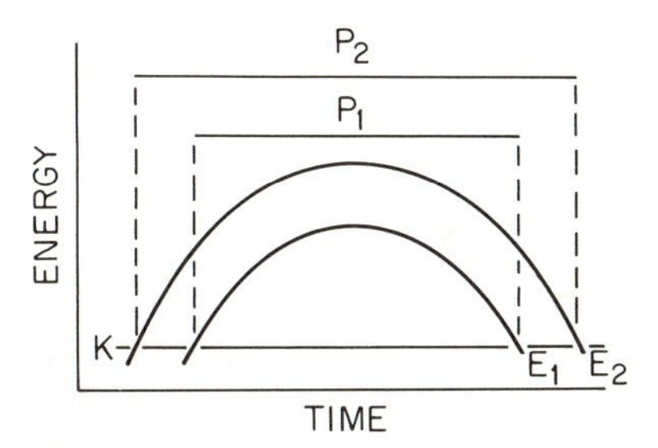

Figure 8.5 Reproduction occurs when the available energy for an individual, E, exceeds the threshold for reproduction, K. P is the period during which $E > K$. E_2 exceeds K for a longer period (P_2) than does E_1 (P_1). This allows two broods in P_2 vs. one for P_1. See also Figure 5.4. (From Brown, 1982a.)

fitness" with multiple clutches, F_m, may be expressed as:

$$F_m = P_{i,j,N}/B_{i,j}. \tag{8}$$

This expression retains dependence on the same factors as equation (5) through the dependence of both $P_{i,j,N}$ and $B_{i,j}$ on the same variables, namely, population density, territory quality, predator pressure, and unit size.

Why Are Some Species Group-territorial and Others Not?

The earlier version of the saturation hypothesis (Brown, 1969a) incorporated the concept of cooperative defense to offset the high intruder pressure and minimize neglect of the young due to aggression by neighboring adults (Hutchinson and MacArthur, 1959). These relationships are made more explicit in equation (5) in which the losses from defense, D, vigilance, V, and parental care, Y, are recouped through aid in defense, A, antipredator activities, S, and in care of young, H. These effects would help to answer the question of what benefits accrue to group-territorial individuals.

They do not help much with another question. Given two species each with a surplus of nonbreeders, why in one species do the surplus individuals join territorial units while in the other species they exist as socially unbonded floaters? Several possibilities exist.

Even with equal population densities and numbers of nonbreeders in the two species, defense costs, D, might differ as a function of physical environment, vegetation, territory size, or other factors. Individuals in a species or habitat with high defense costs might find group territories profitable while those in another species or habitat with low defense costs might not. Defense costs per individual have been estimated for some avian species, but not for any group-territorial species. Indeed, the function relating defense costs to group size and population density seems not to have been studied at all. A difficulty with the cooperative defense hypothesis arises from equation (1). The advantage gained from adding another member to a group of four or five is much less than that gained in units of one or two or even three. This reduces the power of this theory to explain the existence of large groups and suggests that other factors may also be involved.

Similarly, the costs of vigilance, V, and parental care, Y, probably vary with environmental factors, body size, diet, and other factors not considered in the model. Again, however, large group sizes are difficult to explain with models that incorporate effects like that in equation 1.

Susceptibility to Resource Depletion Has Been Overlooked

A factor that has received virtually no attention as a determinant of communal breeding in birds and mammals is diet. The absence of cooperative behavior in nectar-feeding birds and its rarity in frugivores have been noted (Brown, 1978a) but not explained (see Chapter 17).

The influence of diet postulated here concerns the susceptibility of the food resources of a species to depletion, $R_{i,j}$. We are concerned more with the effects of depletion on fitness, $L_{i,j}$, than with the actual amount of energy loss. Effects on fitness will depend on the range of options available for individuals. This is represented in Figure 8.6 by the slope of the curve relating F to Q. A large R has relatively little effect on fitness when the

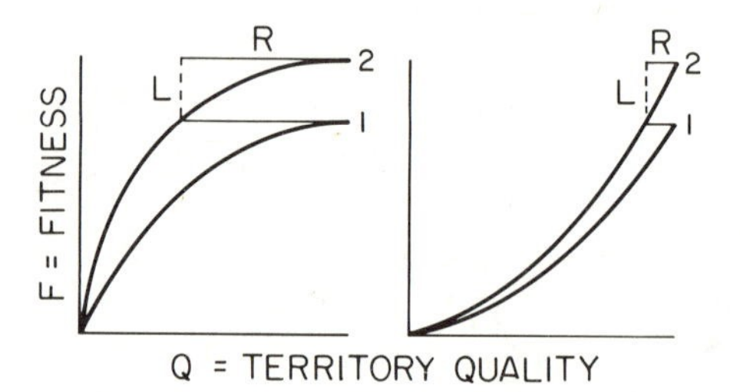

Figure 8.6 The effect of resource depletion on fitness depends on the slope of the fitness function. If shallow (left), a large R may have a relatively small effect. If steep (right), a small R may have a relatively large effect on fitness. (From Brown, 1982a.)

slope among the best territories is shallow as in the left diagram, but would have a drastic effect in the right diagram, where there are few high-quality territories and the slope is steep. Sharing a territory would be favored by smaller R or shallower slope, interdependent factors that together reduce the penalty of resource depletion on fitness. It might be the case that some species are very susceptible to resource depletion, as in Figure 8.4b, and others not, as in Figure 8.4a. Nonbreeders in susceptible species would tend to exist as floaters, while in nonsusceptible species they would tend to join socially bonded, territorial units.

The slope of the function that relates fitness to territory quality may vary in different parts of the curve. A sigmoid shape, as in Figure 8.1, would mean that for similar values of R, high-ranking territories would be less sensitive to resource depletion than middle ranking ones.

Practical Implications of the Model

The central message of this chapter is that individual energy budgets may well be important in individual decisions about joining a territorial unit and admitting or rejecting potential joiners. Energy budgets have long received attention in behavioral ecology (e.g., Orians, 1961) but except for recent work on energetics in social insects (e.g., Heinrich, 1979; Oster and Wilson, 1978) they have been neglected in sociobiology. For communal vertebrates only a few papers have considered time budgets with respect to unit size (Brown and Balda, 1977), although it has long been evident from studies on feeding of young that individuals vary in their energy allocations to different tasks (e.g., Figure 2.1). A pressing problem for communal vertebrates, therefore, is to estimate the relative importance of energy costs for defense, vigilance, and helping ($A_{i,j}$, S_j, and $H_{i,j}$ respectively) in groups of varying size. We do not know the effect of unit size on time-energy budgets for any behavior in communal birds except on the feeding of young (e.g., Brown et al., 1978) and vigilance (Bertram, 1980a, for Ostrich). Unit-size effects on energy phenomena in defense and foraging have received virtually no attention. Many communal species are unsuitable for such work because of the difficulties in observation, but certainly more could be done.

The study of resource depletion has progressed considerably in nectar-feeding birds (e.g., Gill and Wolf, 1975), but this too has not been attempted in communal vertebrates. The problem is tractable. It is possible to estimate time spent foraging by individuals and to compare estimates for individuals in units of different size (e.g., Davies and Houston, 1981; Pulliam and Millikan, 1982; Caraco, 1979b). Perhaps the first study of foraging efficiency in communal birds is that by Rabenold and Christensen (1979), who showed that capture rates for a communal wren were significantly lower in groups than alone. Whether the reduction was caused by resource depletion or another factor remains unknown. On certain occasions, at least, capture rates may be higher in communal groups, such as when a group exploits a rich but transient resource (Brown, 1983b).

There are difficulties in converting effects on time budgets observed in the field into effects on fitness. It would be futile to attempt a general solution to these difficulties here, for each species and behavior will require its own special simplifying assumptions. Time and energy budgets affect both survival and reproduction. Provided something is known about age-specific survival and reproduction and changes in population numbers, it should be possible to evaluate effects of a behavior on fitness by means of its effects on reproductive value.

Several papers have employed optimality approaches to flock size in nonbreeding, nonterritorial groups (Pulliam, 1973, 1976; Caraco, 1979a,b,c). In these approaches, optimal unit size is that which maximizes survival, mainly by appropriate compromises between feeding times and vigilance. Dominants and subordinates may have different optima depending on effects of aggression within flocks on foraging time. The success of these models in predicting individual time budgets and transitions between solitary and social living as a function of temperature, food, cover, and predation risk in field tests has been encouraging for the use of energy and time models in the study of nonbreeding social systems (Caraco, 1979b; Caraco, Martindale and Pulliam, 1980a,b).

A pioneering study on the economics of territoriality with variable unit size by Davies and Houston (1981) on nonbreeding wagtails (*Motacilla alba*) has demonstrated several of the features of the present model, although not for a communal species and not for the breeding season. The addition of a second wagtail to a territory significantly lowered the feeding rate of the owner, an effect that corresponds to $R_{i,2}$. They also demonstrated a reduction in the owner's defense costs due to sharing them with a second bird, an effect corresponding to $A_{i,2}$ in equation (2). The decision to admit a second bird varied with the quality of the territory, as measured by feeding rate, corresponding to Q_i in equation (2). Intruder pressure (the principal cause of $D_{i,N}$) was shown to be linear with respect to Q_i.

To some extent the observations of Davies and Houston bear out the assumption of my 1969 verbal model that group territoriality occurs when the energy profits of group members—both territorial owner and subordinates—are increased through sharing the costs of defense as well as through the benefits of living in a high-quality territory. A major limitation, however, is that in wagtails territorial groups of three or more were not observed under any circumstances. Only high-quality territories could afford dual ownership, but when feeding rates went even higher in warm weather the wagtails became nonterritorial. The reason for this was not investigated, but it might have been caused by a reduction in the profitability of defense due to the ready availability of food without defense, thus bringing $D_{i,N}$ and $A_{i,j}$ to smaller quantities.

In the Galapagos Mockingbird, nonbreeding helpers (mostly young males) were actually more aggressive than male breeders in territorial disputes (Kinnaird and Grant, 1982). By their quantitative observations, Kinnaird and Grant established that "helpers reduce the breeders' energetic costs of territorial defense."

The present model does not deal explicitly with the critical decision in the life of a subordinate nonbreeder, namely, whether to remain as a helper or attempt to become dominant and to breed in a territory of its own. Implicit in this decision, however, is that a favorable energy budget is needed for successful breeding. In this respect, our model is explicit. The model illustrates how several factors interact in an energy budget to determine optimal unit size. For a lone female her decision whether to join two individuals on a good territory or one on a worse territory for breeding may depend on energy. If her energy yield fails to bring her above a threshold for breeding (K in Figure 8.6) she should remain a nonbreeder. For males, an energy threshold for breeding may also be postulated, though the energy would, of course, be used differently.

Ecological Conflict between Dominant and Subordinate

Since the model is for individuals, rather than groups, it may be used to examine conflicts of interest. Conflict of ecological interest between a dominant on the best territory and a subordinate who might join is illustrated in Figure 8.7.

A subordinate individual seeking a territory will have a choice between taking a lower-ranking territory alone, such as Z in Figure 8.7, or sharing the best territory. If the result of sharing is a fitness betwen X and Y for both parties, it should be desirable for both to share, since the fitness of the dominant would be above W and the subordinate above Z. A zone of conflict exists where sharing would raise the subordinate's fitness above Z but would lower the dominant's fitness below W and Y. Under these conditions we should expect the subordinate continually to attempt sharing with the dominant and the latter to refuse. If sharing lowers fitness for

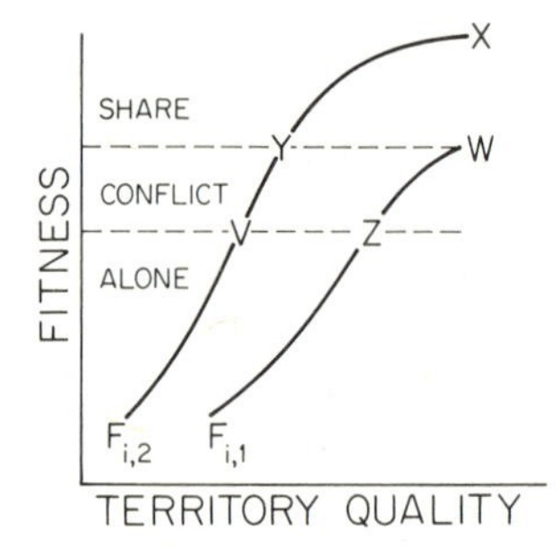

Figure 8.7 Conditions for ecological conflict of interest between a dominant and subordinate. W = dominant alone on best territory. Z = subordinate alone on best available territory. Y = lowest tolerable fitness for sharing by dominant. V = lowest tolerable fitness for sharing by subordinate. The zone of conflict is defined above by the threshold of sociality of the dominant (dashed line through Y and W) and below by the threshold of sociality of the subordinate (dashed line through V and Z). (From Brown, 1985.)

both parties below V, then each would be below its respective sociality threshold for the best territory; each would be better off alone.

In this model both dominant and subordinate may alter their behavior to affect each other's sociality. A subordinate might bring the dominant in the zone of conflict above its sociality threshold (Y) either by lowering costs or raising benefits to the dominant, thereby benefiting the subordinate too by enabling it to share the best territory. A dominant could take advantage of this possibility to raise its own fitness by depressing that of the subordinate. So long as this social suppression maintained the subordinate's fitness above Z, the subordinate should tolerate it. Otherwise, it can leave.

This simple ecological model reveals how competition may persist within a cooperative social relationship. In these cases the fitness axis can be in energy or fitness. The latter can be in direct fitness or inclusive fitness. The addition of indirect fitness would change the costs and benefits to each participant, depending on relatedness, but it would not alter the basic nature of the conflict, which is fundamentally ecological.

Group-size Decisions in a Game-theoretic Context

The preceding treatment employs a simple one-decision-maker model. The model provides a way to think about how an individual should behave if decisions were entirely up to him or her. It allows us to examine the potential relationships among several relevant variables, and it reveals that conflict between individuals with differing optima may ensue. Such models, however, do not reveal the optimal decision when two or more decision-makers are in conflict. Specifically, they do not consider the probabilities of various actions by other individuals nor the consequences of these actions. A simple model that illustrates the potential importance of these considerations was described by Pulliam and Caraco (1984).

For example, consider a situation discussed by Brown (1982a) and Craig (1984). Assume that two groups of group-territorial birds are already at their optimal group size with one-decision-maker models. Each should refuse to admit another individual. If a potential joiner does not join one's own unit as an ally, however, it may join a neighboring unit as a rival or it may leave the area. It may, therefore, be preferable for a dominant to allow the joiner into its group and suffer a small energy loss from depletion of food rather than risk possibly a much greater loss through defense costs or loss of area if the new bird joins a neighboring rival group. The optimal decision depends upon the probability that a neighbor group will recruit the joiner, which, of course depends on its neighbors. Craig has portrayed this situation as a single game of Prisoner's Dilemma (see Chapter 14) and concludes that the optimal solution is to admit the joiner. Pukekos often do admit joiners, and this has negative consequences (loss of space) for rival groups. If the same game is played repeatedly, however, admission might not be the best choice (Axelrod and Hamilton, 1981; see Chapter 14). If the joiner can easily go elsewhere, denial of admission might be a better choice since neighbor groups would be less likely to admit the joiner.

Conclusions

In this chapter we have attempted to view various costs and benefits of communal living in a perspective that is unified around the role of energy and fitness. Certain logical relationships, such as the diminishing return from adding members to a group and the limited amount of food per territory, can be integrated into a comprehensive theory. This theory is not the answer to the problem of communal breeding, but it can lead to answers by suggesting testable hypotheses. An application of the theory described here to the diets of communally breeding birds is developed in Chapter 17.

9 Mutualistic Mating Systems: Polyandry and Uncertain Paternity

The study of mating systems in communal birds has opened new vistas into avian sociobiology. It has extended the classical view through the discovery of mating systems previously unknown in birds, such as cooperative polyandry (Galapagos Hawk) and polygynandry (Pukeko, Acorn Woodpecker), nest-oriented promiscuity (Noisy Miner), and semi-promiscuous monogamy (Ostrich). This chapter and the next explore the implications of these findings for the ecology and evolution of helping.

I suggest that the *sharing of a mate or nest in group-territorial animals provides a widespread mode of origin of helping behavior that is distinctly separate from origins based on the nuclear family.* In mate-sharing and joint-nesting systems breeders typically have no reliable way to distinguish their own young (genetically) from others. These two systems have in common the fact that three or more probable parents have their young in the same nest. Consequently, in their efforts to care for young that have some nonzero probability of being their own they typically also care for young that are not theirs. I refer to such systems as *nest-sharing systems.* In such situations breeders inadvertently become mutual helpers for each other (Skutch, 1961a). Breeding helpers provide one of the classical examples of "helpers-at-the-nest" (see quotation from Skutch in Chapter 1). In this chapter I consider first the various types of mating systems among communal birds, then the consequences of uncertain paternity. The following chapter adds the complication of uncertain maternity.

Mating Systems of Communal Birds

Mating systems are classified traditionally in terms of two factors: (1) the presence or absence of a behavioral *bond* that tends to restrict mating outside the bond, and (2) the presence of one versus two or more members of each sex within the bonded units.

Promiscuity is a mating system without bonds. An individual of either sex is not restricted by conventional bonds from copulating with any available member of the opposite sex within its deme. Promiscuity is most dramatic and conspicuous in lekking species. Lekking does not occur among communal birds, nor does complete promiscuity. Articles claiming promiscuity in Acorn Woodpeckers and some other communal birds have overlooked the term polygynandry (e.g., Stacey, 1979b). Mates are not shared in any useful sense in promiscuity; rather, competition for mates is wide open and all members of one sex in a deme are potentially shared by all members of the other sex. The number of females per male can be very high for some individuals, but this is not a defining feature. It can also be very low.

The four basic mating systems listed in Table 9.1 have in common the importance of a *behavioral bonding* mechanism. Examples of such mechanisms are mate guarding, duetting, chorusing by a communal group (corroboree, Dow, 1971, 1975), and sharing of a territory and group-territorial defense behaviors.

Three of the four entries are well known and need no further introduction. The fourth, *polygynandry*, is common in African Lions, certain primates, Pukekos, and Acorn Woodpeckers, and occurs irregularly in a variety of other communally breeding species. Although this term was not used in the literature on avian mating systems (Emlen and Oring, 1977; Oring, 1982), it has been used in anthropology (Daly and Wilson, 1978). It was introduced to avian sociobiology by Brown (1983a:7). Polygynandry is a mating system in which copulations are shared by two or more males with two or more females within a bonded group such that copulations within the group are the rule and copulations by group members with nonmembers are unusual because of behavioral restrictions arising from the nature of the group. Polygynandry has been described only in communal birds and mammals. Very likely it was overlooked earlier because the mating systems of communal birds and mammals were

Table 9.1

Types of avian mating systems that are based on behavioral bonds, thus excluding intrademic promiscuity.

	Number of Males in Unit	
Number of Females in Unit	1	2+
1	Monogamy	Polyandry
2+	Polygyny	Polygynandry

Table 9.2

Variability in mating systems within a species, as shown by variation in composition of social units in some nonmonogamous species of communally breeding birds. *F* = female. *M* = male. *U* = uncertain sex.

Species	*Number*			*Trios*			*Quartets*			*Larger*			*Percentage of Units*	
Area	*of Units*	*Bachelors*	*Pairs*	*1F*	*2F*	*U*	*1F*	*2 + F*	*U*	*1F*	*2 + F*	*U*	*with > 2 Birds*	*Reference*
Harris' Hawk														
Arizona	50		27	23									46	Mader 1975a,b
Texas	19		18	1									5	Griffin 1976
Galapagos Hawk	31	few	10	13			6			1	1		68	Faaborg et al. 1980
Native Hen	51		26	17	2		4	1		1			49	Ridpath 1972
Dusky Moorhen	48		2	5				19			22		96	Garnett 1980
Pukeko														
stream	13		6										54	Craig 1979
swamp	8		1										88	
Southern Lapwing	20	many	17	2	1								15	Walters 1980
Lonnberg Skua														
New Zealand	50	many	34	16									32	Young 1978
Southern	704	many	700			4							1	
Purple-throated														
Fruit-crow	3			2				1					100	Snow 1971
Dunnock														
Edinburgh	13	3	7	4	1			1					46	Birkhead 1981
Cambridge	25		10	15									60	Davies 1983
Black Tit	19		8	9				1		1			58	Tarboton 1981

not carefully examined until the recent surge of interest in cooperative and altruistic behaviors.

Two bugaboos of such classifications are (1) variability of mating systems within a population and (2) intermediate systems. These will be considered especially in the discussion of Ostriches, Dunnocks and miners. Table 9.2 illustrates the extent of variation found in some simple, nonmonogamous, avian communal breeding systems. Typically, more than one mating type is present in a population. Monogamous social units are often found in the same populations with polyandrous units and sometimes with polygynous units. Some populations of Acorn Woodpeckers (discussed in Chapter 10) include every type of unit shown in Table 9.1. No species listed in Table 9.2 is totally polyandrous, totally polygynous, or totally polygynandrous. Such labels when applied to species are only suggestive and may not even indicate the most common system; they should not be interpreted too strictly.

We shall consider first the simpler systems of mate sharing and then broaden the discussion to include the much more complex polygynandrous forms.

Polyandrous Communal Groups

In certain communally breeding species the helpers are breeding males who share copulations and care of the young with a single female at a single nest. This condition has been termed wife-sharing (Maynard Smith and Ridpath, 1972) or cooperative polyandry (Faaborg and Patterson, 1981). It differs from other forms of polyandry in that the female (1) has only one nest at a time, (2) shares incubation, care of the young, and a territory with the males, and, therefore, (3) does not adopt a male role. Neither sex-role reversal nor reversed sexual dimorphism in plumage or size are found in communal polyandrous species.

The ecological origins of polyandry in communal birds are diverse. Few common threads connect the various cases. Therefore, we shall not attempt any general explanation. Instead we look at various causal factors, each of which is restricted to only a few species.

Richness of Food Supply? Skua

Skuas are large, predatory seabirds that are most abundant around seabird aggregations in the polar regions. In such severe climates they rarely have helpers; however, in a population inhabiting the relatively warm seas near New Zealand helpers occur in roughly a third of the breeding units (Young,

1978). These units are trios, consisting usually of a female and two breeding males, all in adult plumage. Breeding success at one of the New Zealand colonies averaged higher than in the more polar colonies due apparently to the more frequent survival of the smaller chick in a brood of two (1.55 chicks per unit near New Zealand; less than 1 per unit in Antarctic colonies). The food supply at this colony was rich and consisted mainly of nocturnal, small petrels. Here skuas loafed by day and foraged by night. The occurrence of trios was correlated with a less severe climate and a richer food supply for the entire population. Pairs here bred more successfully than pairs in polar regions. Survival of the chicks was so high for pairs that it would be difficult for trios with one female to surpass it. A sample of five trios did not average higher reproductive success than a sample of five pairs. The explanation for the occurrence of trios, therefore, probably resides elsewhere, perhaps in the defense of territories, which remains to be studied. The richness of the food supply for populations with a high frequency of trios might be a permissive factor. Trio territories probably are rich enough that the additional bird does not depress the food in the territory significantly. This amounts to having a flat food-cost function, as in Figure 8.4b. In addition, the relatively high rate of breeding success may affect the population density of skuas and hence the intruder pressure, making it difficult for pairs to maintain occupancy without the aid of another bird. All members of trios shared in defense of territory and nest site.

Variance in Food Intake: Hawks

The breeding unit in most hawks is a mated pair, but in the Galapagos Hawk (deVries, 1973, 1975) and Harris' Hawk (Mader, 1975a,b, 1979; Bednarz, 1985; Dawson and Mannan, 1985) trios constitute roughly a half or more of the units in some populations (Table 9.2). In most cases trios consist of a female and two males. Both males copulated with the female, at least in some trios. The extra males are not known to be related in the Galapagos Hawk, though data on kinship are sparse and inconclusive. In Harris' Hawk some, perhaps most, of the extra birds are young from a previous brood.

The excess of males in the adult population of Arizona Harris' Hawks is conducive to polyandry (31 males to 11 females among adults but 51 males to 47 females in nestlings). In Harris' Hawk, however, recent studies suggest that polyandry might be rare (Bednarz, 1985; Dawson and Mannan, 1985).

In the Galapagos Hawk sex ratio was conducive to polyandry in 1979 (120 males to 83 females) but not in 1977 (119 males to 124 females; Faaborg et al., 1980). Further explanations are needed in any case.

Trios of Harris' Hawks average 1.96 chicks (raised to 28 days) per nest attempt ($n = 23$), compared to 1.30 for pairs ($n = 27$), but the difference was not significant. Clutch sizes were roughly equal for trios and pairs (2.9, trio; 3.0, pair; Mader, 1975a). On a *per male* basis breeding success per nest was lower in trios than pairs in both Harris' Hawks (Mader, 1975a) and Galapagos Hawks (Faaborg et al., 1980:585). Since Harris' Hawks have a long breeding season, much longer than Redtailed Hawks on the same study area in Arizona (Mader, 1978), a more revealing indicator of the relative success of trios and pairs would be a longer-term measure, such as summed reproductive success over a year or more. The species may have three clutches per year (Mader, 1977). Trios were reported at the earliest and two latest nestings recorded by Mader (1978:332), as well as at other late or early nestings (Radke and Klimosewski, 1977; Ellis and Whaley, 1979). Lacking a long-term measure, I agree very tentatively with Mader that the advantage of trios is biologically meaningful, especially for the female of this highly sexually dimorphic hawk. Why?

The males both supplement the female's incubation, feed her, harass potential nest predators, and feed the young, who may be partially dependent for several months. Harris' Hawks are agile and feed on fast-moving prey, mainly Desert Cottontails (*Sylvilagus auduboni*). They hunt cooperatively and share large prey, such as the cottontails. On most days of incubation at one nest watched intensively the female received all her food from her two males.

Compared to pairs, one might expect that trios: (1) could bring nearly twice as much food to the incubating female and bring it with greater consistency and reliability; (2) could bring more food to the young, also with greater consistency and reliability; (3) could provide greater consistency and reliability of food to each adult member on a per capita basis even when not nesting, since large items are shared and hunting is cooperative. These factors should benefit all members of trios provided the addition of a third bird does not depress the food supply too much (i.e., low food-cost function, as in Fig. 8.4b).

All three factors act to *reduce variance* in food intake. The first two also increase the mean. The first two benefit the males by means of the nonzero probability of genetic relatedness of the males to the female or young. The third is independent of relatedness, except as it affects breeding; it depends upon reciprocal aid-giving. When one member of a trio fails to find food on a given day, it can obtain food from a successful member.

The reduction of variance in food intake is especially important when the variance is high. High variance may be expected (1) when the diet consists of a small number of large items, (2) when little time is available each day to hunt, and (3) when periods of food scarcity are unpredictable. All of these conditions are found in the Harris' Hawks of Arizona. In the desert, cottontails are not active in daylight except at dawn and dusk. Therefore, only two brief "windows" of time are avaiable each day for effective hunting (J. C., Bednarz, A.O.U. Meeting, 1984). Cottontail populations are affected by the rains, which vary erratically in Arizona.

The situation resembles an iterated game that is complicated by kinship to the young (see Chapter 14 for a description of a similar game). Cheating by failing to share prey with a trio member has a short-term gain that is the largest payoff available at the time; however, the cheater runs two risks. First, when he needs food, the other male may not share. Second, the breakdown of the sharing mechanism jeopardizes the incubating female and young of the cheater (or of its parents) in the long run.

Trios do not always form. In Texas populations of Harris' Hawks, trios appear to be rare (Griffin, 1976). This suggests that a key element in the above theory is missing in the Texas environment. Further research is needed to test this hypothesis.

The situation in the Galapagos Hawk is similar in many respects to that of the Harris' Hawk. Their life-history features are similar, and trios do a little better at reproduction than pairs. In the Galapagos Hawk, however, variance in the daily food intake appears to me to be less important and the saturation of suitable habitat for breeding territories seems more important than in the Harris' Hawk, in which territorial defense is not obvious. In the Galapagos Hawk the prey are smaller, the hawk less agile, and a large surplus of nonterritorial birds of both sexes is conspicuous.

Brotherly Love among Native Hens

Inclusive fitness theory was first applied quantitatively to communal birds by Maynard Smith and Ridpath (1972), who analyzed the three-year study of Ridpath (1972) on a partially isolated population of Native Hens in Tasmania. These flightless birds inhabit open grassy areas near water and eat green herbage and seeds. Pairs, two-male trios, or occasionally larger units (Table 9.2) live in territories with their young, who remain with them until the next breeding season, at which time they are usually driven out.

The commonest method of formation of a trio was for two brothers to join an unrelated female ($N = 6$). In six of seven other cases of known origin at least two members of the unit were brother or sister to each other. Substitution of a member in a pre-existing group was common; formation of new groups was rare. All members of a unit shared in defense of the territory, nest building, incubation, and care of chicks. Copulations with the female were shared when two males were together.

Interest centers on the question of why trios and larger breeding units occur at all. From the viewpoint of the female there are two advantages of trios over pairs: (1) more eggs are produced, and (2) survival of the young is improved. The data, therefore, suggest that females should encourage polyandry.

From the viewpoint of the males there is an obvious disadvantage, partial loss of paternity, that must be compensated (assuming males have a choice). Four types of compensation seem likely: (1) and (2) as described for females above, (3) kinship with the offspring of a brother or sister, and (4) enhanced probability of obtaining or expanding a territory. The difficulties of males in breeding are exacerbated by the adult sex ratio. Although a sample of 489 birds shot during hunting season yielded a sex ratio insignificantly different from 50:50 (55%), males outnumbered females 14:5 in four clutches from which all eggs survived and in the adult population (111 males, 74 females for the three breeding seasons combined).

In other species, males fight for females on a winner-take-all basis. Why does this not happen in the Native Hen? Risk of injury resulting in lifelong damage is one hypothesis. Risk of losing the female entirely is a second. Neither of these is completely convincing. First, rails and gallinules are no more deadly than most birds, and the loser could probably withdraw before injury after the outcome became obvious. Second, the cooperating males tend to come from the same family, hence know each other well enough to predict the outcome. Therefore, we must revise the question to read, "Why does the dominant not drive out the subordinate?"

Maynard Smith and Ridpath formulated this question mathematically, but they did not answer it. They assumed that two types of individual might occur as follows:

D: If dominant, *D* attempts to drive out rival males by fighting.
d: If attacked, *d* can win a fight if dominant, but it does not provoke a fight.

Now define fitness of males as follows:

P = number offspring in a pair over a lifetime;
T = number of offspring in a trio with another male over a lifetime;
L = number of offspring of a loser that must leave his trio, over a lifetime.

If we assume that males already know the potential outcome of a fight for exclusive possession of their female, then it is reasonable to assume that D will fight only when D is sure of winning. When D is subordinate it will share the female with d equally. Hence we need to examine only the case in which D could win if it fought. In this case D would be selected if a paired male has greater lifetime reproductive success than a male who shares copulations equally in a trio:

$$P > T/2.$$

Maynard Smith and Ridpath estimated lifetime reproductive success for T and P from the data in Table 9.3 using the observation that mature units were 1.6 as common as first-year units. They argued that lifetime reproductive success could be estimated using the relative frequencies of mature and first-year groups. Over three years this ratio was 1.6:1.0 and ranged from 0.9 to 4.3. Therefore,

$P = 1.1 + 1.6(5.5) = 9.9$ surviving young;
$T = 3.1 + 1.6(6.5) = 13.5$ surviving young;
$T/2 = 6.75$ surviving young.

Since P is greater than $T/2$, D should be selected; but the prevalence of trios shows that D has not been selected. Therefore, this model is inadequate.

The authors felt that the inclusion of indirect fitness would improve their model. Consequently, they added terms for the direct fitness of the brother weighted by $r_{D,d}$, the relatedness of D to the losing d type. D is selected if

$$P + r_{D,d}L > T/2 + r_{D,d}T/2.$$

When $r = 1/2$, D is selected if

$$P > 3T/4 - L/2.$$

If losers fare badly (L near zero) this is a more stringent criterion for D and, therefore, more conducive to sharing than $P > T/2$. If losers could

Table 9.3

Comparison of trios and pairs of the Native Hen. (From Maynard Smith and Ridpath, 1972).

	Mature Groups		*First-year Groups*	
Breeding Unit	*Number of Units*	*Surviving Young*	*Number of Units*	*Surviving Young*
Pair	22	5.5	15	1.1
Trio	24	6.5	7	3.1

easily re-establish themselves into another trio ($L = T/2$) then monopolization of the female by D would be easier, the criterion becoming as before $P > T/2$. Ease of entering into a trio would diminish, however, as P increased, thus forcing D's criterion toward the easier version for aggression.

The fate of losers is heavily influenced by the sex ratio. If the number of breeding females equaled or exceeded the number of males, then all males could pair, and none should share, the criterion for D being simply $P > T/2$.

We have considered the case in which the males in a trio can predict the outcome of an attempt at monopolization of the female with complete certainty. Now let us suppose that the males *cannot predict* the outcome. We assume now that D males always fight but win only half the time. The criterion for D to be selected over the sharing genotype if relatedness is ignored would be $P > T - L$. This is a more stringent criterion for aggression than $P > T/2$ if the cost of losing is high (L near zero), but it reverts to $P > T/2$ when $L = T/2$. If males cannot predict the outcome and win only half the fights, relatedness does not alter the criterion, which remains $P > T - L$.

Maynard Smith and Ridpath concluded as follows: "If the above analysis is correct, wife sharing in the native hen is an example of kin selection, whereby the increase in frequency of a gene depends on . . . relatives of individuals with that gene." This statement is deceptive. In my view, their paper does *not* justify the conclusion that indirect selection has been a critical factor in the evolution of sharing or even that it has been important. It does justify the speculation that indirect selection *might* have been involved, and it shows that two important conditions have been met, close kinship and $T > P$.

What then do we need to carry the analysis further? For all the criteria discussed, the value of L is critical. This is the only variable whose value was not estimated. L reflects the difficulty of obtaining a female and a territory. Thus L should be heavily influenced by the sex ratio and by territoriality in the conventional sense (Brown, 1975a). In fact, Ridpath makes it clear that the sex ratio is skewed from hatching (probably) and among breeding adults on the study area (certainly) and that territoriality is conspicuous. Territorial defense was vigorous. He wrote of the young: "Only a few succeeded in the face of the opposition from surrounding groups. Territorial behaviour greatly reduced the density of the population during the breeding season" (1972:53). These data suggest that L is very low, probably less than $P/2$ and less than $T/4$, but we need actual field estimates of L over a series of years. Specifically, what fraction of birds entering the breeding season (10–11 months of age) acquire a mate in their first year and what are the life table statistics (l_x, m_x, see Chapter 4) of successful and unsuccessful males?

The authors did not consider the hypothesis of habitat saturation, which had been mentioned by Selander (1964) and developed more generally by Brown (1969a). Since a female should prefer polyandry to monogamy she might encourage both males, making it difficult for a dominant to completely exclude subordinates from copulations.

This study is consistent with the following effects:

(1) A strong effect of sex ratio in causing low L;
(2) A strong effect of territoriality in causing low L;
(3) A conducive or even an important effect of kinship in ameliorating the criterion for sharing.

Although Ridpath's field study on the Native Hen was impressive and the analysis with Maynard Smith was thought-provoking, I feel that re-examination of this case in view of more recent work on polyandrous systems would be justified. In particular, the following studies on Dunnocks suggest new insights.

The "Undefendability" of Female Dunnocks

Close scrutiny of the Dunnock by a series of workers has led to several new insights into the sociobiology of helping behavior. These insights bear on size of food items, cover, and female infidelity. They enable the rejection of various hypotheses for helping in this species.

The Dunnock is one of twelve species in a small family (Prunellidae) of inconspicuous, little brown birds that inhabit scrubby vegetation across Eurasia. Helping behavior was first described in the Prunellidae in the Alpine Accentor (Dyrcz, 1977), followed shortly by studies of helping in the congeneric Dunnock (Birkhead, 1981; Snow and Snow, 1982). These studies laid the groundwork for systematic observations, experiments, and modeling (Davies, 1983, 1985, 1986; Davies and Houston, 1986; Davies and Lundberg, 1984; Houston and Davies, 1985). The following account is based on these studies.

Contrary to previously prevailing opinion, Dunnocks frequently breed in social units larger than a simple monogamous, mated pair. This could not have been realized without detailed studies of color-banded birds. Some idea of the variability found in breeding units of Dunnocks is conveyed in Table 9.2. In Edinburgh, U.K., pairs and larger groups up to four were equally common; in Cambridge, U.K., trios and larger groups predominated over pairs. Most units larger than a pair were biandrous trios; however, polygyny occurred once naturally, as did polygynandry. The latter was also created by an ecological experiment.

In the social structure of passerine birds much depends upon the manner of establishment of territories. In most migrant species the males arrive early, attempt to monopolize critical resources, and await the females. In the British Isles Dunnocks spend the winter on the breeding grounds. Groups of fairly constant composition occur around good breeding sites all year. In winter and spring readjustments of home ranges and group composition occur. Established birds of both sexes breed on almost exactly the same territories from year to year. Tactics of males, therefore, begin with residence in a group and involve attempts to maintain a holding or to found and enlarge a new one against continual opposition. Sometimes severe compromises need to be made and territories are shared (Fig. 9.1b,d).

Two female Dunnocks do not share their foraging areas with each other even if these areas lie within their male's territory (Fig. 9.1c,d). It is only the males that are forced to compromise their sexual interests. This is consistent with the excess of males per female in the population (1.13,

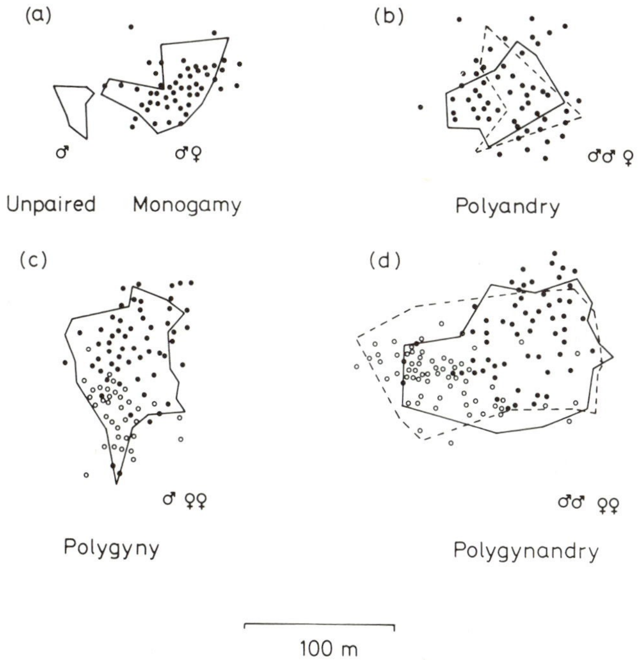

Figure 9.1 Examples of the different mating combinations in the Dunnock. Male song polygons are shown by dashed and solid lines; female ranges, by points (open and closed circles) representing a first sighting during a transect around the study area. Female ranges are largely exclusive; but when two males share one or two females, their song polygons overlap. (From Davies and Lundberg, 1984.)

1.48, 1.26 males per female for 1981–83), which stems from differential winter mortality.

In typical group territoriality all members of a group share all parts of the group territory. Dunnocks are a clear exception to this pattern.

Helping in Dunnocks, as in all the species considered so far in this chapter, is the joint product of constrained sexual selection and direct kin selection. As these terms are often misused, let me clarify (1) that sexual selection is selection based on competitive mating, and (2) that care of offspring that are likely to be one's own is kin selected (Maynard Smith, 1964; see quotation in Chapter 4). The remainder of this treatment develops these two themes.

First, we consider the question of mate sharing. In most passerine species contests over females are resolved on a winner-take-all basis. Are there ecological factors that mediate against this in Dunnocks? Size of territory appears to be one of the important variables for answering this question. The most successful males have a territory that is large enough and rich enough for one or more females but small enough to be economically defendable. As shown in Figure 9.2 the success of a male is correlated with the size of his territory. More commonly, two small male territories may be included in one female territory. In such cases the males merged their territories and ended up establishing a dominance relationship and

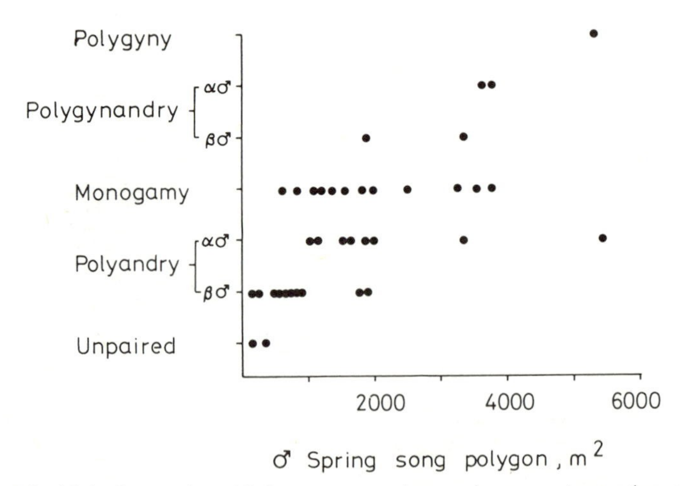

Figure 9.2 Male Dunnocks with larger song polygons have greater mating success. Spearman rank correlation, corrected for ties, = 0.641, $P < 0.01$. The different mating combinations have been ranked up the *y* axis in increasing order of male mating success. Spring song polygons were measured in February and early March before any coalescence of male territories had occurred. (From Davies and Lundberg, 1984.)

sharing the female. Trios formed by coalescence of formerly distinct male territories in fourteen of sixteen cases, males adjusting their territories to the female's movements, rather than vice versa.

Quality of territory is even more important. Dunnocks forage for tiny seeds and invertebrates on the ground in dense vegetation. Suitable vegetation on Davies's study area (a botanical garden) was patchily dispersed. Range size of females was inversely correlated with range quality, as judged by percentage of the range with "good feeding habitat." It appeared, therefore, that the number of males per female depended on the size of a female's home range (relative to a male's), with larger sizes averaging lower quality than smaller ones.

Rephrasing these observations in terms of a hypothesis, the mating system was dependent on quality of the female's range. A testable prediction is that if female ranges could be made richer, they would contract in size, with the result that the number of monogamous pairs would increase relative to the number of biandrous pairs, in the affected area. Davies and Lundberg (1984) tested the prediction by putting extra food into the home ranges of females. The results were as follows: (1) females spent 36% of their foraging time on the provisioned food; (2) home ranges with feeders were smaller than those without for females but not for males; (3) on average, males on feeder territories had more females per male than did control males.

The role of the female Dunnock is critical. The nestlings of females with two males were fed more frequently and were heavier on average at all the normal brood sizes than young of monogamous females. This puts the female in conflict with the males. *She* profits more in polyandry; *he*, in monogamy. The result is sexual strife, infidelity, intense mate guarding, and uncertainty of paternity. Females actively encourage copulations from the beta male and the alpha male. The alpha male guards the female assiduously, but due to the dense cover he cannot easily find or drive out the beta male, nor can he completely prevent copulations by the beta male. Thus, the alpha male does not choose polyandry; *he chooses monogamy and gets polyandry.*

Beta males will not feed the nestlings unless they have copulated with the female. Even then they feed less than the alpha male (Houston and Davies, 1985). This discrimination represents a crude form of kin recognition based on a simple rule of thumb, such as "feed young of a particular female if I have copulated with her, but not if I have not, and at a rate that reflects my chances of being the father." There is also a suspiciously high frequency of failure among nests of females not known to have copulated with the beta male; however, causation by the beta male has not been confirmed.

Copulation does not, of course, guarantee paternity. Before copulating, males peck the cloaca of the female, causing her "to eject a small pale mass" from which sperm have been recovered (Davies, 1983). This mechanism should benefit the alpha male most since he does 60% of the copulations when they are shared. We do not know how effective it is.

The rate of feeding of nestlings is sensitive to the number of feeders. This is not true in several other species of communally breeding birds (Chapter 11). The rate may reach 27 feeds per brood per hour for trios, which is rather high for small passerines. Could it be that the system depends on the *smallness of the food items* that are utilized by Dunnocks? Small items tend to require many trips to the nest; large items could be brought less frequently. Small items might be accumulated before a trip, but this might result in some items being lost by slipping out of the bill. In the dense cover where Dunnocks breed, frequent visits to the nest might not draw the attention of predators to the nest as much as they would in more open habitats.

We may now generalize these findings from the Dunnock to consider various hypotheses for the evolution of helping and communal breeding in mate-sharing systems. The following alternative hypotheses can be rejected:

(1) Males choose to be bigamous in a good territory in preference to monogamy in a bad territory (the male version of the polygyny threshold model). False. The male is not choosing polyandry; it is forced on him by circumstances.

(2) Males choose to share the female because their reproductive success is higher if they share territorial defense (Stacey, 1979b). False. The male's gain in defense is less than his loss of paternity.

(3) Males choose to share the female because in such a food-poor environment a helper is needed to feed the young (Gowaty, 1981). False. Males do better with monogamy.

(4) Males choose to share because they are related (Maynard Smith and Ridpath, 1972). False. The males were not related to each other.

(5) It is advantageous to males to share to reduce the variance in the daily food intake (this chapter and Chapter 14). Unlikely, because of small item size.

(6) Males by feeding each other's young are engaging in some kind of score-keeping reciprocity, such as Tit-for-tat as envisioned by Ligon (1983). False, as shown by Davies (1986). Males do not share the female willingly. They share feeding of the young because of a nonzero probability of kinship with the young.

(7) Males help in order to acquire a territory in the future (Woofenden and Fitzpatrick, 1978). False. Most sharing males already have a territory before sharing.

What is the optimal feeding effort for an alpha or beta male or a female? Optimality models, such as that of Houston and Davies (1985) combined with controlled field experiments (not just "natural experiments") are needed. Houston and Davies have constructed a model based on parental confidence, clutch size, effort of others, survival of feeder, and survival of young. The model has been designed to fit current knowledge, which it does. Further testing is desirable.

More questions remain. Why don't males refuse to feed the young altogether, as in many promiscuous species whose confidence in paternity is sometimes higher than an alpha Dunnock's? In brief, males who default on feedings would lower the total amount of food brought to the young, causing a reduction in success of the brood and of their fitness. If this were not true it would pay a male to cheat. Other males do not compensate entirely for the loss.

What hypotheses have not been rejected? One reason for this chapter is to spotlight the importance of shared paternity as an evolutionary pathway to helping behavior. Although a nonzero probability of kinship with the young explains in part the sharing of feedings, it does not explain the sharing of the female. In Dunnocks wife-sharing seems to be based on the density of the cover; the need for frequent feeding of the nestlings, perhaps due to small item size, resulting in advantage to the female from extra males; and the inability of one male to monopolize a female in the face of her efforts to solicit other males.

In helper systems that are based on shared paternity, the total feeding rate of a brood of nestlings appears not to be completely compensated for the loss of a feeder. When a feeder is lost, the total feeding rate drops. In contrast, in helper systems not based on shared paternity, compensation may be better and perhaps even complete under normal conditions. For example, Dunnocks, in which helping is based on shared paternity, do *not* compensate fully. Grey-crowned babblers, in which helping is probably not based on shared paternity, *do* compensate fully—at least in a productive year (Brown et al., 1978).

Female Miners Sell Sex for Paternal Care

The Noisy Miner of Australia goes beyond polyandry toward a kind of *nest-oriented promiscuity.* Copulation as a means of securing paternal care at the nest has been carried about as far as it can go in the Noisy Miner. The miners are a genus of three species of small (60 g) arboreal insectivorous honeyeaters (Meliphagidae). The Noisy Miner was among the first species of communally breeding birds to be studied with colored leg bands (Dow, 1970). Despite much study, its social organization remains an enigma full of fascinating evolutionary problems. It is unique among communal birds and defies categorization.

Female Noisy Miners live in home ranges that in the breeding season are mainly mutually exclusive (Fig. 9.4). Home ranges of breeding males, in contrast, overlap broadly with each other and those of several females (Fig. 9.3; Dow, 1979a). The birds are typically seen in groups, but their members clearly do not all have the same home range (Figs. 9.3, 9.4) during the breeding season, as would group-territorial species. A group territory in relation to reproduction cannot be identified in Noisy Miners. Nests are defended from predators by groups of males; however, males differ in which nests they frequent.

Only one female incubates the eggs at a given nest (Dow, 1978a,b). Care of her young is provided by herself and up to 22 males (usually 9 males; Dow, 1979a). Females "almost never visited the nests of other females and only rarely fed their fledged offspring" (Dow, 1979a).

In a three-year study Dow (1978a) took notes on 134 copulations and observed many more. This contrasts with the 6 copulations seen by Rowley

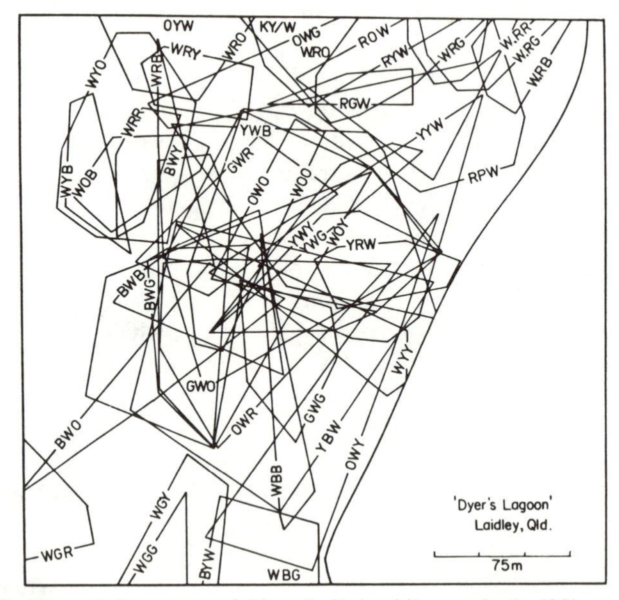

Figure 9.3 The activity spaces of 40 male Noisy Miners wholly (28) or partly (12) resident in the study area from January–May 1972. Four others, birds which appeared only sporadically, are shown without boundaries near the edge of the area. Each maximum polygon (activity space) was constructed from the points at which that bird was recorded. (From Dow, 1979b.)

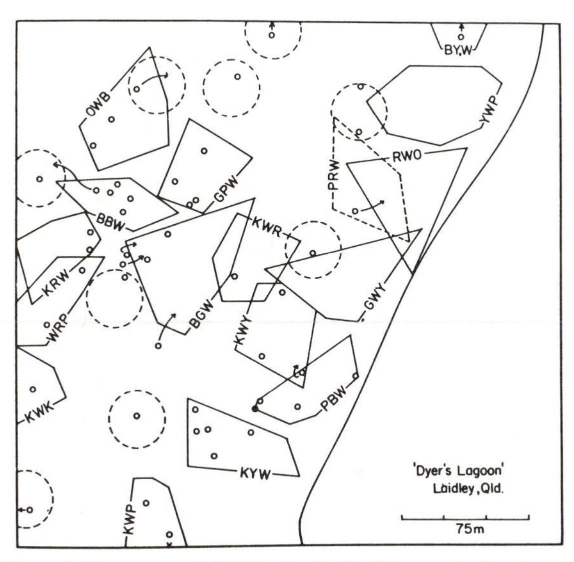

Figure 9.4 The activity spaces of 17 female Noisy Miners wholly or partly resident in the study area from January–July 1972. Dashed circles show the general location of unbanded females during the nesting season of 1972. Successive nest locations (sites selected after activity spaces mapped) are shown by small circles. The dashed range is of a female (PRW) banded after KWR and GWY had been killed and while RWD was temporarily absent. Arrows indicate the female to which a nest belonged. (From Dow, 1979b.)

(1965a) in a five-year study of the Superb Blue Wren and two copulations seen by Woolfenden (1975) in five years of study of the Scrub Jay. No other communal bird rivals the Noisy Miner in frequency of observed copulations.

Copulations were not constrained by social bonds, but "the dominant male in a coalition near a nest appeared to achieve most of the copulations" (Dow, 1978a: 164). Females were observed copulating with more than one male on the same day. Likewise, males often attempted copulation with more than one female on one day. Copulations occurred at all times of day and in every month, though they were more frequent in July during nesting. "Females copulated in all stages of nesting as well as outside the breeding season" (Dow, 1978a: 168). During nesting "groups of up to 30 birds gathered in the canopies of a few adjacent trees," and many copulations were seen. "Violent attacks and chases occurred on all sides. Sometimes feathers were torn out in flight chases. Fights ensued and scuffling birds dropped, claws locked, to the ground. I have seen as many as seven duos of fighting miners drop around me in ten minutes" (Dow, 1978a: 168).

Usually such violent activity "occurred near the site of a nest under construction."

Unfortunately, data on reproductive success of females as a function of the number of males caring for their young are not available. There are, however, two reasons why the number of males may be important. First, most of the feedings are by males (Dow, 1979a). Second, the males are very aggressive toward birds and other animals dangerous to the young, as well as to harmless species (Dow, 1977).

With regard to the feeding of the young, miners have special features that are suggestive of a strong reliance on a large number of feeders. The rate at which the young are fed in terms of visits per hour is one of the wonders of the bird world. Feeding rates of nestlings by males reached 55/hr and of fledglings, 80/hr (Dow, 1979a). The maximum rates at four nests were 34, 44, 49, and 71 visits/hr when all visitors of both sexes were included (Dow, 1970). Few passerine species exceed 20 visits/hr (Nice, 1943, Moreau, 1947).

Why this high rate? Perhaps it is because miners are canopy foliage-gleaners and feed on very small prey (Dow, 1978a:181). Very small items, such as scale insects, appeared to account for a high proportion of the diet of nestlings, and very few were brought at a trip. In any case, females act as if it were important to them that males know the location of their nests; they perform a head-up flight display on the way to their nest when building.

"Copulation appears to be a means by which females maintain bonds with males, because Sexual Driving and copulation can be observed at any time of year. . . . A female's participation in such sexual activity might relate to the number of males visiting her nest and later feeding her young. . . . The degree of participation by a male in nest affairs could well be related to the probability of his genetic investment in its contents, because males with an apparently higher probability show more defense of the nest and less interruption of the incubating female" (Dow, 1978a:182).

Why do female Noisy Miners employ this method to an extreme while other species do not? Several levels of explanation may be needed. To begin, the size of food items delivered to the young is tiny. The required number of feedings per hour, f, of items of caloric value, i, for a brood of biomass b grams needing c calories per gram, assuming one item per trip would be:

$$\text{Required feeding rate} = f = bc/i.$$

Therefore, the required feeding rate becomes higher at an increasing rate as item size (or calories delivered per trip), i, diminishes. If many trips are needed, the energetic cost of delivery may become relatively more important than the energy content of the item.

Given a small item size, it may be extremely difficult for a female with only one male to feed the brood. She would then profit by extra feeders, most of which must be males since most females have their own nests. For simplicity assume that a female with one male can produce G young but that with infinite help she could produce an additional R young, constrained by clutch size. If the number of additional males is m, then a simple function expressing a female's average number of young per nest, Y, is:

$$\text{Number of young per nest} = Y = G + R\,(1 - e^{-sm}),$$

when s is a scaling factor that affects the shape of the curve.

The female's game then is to attract males. She is aided by the sex ratio, which was 2.2 and 3.3 males to females in one population in two years (Dow, 1973, 1978b). Thus, females competing for surplus males can expect 1.2 to 2.3 "additional" males per female. Females would also be competing for "each other's" males, even under a more even sex ratio. If females are consistently successful at attracting extra males—and it appears that they are—then they should lay a larger clutch to take full advantage of the maximum possible value of R. One might predict, therefore, that clutch size would be higher in populations with a higher sex ratio. By the same reasoning, rates of feeding young should also reflect the sex ratio. If so, experimental evening of the sex ratio by removal of extra males should lower reproductive success per nest.

Since this system is influenced by size of food items delivered to nestlings, the determinants of item size are of interest. Several factors may be important.

(1) Small nestlings may require small items, especially if adults have trouble dividing large items into pieces, as in most passerines.

(2) The species may be adapted to small item sizes in its feeding ecology, and this may be true for miners generally. It is known, however, that miners do consume some large items, such as large caterpillars, moths, mantids, and even a tree frog (Dow, 1977).

(3) As suggested by Dow, adults living at high densities may consume the larger items themselves leaving only the smaller items for nestlings. This could be tested by comparing item sizes and feeding rates in high and low density areas with controls for item size abundance.

(4) It might benefit the males socially to make many visits. The sight of a male feeding the young might increase a male's chances of paternity in the future somehow, either by influencing the female or other males. This could be done most cheaply by bringing the smallest possible items.

(5) There might also be some value in developing a social relationship with the young that might benefit the feeder's ability to escape predation or get a mate in the future.

The behavior of males is also puzzling. Given their relative promiscuity, why do they feed the young at all? At the other extreme, why do not individual males monopolize one or more females?

It is easy to understand why a male would feed his own young or young that were probably his, but the large number of males at some miner nests strains our credulity. There are only 2–4 eggs in a nest, yet up to 22 males may feed them. Have they all copulated with the female? Obviously, most males are not the father. Furthermore, it would seem that no male could be sure of being the father. Worse yet, even the probability of being a father of any young at such a nest must be low. The patterns of visits by males to nests are complicated. The mean number of nests visited by 33 marked males was 5.2 (Dow, 1979a). Some males were seen at only one nest; others were seen at fourteen. Fidelity to a single nest was greater than it might seem from these data. In addition to the males who visited only one nest, others concentrated their attention at one nest. On the other hand, some males distributed their efforts rather evenly.

A possibility in such systems is that of alternative male strategies. A dominant male appears to get most of the copulations at each nest. The system may, therefore, approach monogamy for dominant males. Subordinate males might do better to spread their efforts more evenly among females and nests. With a highly skewed sex ratio or if males are dominant at more than one nest, the subordinates constitute a large fraction of the population. This would make exclusion of rival males more difficult for a dominant male. If several subordinates approached a receptive female together, one or more could copulate while the dominant male chased the others. On the other hand, a dominant might tolerate copulations of "his" female with other males if they were outside of her fertile period and resulted in more alloparental care of the young.

Conclusions

Polyandrous communal birds rarely form large, close-knit groups. Trios are common, but not quartets. Their typically small group size sets the polyandrous species apart from species in which communality is based originally on nuclear families. In the latter species, groups may be large because young from one to several broods stay in the group one to three years or longer before dispersing. Of course, the two modes of origin may be combined (as in Acorn Woodpeckers and Pukekos, Chapter 10), which also leads to larger groups; and polyandry may lead to larger groups through polygynandry. Nevertheless, we may ask the rhetorical question, why are polyandrous units so small?

The answer appears to lie in simple arithmetic. Because the chances of being the genetic father rapidly diminish as more males are added, the addition of males quickly becomes unprofitable for the males.

Genetically, the polyandrous species of communal birds have one feature in common. The males share a nonzero probability of kinship with the young they care for. As in all explanations involving kin selection, relatedness is not by itself a sufficient answer to the problem of cooperation or mutualism. Ecological explanations are also needed.

The ecological problem boils down to one of costs and benefits to the dominant male. Why does the cost of excluding subordinate males exceed the benefits—as we presume it does? After our examination of only a small number of species, a plethora of suggested ecological factors has appeared: environmental richness, territorial defense and habitat saturation, territory heterogeneity, sex ratio, size of prey, foraging habits, type of nesting habitat, patchiness of nesting habitat, size of female's home range. The list is likely to be extended. Brotherhood may facilitate polyandry, but more cases have been found in which cooperating males are thought to be unrelated than related. It seems that the only appropriate conclusion to this chapter is "more work is needed."

10 Mutualistic Mating Systems: Joint Nesting and Uncertain Maternity

Mate sharing in polyandrous systems may seem complex, but it is relatively simple compared to the systems considered in this chapter. When a second female is added to a polyandrous unit, polygynandry occurs, as in the Acorn Woodpecker and Pukeko. This possibility raises a new set of questions concerning the female in particular. In addition to strict polygynandry, other unusual mating systems also involve two or more males copulating with two or more females, as in miners and ostriches. Even monogamous species may share a nest, leading to complex decisions, as in Groove-billed Anis. In this chapter we survey a variety of social systems that share the feature of multiple females laying in a single nest, known as joint nesting. We begin with a species whose mating systems are extemely variable, forming a natural bridge from the preceding chapter to this one.

Crossing the Polygynandry Threshold with the Acorn Woodpecker

Mating systems in the Acorn Woodpecker run from monogamous pairs through polyandry and polygyny to polygynandry (Koenig et al., 1984), thus providing an opportunity to seek environmental determinants of mating systems and their associated cooperative and selfish behaviors. The species is notable for its conspicuous linkage with a simple food resource, namely, acorns. These woodpeckers store acorns and even pinyon nuts (Stacey and Jansma, 1977), not secretly in the ground as do many jays, but openly concentrated in specially constructed, acorn-sized holes, usually in a prominent, large tree branch or trunk (Michael, 1926; Ritter, 1938; MacRoberts and MacRoberts, 1976; Gutierrez and Koenig, 1978). Similar storage sites and behavior are found at most study localities in the United States, as well as some in Central America (Eisenmann, 1946; Skutch, 1969b; Koenig and Williams, 1979; Stacey, 1981). These sites are commonly referred to as granaries. The birds depend upon these stores through the winter and vigorously defend them.

The transition from simple, breeding pairs to polyandrous trios can be studied in the oak zone of Arizona and New Mexico, where both mating systems are common (Table 10.1). The prevalent mating system may change from one year to the next at one locality. Most breeding units in the Chiricahua Mountains in 1977 were pairs; but in 1976 they were mainly polyandrous (Trail, 1980). Trail suggested that "the size and reliability of acorn crops control the composition of acorn woodpecker social units." Following a poor crop of acorns in 1976, the home ranges of breeding units in 1977 more than doubled, and the number of units with apparently "nonbreeding" adults decreased from 90% to 20%. Once green acorns became harvestable in 1977 the birds no longer foraged as widely. The acorn crop in 1976 was a quarter that of 1977, but unfortunately that of 1975 was unknown.

Another type of social plasticity in response to acorn availability was suggested by Stacey and Bock (1978) for a population next to the Huachuca Mountains. At the lower extremity of the ecological range of oaks and acorn woodpeckers, the birds lacked storage trees, did not store acorns in granaries, and *left the area in fall* just after the acorn crop was exhausted, soon after its production. During the winter of 1976–77, only one of twelve summer territories remained occupied. Interestingly, 1976 was a poor year for acorn production in the Chiricahua Mountains, not far away (Trail, 1980). Some social units which were more in the center of the ecological range in the Huachuca Mountains stayed for the winter. In one census all of the five wintering groups had large storage trees or an equivalent, which were used in winter.

Further influence of acorn abundance and predictability on social structure in the Acorn Woodpecker has been inferred from a comparison of a population with large unit sizes (California) to one with small units (New

Table 10.1

Number of adults in reproductive units of Acorn Woodpeckers in two populations (Stacey, 1979a; MacRoberts and MacRoberts, 1976).

	New Mexico	*California*
Number of units	44	41
Average size of units	3.0*	4.8*
Range of size	2–5	2–12
Percentage larger than a pair	59%	71%

* $P < 0.001$, t-test.

Mexico; Table 10.1). The California population had a higher average unit size, had the largest units by far, and contained the highest percentage of groups larger than a pair. Unlike the New Mexico population, in the California population vacant, previously occupied territories were absent and young of the year tended to remain in their natal unit. These results were rightly interpreted by Stacey (1979a) as strongly supporting the hypothesis of Selander (1964) and Brown (1969a, 1974) known as habitat saturation (see Chapter 5). The lesson from the papers of Stacey (1979a), Trail (1980), and Roberts (1979) is that the California populations enjoy a more reliable supply of acorns than the Arizona-New Mexico populations (Bock and Bock, 1974) and a richer one. The oak species are more numerous and the acorns are larger in California. The result is that unit sizes are larger in California while the number of territories per mile is similar. Therefore, California units are probably larger because overwintering is easier there. Not only does the food supply facilitate overwintering, but the climate in winter is also warmer.

These studies indicate that the environmental determinants of the ability to overwinter on the breeding territory modulate group size in Acorn Woodpeckers. What they fail to explain, however, is why Acorn Woodpeckers have the capacity to respond to a rich environment by forming persistent, territorial groups larger than a pair and why most members, especially the extra males in trios, feed the young.

With regard to the extra males, a breakthrough was achieved by Stacey (1979b, 1982). He observed in one two-male trio that both males copulated with the female, thus replicating earlier reports (Ritter, 1938; MacRoberts and MacRoberts, 1976). Extensive behavioral observations on the California and New Mexico populations, together with scanty electrophoretic data, have confirmed this finding (Mumme et al., 1983b, 1985; Joste et al., 1985). More important, Stacey reported that adults who had not participated in copulation or been present during the period of female receptivity did not feed the young (4 females, 1 male). Feeding of young was dependent upon a nonzero probability of parenthood or, in other words, on *kinship with the young*. It was, therefore, a case of kin selection in which the direct rather than indirect component of fitness was crucial. Stacey's (1979b, 1982) papers were important not just because of his observations of nonmonogamous copulations but because of his realization of the importance of the observations for understanding helping in Acorn Woodpeckers and other communally breeding birds. Before these papers the possibility that helping could arise from mate-sharing was not widely appreciated.

To understand nonmonogamous mating systems, such as in the Acorn Woodpecker, we need data for each sex separately. The literature before 1983 largely failed to supply them (e.g., Stacey, 1979a,b; Koenig, 1981a,b), but a good start has now been made by these same workers.

A conflict of interest may exist between males and females in some situations, much as in the Dunnock but with a new wrinkle. In New Mexico, females in trios produced more young than those in pairs (3.16 versus 1.26 fledglings/year respectively; Stacey, 1979b). In California, females also produced more young when more males were present, as shown in Figure 10.1. It is not clear whether this gain was caused by the males' behavior or by some other correlate of number of males, such as territory quality. This is a crucial point since it may control the female's willingness to share herself.

In contrast to females, males either break even or lose by adding males to a pair. In California, males on average lose from polyandry (Fig. 10.1). In New Mexico, male reproductive success was 1.10 fledglings per male in pairs and 1.12 in trios (Stacey, pers. comm.). Although there is a weak hint that dominance might increase the variance in reproductive success in trios over that with evenly shared paternity (Joste et al., 1982), in none of eleven trios studied later in California did the two males differ significantly in the intensity of their "guarding" (Mumme et al., 1983b). A male on seeing a rival mount the female would instantly fly at the couple "pecking and shoving the male off the female." Males in New Mexico have been reported to share the female more peacefully (Stacey, 1979b:61).

An individual male then either loses in a trio on average (California) or could lose through dominance if he were not assiduous in guarding his "right to copulate." Why do males share if some of them at least could do better in a pair?

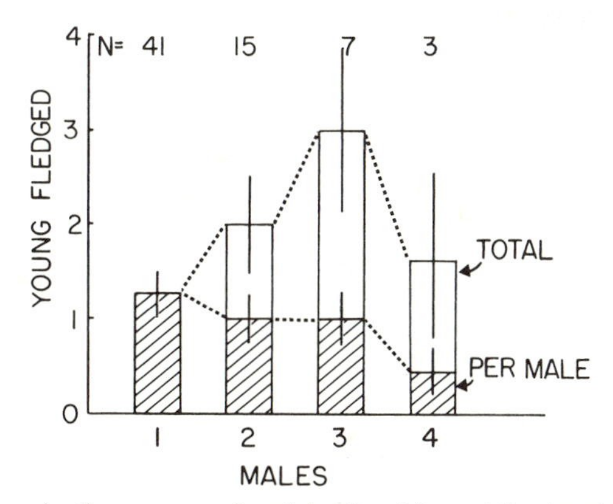

Figure 10.1 Reproductive success of social units of Acorn Woodpeckers with 1 breeding female, no nonbreeding helpers, and 1–4 core males. Total height of bars is the total number of young fledged per group (number per female); hatched portion is number of young fledged per male, assuming equality of individual mating success. Bars indicate means; centerlines indicate standard errors. (From Koenig et al., 1983.)

In the first place, it may be difficult for any male completely to prevent a female bent on sharing herself from so doing. To attempt monopolization of her might result in a greater loss for a male than mere sharing of paternity. Monopolization might entail severe risk of injury through fighting or predation. To make a long story short (skipping the seven hypotheses discussed for the Dunnock), Stacey and Koenig (1984) favor the idea that male losses of paternity in trios are compensated by enhanced survival. In California, males or females that are the only member of their sex in their group survive at the rate of 0.70 annually. Single females breeding polyandrously survive at 0.79. Males sharing a female survive at 0.86. In New Mexico, the pattern was similar although the rates were lower (0.47 for both sexes in pairs; 0.65 for larger groups). Expected lifetime reproductive success is thought to be comparable for males across group sizes when there is only one female per group. Why survival is worse in pairs than larger groups in Acorn Woodpeckers is unknown. It remains a critical problem for understanding why Acorn Woodpeckers breed in groups. Do territories with nonmonogamous units have better food resources for overwintering, thus causing better survival?

Having gone from a social structure based on pairs to one based on polyandrous groups, the Acorn Woodpecker is positioned to cross the polygynandry threshold. Polygynandry is achieved when a female joins a polyandrous group. It has not been reported from Arizona and New Mexico. In California, however, "26 to 37 percent of all nests are the product of two or more females and at least 41 to 54 percent of all breeding females" share a nest (Koenig and Pitelka, 1979; Koenig et al., 1984).

What factors favor polygynandry? Again we must examine the consequences for each sex separately. One would normally expect that adding a second female would double the number of eggs laid and thereby increase the number of fledglings per nest. Instead, as shown in Figure 10.2, the number of fledglings per group declined when a second female was present. Reproduction per female was roughly halved.

One cause of this reduction would seem to be the eating of each other's eggs by females who lay in the same nest (Mumme et al., 1983a). At 15 nests with 2 females 52 eggs were destroyed, 94% by breeding females. Disregarding abnormal eggs, 94% of eggs removed were taken by females other than the layer. Destruction of eggs, or infanticide, is discussed further in Chapter 17. A second cause of the reduction is a lowered rate of hatching of normal eggs in two-female nests (79%) compared to one-female nests (88%; Koenig et al., 1983). Finally, the considerable amount of brood loss during the nestling phase suggests that there may be a limit to the number of young that can be adequately fed by such groups. The reduction caused by additional females is still poorly understood.

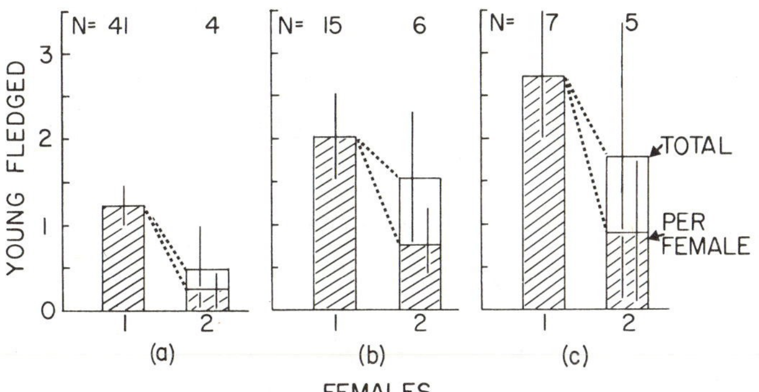

Figure 10:2 Reproductive success of social units of the Acorn Woodpecker with (a) 1 breeding male, (b) 2 breeding males, and (c) 3 breeding males; no nonbreeding helpers; and 1 or 2 core females. Total height of bars is the total number of young fledged per group; the hatched portion is number of young fledged per female. Bars indicate means; centerlines indicate standard errors. (From Koenig et al., 1983.)

Also poorly understood is why one female joins another when she would be better off in a pair. The California authors (Koenig, 1981a; Koenig et al., 1983) have suggested a variation of the habitat saturation theory. Especially in California, options for dispersal to a vacant and suitable territory are reduced by the fact that units control virtually all suitable territories all year. They invoke granaries as an important agent. While not disagreeing that energetic benefits accrue from a central storage site, I would shift the emphasis from the specific (conspicuous, concentrated storage sites) to the general. I suspect that the important factor is the ability based on stored food *to overwinter on the territory*; this is done in other group-territorial species without concentration of food in conspicuous storage sites. Even Acorn Woodpeckers can store food in scattered locations (pers. observ.), and jays in the same areas certainly do. It may be that a central storage site is more economically defended than are scattered sites; but this aspect has not been studied nor is it as important as the presence of stored food *per se*, regardless of where it is stored.

Female Acorn Woodpeckers have provided yet another example of density-dependent reproductive success, in which increasing competition leads to decreasing success. A simple *a posteriori* model that describes some features of female behavior can be generated using these assumptions:

(1) The number of suitable territories is limited.
(2) A range of territory qualities exists. For simplicity assume only two, rich (R) and moderate (M).
(3) Reproductive success of females is lower with two females than one in any territory.

(4) Reproductive success, RS, in the various combinations is ordered as follows:

$$RS_{R,1} > RS_{M,1} > RS_{R,2} > RS_{M,2} > RS_{NB},$$

where RS_{NB} is the success of a nonbreeder and 1 and 2 designate the number of females.

The model allows the following predictions:

(a) Females will breed singly in the best territories until all are occupied (saturation).
(b) Thereafter, females will breed singly in moderate territories until they are saturated.
(c) Thereafter, females will share rich territories until saturated.
(d) Thereafter, females will share moderate territories until saturated.
(e) Thereafter, nonbreeding females will help until they can occupy a nearby territory as a breeder under more favorable conditions.
(f) A state may exist in which females in pairs have higher RS than females in two-female groups.
(g) As territories fill up in the expected order, the average reproductive success per eligible female should decline.

In this section I have discussed why breeders help each other in various combinations. The reasons why nonbreeders exist and why they help have not been discussed. Suffice it to say here that indirect fitness is implicated and that no known theory based on direct selection is very satisfactory (Stacey, 1979b; Koenig et al., 1983). Little has been said about long-distance dispersal in Acorn Woodpeckers; however, the general pattern is consistent with a version of habitat-saturation theory. Where vacant territories and mates are not saturated because overwintering conditions are poor, as in Arizona, dispersal over presumed long distances occurs (Stacey and Bock, 1978). In contrast, where saturation is extreme and overwintering easy, as in California, dispersal is conservative, struggles over the filling of breeding vacancies are intense (Koenig, 1981a), territorial inheritance is common, and mating is constrained by avoidance of pairs with close relatives (Koenig and Pitelka, 1979). Much has been learned about Acorn Woodpeckers since the inception of the era of studies based on color-banding; however, I agree with the various authors who claim that several key elements of the explanation are still missing.

Polygyny in Magpie Geese

Among communally breeding birds, polygyny is rare, and helping in these cases is atypical of communal birds generally. In most polygynous species females lay their eggs in their own, separate nests, then incubate and care for only their own eggs and young. In the species discussed here two or more females of a single male may lay in the same nest.

The Australian Magpie Goose has been studied in tropical marshes, where it nests in two or three feet of water in social units composed of a male with one or two females, trios being more common than pairs (Frith and Davies, 1961). All members of a trio share nest building, incubation, and care of young. Unlike other ducks and geese, Magpie Geese "directly feed the goslings in a bill-to-bill fashion" from the first day off the nest to five weeks of age (Johnsgard, 1961). The young also graze and pick up food independently. Defense of the brood by the parents is conspicuous. The costs and benefits for females in a trio have not been systematically compared to those in pairs; however, Davies and Frith (1964) observed that during high temperatures (e.g., 120°F or more) each member of the trio could reduce its time shading the eggs, thus minimizing desertions and consequent losses of eggs.

Semipromiscuous Monogamy in Ostriches

Ostriches have an unusual system of mating, which combines elements of monogamy, polygyny, and promiscuity. Although males mate with several females and vice versa, a pair bond is still apparent. Males defend rather loose territories and construct simple nests in which a dominant or major hen lays. She then incubates the eggs almost exclusively, and the "pair" leads and guards the brood (Sauer and Sauer, 1966; Bertram, 1980b). The male fathers most of the eggs in his nest and the major hen is the mother of a larger than even share of them. These two conditions make the pair bond possible in a sort of "open marriage."

Two complications give rise to situations in which alloparenting may be said to occur. The first is basically intraspecific brood parasitism. In addition to the eggs of the major hens, it is common for two or three minor hens to lay in the same nest. Some of these minor-hen eggs are rolled out of the nest by the major hen, but she probably still ends up caring for many young that are not her own (Bertram, 1979). Minor hens are usually not allowed to incubate. They might roll out the major hen's eggs.

The second situation arises when a stepmother replaces a major hen who has gone elsewhere. In both cases the females are likely to be related to some of the young. The father too is unlikely to be related to all the young.

The advantage to the major hen has been attributed to dilution of risk to the young. If a predator kills a fixed number from a brood, the chance that the survivors include young of the major hen is improved if other young are in the brood.

Gang Warfare among the Gallinules

Helping behavior and communal breeding have been reported for several members of the Rallidae, or rails and gallinules (Table 2.2). Aside from the Native Hen, discussed in Chapter 9, the most extensively studied rallid has been the Pukeko, a subspecies of the Purple Swamp-hen confined to New Zealand.

The social structure of a population of Pukekos varied with habitat (Craig, 1979). When inhabiting stream-pasture habitat, the social units were often pairs. Larger units predominated in swamp-lake habitat. Unit sizes are shown in Table 9.2.

Unit size depended upon the defendability of the territory in two ways. Along streams, the borders needing defense were short (40–80 meters), being confined mainly to the upstream and downstream parts of the creek. In addition, neighboring units were usually small. About 22% of units were pairs; 78% were larger than pairs.

In swamps, the borders were longer (120–320 meters), and the neighboring units were generally larger than pairs. To make matters even worse for defenders of swamp-lake habitats, a large reservoir of nonterritorial birds formed there each summer. Thus owners of swamp-lake territories had to defend long borders against large established neighboring groups and a variable number of nonterritorial birds attempting to enter established groups or to displace them (Craig, 1979).

All members of a unit shared the group territory. Units of four or larger had at least two females and were polygynandrous. Copulations occurred between all adult members of a group, but male-female copulations were the most frequent (Craig, 1980a,b). All group members aided in care of chicks. Parental efforts were not clearly related to copulation frequency in males or egg contribution in females (Craig and Jamieson, 1985). Males who performed more copulations incubated less but tended chicks more.

Territory size was sensitive to the size of the group, to the number able to participate in defense, and to the presence of a large male (Craig, 1976, 1979). Territories enlarged when a new male joined and shrank when a bird was incubating, hence unable to defend. This sensitivity to group size was found also by Woolfenden and Fitzpatrick (1978) for Florida Scrub Jays. It is consistent with some ideas of mine (Brown, 1969a) developed more fully in 1982 (Chapter 8). This situation has been viewed as a game by Brown (1982a) and Craig (1984).

Why does the dominant male Pukeko not drive out his subordinate males? Does the female profit from adding a second male? The data were not presented in a way that enabled these questions to be answered unambiguously. Craig's interpretation is that males are added as needed to defend the territory, even if reproductive success per male is reduced. On some occasions a male was admitted when needed for defense, then excluded when the female became receptive.

Why would an established female admit another? Since overall reproductive success per unit was greater in pairs than in larger units, females in polygynandrous units did less than half as well as females in pairs (Craig, 1980a). This difference cannot be attributed entirely to the size and composition of the groups. Units with two or more females suffered more losses due to lowering of water level. They had a higher predation rate from harriers and mustelids. Their boundaries were longer and less defensible. And their neighboring units were larger, thus harder to keep out. Unqualified conclusions cannot be drawn from this study, but it seems that nonmonogamous units in Pukekos form where large populations exist, and where boundaries are long, leading to strong intruder pressure and making some territories economically undefendable by pairs. In the most desirable territories, admission of more birds to maintain ownership may become a necessity, in the manner of Chapter 8, but with the added cost to males of mate sharing.

In the Purple Gallinule more than two adults may be in a social unit; and, although monogamy seems likely, polyandry is not precluded (Krekorian, 1978). In the Dusky Moorhen, on the other hand, most breeding was done in polyandrous or polygynandrous units, somewhat like the Pukeko (Garnett, 1980).

Joint Nesting with Monogamy in Anis

Joint nesting is not necessarily linked with the kinds of nonmonogamous mating that we have seen in the species previously considered (Chapters 9 and 10). Monogamy provides similar uncertainty of paternity and maternity, as long as two or more females lay in the same nest and their eggs cannot be distinguished. The only commonly joint-nesting species thought to be monogamous are the anis. The three species of anis, together with the Guira, comprise a New World subfamily (Crotophaginae) of the worldwide cuckoo family (Cuculidae). The anis were among the first birds to be studied with respect to communal breeding. Pioneering field studies were done on three species of the subfamily by Davis (1940a,b, 1941, 1942), supplemented by Koester (1971). Skutch (1954, 1959) added observations on a fourth species. More recently detailed studies of Vehrencamp (1977, 1978; Vehrencamp et al., 1986a,b) on color-banded populations have clarified the ecological selection pressures on these birds and revealed their complexity.

Since most species of cuckoos are not social, Davis (1942) argued that the least social of the species of Crotophaginae, namely *Guira guira*, was the most primitive of the group. He therefore looked to *Guira* for clues to the evolution of the more elaborate societies of the three species of *Crotophaga*. *Guira* inhabits dry, grassy plains with widely scattered trees in Argentina and Brazil. It roosts by night in flocks in trees. Since it also nests in trees, there may be a shortage of nest sites, and this may be conducive to colonial or communal nesting. There may be a mild shortage of nest sites also in the original habitat of the species of *Crotophaga*, which was probably marsh and wetlands. Of course, patchiness of nesting habitat is not a sufficient explanation by itself, as several other species using the same habitat have not become communal. The smaller anis now inhabit pastures and cultivated land that was unavailable when forested. Davis's (1942) answer to the question "Why do anis breed in groups?" was that there is a shortage of nest sites.

A different hypothesis is that reproductive success for various reasons increases with group size to an asymptote. Vehrencamp (1978) has examined reproductive success as a function of unit size in the Groove-billed Ani. These birds breed when one year old. Therefore, units were composed mainly of breeding pairs with relatively few nonbreeders, which were apparently not considered in Vehrencamp's analyses. Annual reproductive success per female per year did not differ significantly as a function of number of laying females per unit, except that the rare unusually large units (4+ in marsh; 3 in pasture) did worse than other unit sizes. We can therefore reject the hypothesis that pairs of anis share nests because in doing so they increase their annual reproductive success. It is surprising that females did not do worse in multifemale units than in pairs, because they removed each others' eggs from the joint nest until all females were laying; however, these losses were compensated by laying more eggs (Vehrencamp, 1977).

Survival appears to play a key role in the ecological bases of social nesting in anis (Vehrencamp, 1978; Vehrencamp et al., 1986a). Only the alpha male assumes the risky job of incubating the clutch at night. Other males and the females alternate incubating during daytime. Nocturnally incubating males had a lower probability of survival than nonnocturnal incubators. Nest predation occurred principally at night and involved a variety of mammals and snakes. For females, survival increased with unit size, especially with the addition of a second female to a pair. Females in pairs averaged 50% of diurnal incubation, but in larger groups females spent 17% or less of their day time incubating. Thus, mortality increased with time on the nest.

I suggest that joint nesting in Crotophaginae may in some instances involve *parental facilitation* (Chapter 7). A male in a pair who lets his son remain on the natal territory into the breeding season to breed with an immigrant female suffers no greater risk himself and does not lower his own reproductive success, but he spares his son the risks of dispersal plus the risk to his son of nocturnal incubation if the son were to breed in a pair. The father and mother benefit by having more grandoffspring through their son. The mother would allow an added female to mate with her son because a second female would increase the mother's grandoffspring through her son without harming her own reproductive success. The son benefits by living longer and by breeding on a territory known to be productive of young. All breeders benefit by reducing the risks associated with daytime incubation and, consequently, by having more time for foraging and antipredator behaviors. In this species, 19% of surviving yearling sons commonly breed on their natal territory; but daughters, rarely. The advantage to the parents of retaining a daughter instead of a son would be less because females do not incubate at night. Since 42% of yearling males join a new (nonparental) group, factors other than parental facilitation must be important (Vehrencamp, pers. comm.). In going from two pairs to three, only one of the fathers would reap this advantage for his son; however, neither father would know which males were his sons so the advantage would be halved on average. Units of four or more pairs are rather rare. This unit size seems to represent the point of diminishing returns for anis. Multifemale groups tend to form mainly in territories with a good supply of food in the dry, nonbreeding season. Food in the breeding season is not correlated with group size, hence probably not limiting for group size or reproductive success.

Why are anis monogamous? Polyandry, polygyny, and nonmonogamous copulations are known for the small anis (Davis, 1940a; Koester, 1971), but monogamy appears to be the predominant system, even in multifemale groups (Davis, 1940a; Koester, 1971; Vehrencamp, 1978). Little attention has been given to this problem. The behavior of the female may be critical, as in the Dunnock. If a female were to choose polygamy, she could probably not be prevented from doing so by a subordinate male. If a joint-nesting female chose monogamy with a subordinate male, it would be difficult for the dominant male to prevent it because he would have to give up guarding his first female to court the second. Under polygamy a male who had no confidence of paternity might leave the group, thus lowering the expectation of survival for all females in the group. The synchrony of laying in joint-nesting anis makes possible the above scenario, which benefits all the females and the subordinate males at the hypothetical expense of the dominant male.

Monogamy among joint-nesting species is the exception rather than the rule. Could it be that monogamy in anis hinges on the ability of females to improve their chances of survival by allowing more males in the group?

If so, then the presumed absence of this opportunity in other joint nesters might partially explain their lack of monogamy. It is also clear that a dominant male would lose the advantage of facilitating reproduction by his son in exchange for taking the son's female.

Kin Selection in Nest-sharing Systems

Kin selection has two forms, direct and indirect. Direct kin selection is characteristic of both mate-sharing and monogamous joint-nesting systems, to which I refer collectively as nest-sharing systems since both employ only one nest for all males and females in the unit. Indirect kin selection operates in prolonged nuclear-family systems (Fig. 10.3). It is illuminating to compare how kin selection operates in the two systems. Two aspects of kin selection are clarified by such comparisons. First, the minimum degree of confidence in relatedness that is sufficient to motivate kin-care can be estimated. Second, the importance of rules-of-thumb as criteria for decisions about kin-care emerges.

In polyandrous and joint-nesting situations some breeders are uncertain whether they are the genetic parents of a particular young individual. For the sake of comparison it is useful in such cases to employ an average coefficient of relatedness that is the product of the relatedness to one's

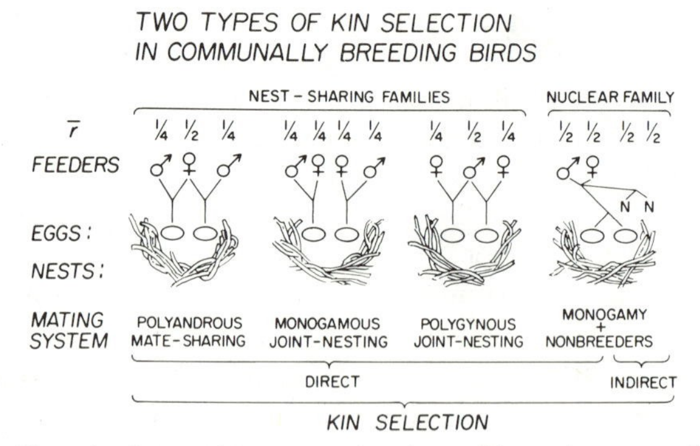

Figure 10.3 Kin selection and average relatedness ($\bar{r}$) to the young in the nest in communally breeding birds. Each feeder, whether a mother, a father, or a nonbreeder (*N*), is related by its own $\bar{r}$ to the pooled young (represented by eggs). Parents are related by direct kinship (to some or all of the young); nonbreeders, by indirect kinship. Each feeder that is not a parent of one or more of the young is a helper to those young. The first three systems have breeding helpers; the last, nonbreeding helpers. In breeding helpers $\bar{r}$ depends on the probability of parenthood, which is assumed here to be 1/2 when it is not obviously 0 or 1.

own young, $r = 1/2$, and a coefficient of parenthood, P, which is simply a binomial probability that the potential parent is the genetic parent of a specified individual. Thus, average relatedness $= \bar{r} = rP$.

It is common for two males to share one female. In this case $\bar{r} = 1/4$. In Dunnocks and perhaps other multimale species, paternity may not always be shared equally, so the subordinate male has an even lower $\bar{r}$. In Groove-billed Anis up to four females may lay in the same nest. Although maternity tends to be evened out somewhat, destruction of eggs by the last-to-lay female (Chapter 16) means that some females will experience $\bar{r} < 1/8$. Similar situations may occur in Acorn Woodpecker units composed of one female and four males.

It is readily apparent from these facts that there is nothing sacred about $\bar{r} = 1/2$. Much lower values of $\bar{r}$ are sufficient to justify parent-like behavior, at least in the contexts of joint nesting and mate sharing.

Average relatedness is not the only factor involved. As in nuclear-family systems, the future-direct component of fitness might also be involved. The various researchers of mate-sharing and joint-nesting species have given little attention to such possibilities. They will be discussed in Chapter 14.

From the examples in these two chapters on mutualistic mating systems we may deduce that absolute certainty of paternity or maternity or relatedness in general is definitely not needed to motivate the care of young birds. It is probably necessary in most of these cases, however, for the average relatedness to be 1/8 to 1/4 or more.

How do the birds know their average relatedness? Of course, they do not know it; they use instead a "rule-of-thumb" that on average yields good results. For example, for a male, having copulated with the female is a rule-of-thumb that might lead him to channel more energy toward care of the female's young than if he had not copulated with her. For a female, having laid some eggs into a nest also containing eggs of other females appears to be an adequate rule-of-thumb to motivate care of the young even though she cannot be certain of caring for her own young exclusively. The main points are: (1) *birds use rules-of-thumb that do not require perfect kin-recognition*, (2) *these rules are correlated with relatedness*, and (3) *the rules and the relatedness are probabilistic, not certain.*

Conclusions

Looking back at our survey of mating systems in this and the previous chapter, four important kinds of progress impress me.

(1) The diversity of known mating systems has been considerably extended by the study of communal birds. Most prominent has been the

discovery and study of *polyandrous* communal groups (cooperative polyandry). *Polygynandry* represents a mating system not differentiated under that name from promiscuity in birds until 1983. It had simply been overlooked. The unusual mating-and-care systems of Noisy Miners, Ostriches, and Magpie Geese have also broadened our perspective on avian mating systems.

(2) The role of kin selection in the species covered in these two chapters has been misinterpreted by earlier authors. Helping in all of these mate-sharing or joint-nesting systems depends to some degree on the probability that the helper is the parent. Since the probability is a variable rather than a certainty, breeders are led by chance to care for young not their own. Although most authors working on these species have argued against "kin selection," they have been confused about the meaning of this term. We can probably reject indirect (kin) selection in most of these cases—and this is probably what the authors had in mind—but direct (kin) selection is obviously involved since the parent-offspring relationship seems to be crucial in all the cases discussed. More important, however, than dispelling the semantic confusion that previous misusage has generated, is the realization that the *average degree of relatedness needed to justify parent-like behavior can be much lower than previously realized*—probably lower than in singular-breeding helper systems.

(3) The *strategies of members of one sex are strongly influenced by strategies of the other sex* and in some cases, at least, can only be understood by considering the other sex. An analysis in terms of sex (and dominance within a sex) is, therefore, preferable to one based on per capita fitness (lumping males together with females, breeders together with nonbreeders) or on reproductive success per group (which is appropriate for nuclear families but not for mate-sharing or nest-sharing groups).

(4) This chapter and the preceding one illustrate a *distinctly different origin of helping from that in nuclear-family systems.* Therefore, regular helping can be said to have evolved in two ways, one based upon nest sharing with breeding helpers (joint nesting and mate sharing), and the other on the nuclear family with nonbreeding helpers. Separate theories are needed for each type of origin. Simply to point out that theories intended for nuclear-family systems do not work well for nest-sharing systems does not take us far. It is necessary to develop new and separate theories for nest-sharing systems that take into account the peculiarities of such systems. The work on the Dunnock provides a good example.

11 Does Helping Really Benefit the Helped?

It is generally understood that theories of helping which invoke indirect selection (Lack, 1968; Brown, 1969a, 1974; Emlen, 1982b; Maynard Smith and Ridpath, 1972; Ricklefs, 1975; Vehrencamp, 1979, 1980) require the recipient to benefit from the alloparental care provided by the helper. Early efforts to refute these theories concentrated on denying or questioning such benefits (Zahavi, 1974; Gaston, 1973, 1978b; Ligon and Ligon, 1979). Some people may have assumed that such an argument against indirect selection is, therefore, an argument in favor of an alternative hypothesis, reciprocity or mutualism. This is actually incorrect. Any form of mutualism involving helpers, including most types of reciprocity, must by definition also involve measurable benefit to the fitness of the individual receiving the help. Mutualism is, of course, defined in terms of effects on fitness.

It is interesting to observe the changes in thinking on this issue that occurred in the 1970s and 1980s. At first it was widely assumed that the net effect of nonbreeding alloparents on the breeders was positive. After all, this interpretation was in line with Rowley's (1965a) discovery that the presence of a helper was associated with an advantage in reproductive success; and it appealed to inclusive-fitness theorists.

Then apparently in response to Zahavi's (1974) speculation that helpers are hinderers, a wave of skepticism followed. Some studies failed to find the association found by Rowley. Others established the plausibility of rival hypotheses to explain the observed positive correlation between reproductive success and number of helpers. At the 1974 International Ornithological Congress roughly half of the researchers on communal birds indicated their doubts about the hypothesis that recipients benefit (pers. observ.). Critical data, however, were missing then.

An experiment in the field with the controversial variables controlled was needed. The results of such an experiment supported the hypothesis of recipient benefit (Brown et al., 1982b). By the early 1980s the hinderer hypothesis was unpopular and today the benefit hypothesis is largely unchallenged, although it should be. One experiment is not enough.

This chapter reviews and documents these issues and events. We consider how much benefit is conferred by a helper, if any, and the mechanism involved. In this chapter we limit the discussion to nonbreeding helpers. Breeding, or mutual, helpers are discussed in Chapters 9 and 10.

If helpers bring food to the young and otherwise participate in activities presumably beneficial to the parents and their young, is it really necessary to ask whether the helpers benefit the fitness of the recipients? Yes: if we intend to test theories about helping and natural selection, we must estimate the *amount* of benefit to the direct fitness of the recipients. This is done by estimating the components of the recipient's lifetime reproductive success. A convenient measure is reproductive success per year, but annual survival of the recipient may also be measured in long-term studies. Bringing food to the young and other forms of alloparental effort do not necessarily cause *measurable* benefits to fitness of parents or young. Therefore, the null hypothesis tested in this chapter is that helpers do not measurably increase lifetime reproductive success of recipients. Lack of a measurable effect on direct fitness is grounds for skepticism about any selection hypothesis.

Effects on Reproductive Success

Singular Breeders: A Simple Correlation. The first study to demonstrate a positive correlation of reproductive success with number of helpers was that of Rowley (1965a) on the Superb Blue Wren. These Australian "wrens" (now called fairy wrens; not closely related to Northern Hemisphere wrens) nested either in pairs or trios, the latter consisting of a pair plus an extra, or supernumerary, male. All three birds fed the young. Rowley found when he added up all the broods produced in a breeding season in Canberra, that trios outproduced pairs. He obtained a similar though less decisive result with the Splendid Wren in Western Australia (Rowley, 1981a).

The second study to show increased reproductive success associated with the presence of nonbreeding helpers was that of Parry (1973) on Kookaburras in eastern Australia. Over the years since these early studies, many workers have demonstrated benefits to breeders from helpers (Table 11.1). The statistical effect of helpers is clearest in singular breeders with few helpers. In such cases one can neglect the number of helpers above 1 and simply compare units with and without helpers. In such cases the augmentation of reproductive success in a season with a helper versus without is usually distinct.

In singular-breeding species with more than one or two helpers, a comparison of reproductive success with and without helpers is inappropriate because few units have no helpers. Instead, the correlation between reproductive success and number of nonbreeding helpers has been exam-

Table 11.1

Effects of nonbreeding helpers on reproduction in singular-breeding social units, for species allowing a comparison between units with and without helpers. Sample size in parentheses. ***$P < 0.001$. **$P < 0.01$. *$P < 0.05$. S = significant, NS = not significant. Significance given is that reported in reference. fl/n = fledglings per nest. juv/yr = juveniles per year. ind/yr = independent young per year. yng/unit = young per unit. yr/n = yearlings per nest.

Species	*With Help*	*Without Help*	*Reference*
Hoatzin			Strahl 1985
Island	1.2–2.5 (21)	1.0 (4) ind/yr *	
Nonisland	0.6–1.8 (57)	0.6 (58) ind/yr *	
Pied Kingfisher			
Lake Victoria	3.6–4.7 (18)	1.8 (14) fl/n **	Reyer 1980
Lake Naivasha	4.3 (4)	3.7 (9) fl/n NS	
Kookaburra	2.3 (10)	1.2 (9) fl/yr S	Parry 1973
Red-throated Bee-eater	2.7 (22)	2.0 (56) fl/n **	Dyer & Fry 1980
White-fronted Bee-eater	1.1–1.3 (74)	0.7 (93) fl/n S	Emlen 1978
Bicolored Wren			
shrub woodland	1.40 (10)	0.40 (35) juv/yr ***	Austad & Rabenold 1985
palm savanna	1.14–1.33 (17)	0.46 (13)	
Galapagos Mockingbird	2.5 (24)	1.6 (43) fl/n **	Kinnaird & Grant 1982
Black-capped Donacobius	1.2 (5)	0.4 (10) fl/n NS	Kiltie & Fitzpatrick 1984
Superb Blue Wren	2.8 (12)	1.5 (16) ind/yr S	Rowley 1965
Buff-rumped Thornbill	0.50–0.75 (6)	0.35 (14) yng/unit	Bell 1983b
Brown-and-yellow Marshbird	2.0 (11)	1.3 (4) fl/n	Orians et al. 1977b
Scrub Jay	1.5 (59)	0.6 (41) ind/yr S	Woolfenden 1975
Beechey Jay	1.84 (16)	1.19(5) yr/n $P = 0.06$	Raitt et al. 1984

ined in several species. For example, in the Grey-crowned Babbler social units raised 0.46 fledglings per helper in one year above what unaided parents achieved (Brown and Brown, 1981b). Helpers in the White-browed Sparrow-weaver were associated with 0.30, 0.47, and 0.54 independent juveniles per helper in three successive years (Lewis, 1981). Somewhat lower values were obtained in the Hoatzin: 0.29 and 0.26 fledglings per helper per year for island and nonisland nest sites (Strahl, 1985).

In some singular-breeding species with more than one or two helpers it is possible to examine the shape of the curve that relates reproductive success in a unit to the number of helpers in the unit. The curve is not always linear. In the Stripe-backed Wren studied in Venezuela by Rabenold (1984, 1985) a *step function* was observed (Fig. 11.1). One helper added to a pair made little difference; but a second helper made a big difference, while further helpers added but little. In the Bicolored Wren,

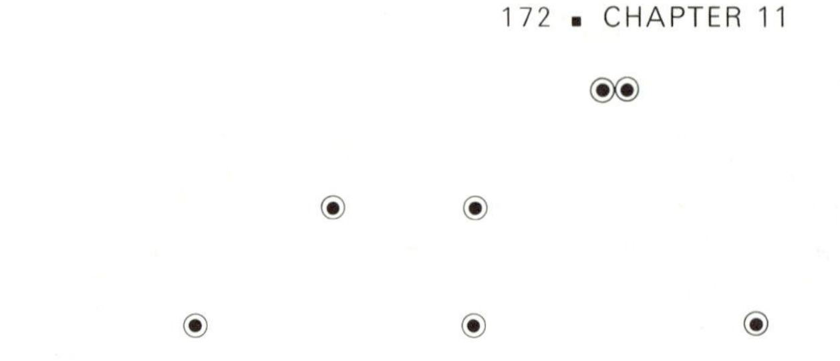

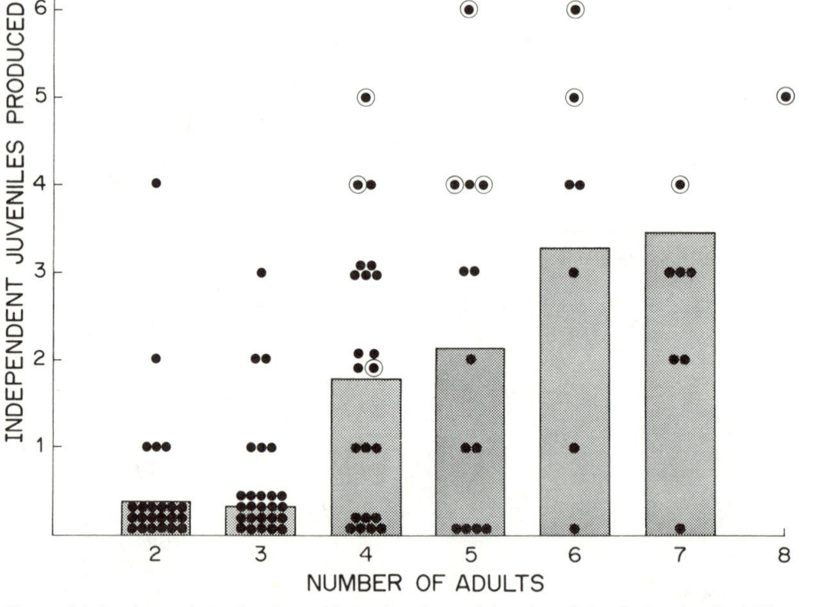

Figure 11.1 Annual production of juveniles in social units of the Stripe-backed Wren. Each point represents the number of independent juveniles raised in a year by the breeding pair plus their nonbreeding helpers. Circled points indicate two broods per year. Bars show the means for each unit-size class. (From Rabenold, 1984.)

which was studied in the same area, a single helper tripled reproductive success per group; but additional helpers were essentially without effect (Austad and Rabenold, 1985). In the species named above the correlations were all significant. Some species in which significance was not attained are discussed below under "Controversies."

Plural Breeders. In species with a variable number of breeding females per social unit it is considerably more difficult to estimate the contribution of helpers. The first difficulty is that there are two kinds of helpers, nonbreeding helpers and breeding helpers, and the two are not always easily distinguished, especially if they are males. Second, because the units and their territories tend to be larger, fewer units can be studied while more are needed to control relevant variables. For example, when the number of breeding females is a variable, more units must be studied to control this variable, since the number of eggs laid tends to go up with the number of laying females. For these and other reasons, with exception of the Acorn Woodpecker (Chapter 10), data for plural breeders are either absent, rare, or based on small samples.

Controversies

The early results showing an apparently beneficial effect of nonbreeding helpers, under circumstances where close relatedness was likely, agreed with the idea that indirect kin selection was involved. There were, however, two types of objections. One group argued that in too many species the expected positive correlation could not be demonstrated (Zahavi, 1974; Gaston, 1978b, Ligon and Ligon, 1979). The other group pointed out that even if a positive correlation between number of helpers and unit reproductive success were found, a cause-effect relationship was unproven because other positively correlated variables were not properly controlled.

Insignificant Correlations. A few researchers have failed to find a beneficial effect of helpers. These cases also deserve careful consideration.

In a study of the Long-tailed Tit in England, Gaston (1973) found that helpers, which were mainly failed breeders, did not add significantly to reproductive success. Unfortunately, however, helpers were uncommon in his population and his sample size was too small to be convincing (3 unaided pairs; 6 pairs with supernumeraries). In this species helpers "arrive" when the young are most in need of food. Thus the social system is not exactly comparable with the nuclear-family species that are characteristic of singular breeding. Could the helpers in this species be more interested in mating than the survival of the young, as in Pied Kingfishers (see Chapter 13)?

Raitt et al. (1984) in a study of the Beechey Jay in Mexico found no significant difference in the number of fledglings per nest comparing nests with and without helpers; however, they found few units without helpers, and these were on average less productive. Helpers did bring food and guard the nest.

The Arabian Babbler has been studied in the Negev Desert of Israel by Zahavi (1974). He reported that "large groups did not raise more young than smaller groups," and he emphasized the ways in which helpers could be detrimental. His data, however, on re-examination showed a correlation of +0.35 ($P = 0.045$ with a one-tailed test; Brown, 1975b). As often happens there were few of the less productive, smaller units in his sample, and it could be argued that a larger number of units was needed for an adequate test. In the subsequent decade Zahavi has not backed up his original claim with confirmatory data.

Gaston carried out an extensive study on the Jungle Babbler near New Delhi, India, in various habitats. His sample sizes were small and his correlations were often lacking in significance. A conspicuous positive correlation between group size and reproductive success was complicated by positive correlation of these two variables with a third, habitat quality.

The Green Woodhoopoe is a singular-breeding species with a single breeding pair accompanied usually by 1–7 nonbreeding helpers. A population of 22–24 social units was studied in acacia woodland in Kenya (Ligon and Ligon, 1979; Ligon, 1981a). Some groups bred successfully as many as two or rarely three times in a year. Combining all broods of a unit per year, Ligon (1981a:242) summarized the results as follows: "Our data over three years do not convincingly demonstrate that more helpers per flock yield more surviving young woodhoopoes." If anything, they said, there was an inverse relationship. A significant positive effect of helpers on recipient lifetime reproductive success was not found.

These data provide the most convincing case available for "no benefit." There are several other possible explanations, however, that allow the possibility of an undetected benefit to recipient lifetime reproductive success, as follows:

(1) The results were due to chance, small sample size, and high variance.
(2) These three years were aberrant in some unknown way.
(3) The effect of heavy and erratic predation makes the detection of a helper effect very difficult in this species. These birds did have a rather low survival rate compared to other communal breeders (Table 3.2).
(4) The researchers measured the wrong component of recipient fitness. There could have been effects on survival of recipient breeders or on their long-term reproductive success.

None of these alternative hypotheses is convincing without more data.

In 1983 the Ligons apparently reversed their position. Justifying their interpretation of helping as reciprocity, they wrote as follows: "In green woodhoopoes nonbreeding helpers generally provide a large proportion of the food brought to nestlings. . . . It appears then that helpers provide significant aid to the breeders and their offspring." No new data on this matter were presented nor was any explanation provided for the apparent disagreement between this statement claiming benefit and the earlier statements denying benefit. As noted in the introduction to this chapter, the mere provision of food or other care by a helper does not demonstrate that the fitness of the recipients benefits from it in the sense required for reciprocity (measurable net effects on direct or inclusive fitness). For further discussion of this point see Chapter 14.

In the Pukeko large groups were not more productive than pairs, and this finding has been used to question the positive effect of helpers (Craig,

1979). As discussed in Chapter 10, however, the comparison was not made within a single population; so the habitat differences were not controlled. In fact, the habitat differences were extreme. Pairs occupied mainly stream habitats with short, easily defended borders. Larger units occupied mainly swamp habitat with long borders and higher intruder pressure. Other data for this species reveal the beneficial effect of nonbreeding helpers in territory defense.

Confounding Variables: Territory Quality. In the second type of objection to the hypothesis that helpers benefit recipients, the positive correlation between number of helpers and reproductive success of recipients was not denied. Instead it was interpreted as being no more than a by-product of some other causal relationship and not a cause-effect relationship itself. If a third variable, *C*, causes *H* and *R* to be positively correlated with *C*, then *H* and *R* are likely to be correlated with each other even though there may be no cause-effect relationship between *H* and *R*. This condition is called colinearity.

For example, if good territories consistently allow a pair of breeders to produce more young than poor territories, and if these young all stay with their parents on their natal territory for a year, then good territories will have larger groups than poor territories. This effect alone would provide a positive correlation between number of retained young, which are potential helpers, and unit reproductive success, even if the helpers were completely without effect. Although Lack (1968) mentioned this problem, as did Woolfenden (1975), neither worker published relevant data. In 1977 Balda and Brown demonstrated a positive correlation between vegetation quality and group size *within* a population of the Hall's Babbler, a species that is probably group-territorial (Balda and Brown, 1977). Similarly, Lewis (1981) found a positive correlation between amount of vegetative cover used for foraging, group size, and reproductive success per group within a population of White-browed Sparrow-weavers in Zambia. Curiously, the correlations with estimates of food, namely, the density of seeds and grasshoppers, were not significant. In the Grey-crowned Babbler weak positive correlations were found between certain measures of vegetation, group size, and fledglings produced per social unit (Brown and Brown, 1981b; Brown et al., 1983). Stronger correlations between group size and territory quality have been found in the Superb Blue Wren (Nias, 1984).

Predators also affect territory quality. In the Bicolored Wren, experimental exclusion of tree-climbing predators by metal collars around the trunks of nest trees resulted in a significant increase in reproductive success compared with unprotected control nests (Austad and Rabenold, 1985). Hoatzins are subject to predation at the nest by Wedge-capped Capuchins

(*Cebus olivaceus*). Nests on islands had greater reproductive success than other nests, as well as having more helpers (Strahl, 1985). By creating artificial islands, Strahl demonstrated experimentally the protective effect of insular location. This finding suggested that the positive correlation between number of helpers and reproductive success might not have been a cause-effect relationship; however, a positive correlation remained when habitat (island versus nonisland) was controlled by limiting the analysis to within-habitat.

In the Acorn Woodpecker, acorn abundance was predicted to be correlated with reproductive success and group size. Although correlations between populations are impressive, correlations within a population have been weak (Trail, 1980, Koenig, 1981b). Variation in group size was also correlated with territory quality qualitatively in the Jungle Babbler (Gaston, 1978b) and quantitatively in the Superb Blue Wren (Nias, 1984).

These studies provide sufficient evidence for us to use caution in our interpretation of the positive correlation between helpers and reproduc-

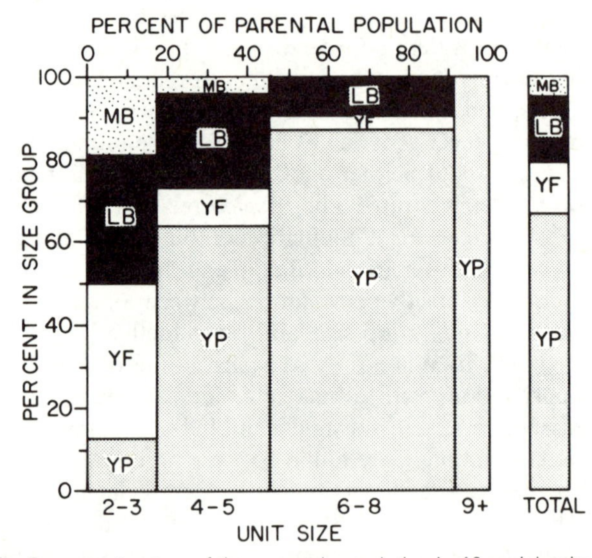

Figure 11.2 Eye-age structure of the parental population in 46 social units of the Grey-crowned Babbler. Breeding males and females are combined to comprise a sample of 92 birds. Social units have been combined according to their sizes early in the breeding season. The distance along the abscissa of a size group is proportional to the number of units in that group. Distance along the ordinate is proportional to the number of birds in each eye-age category. Abbreviations for iris colors: MB, medium brown; LB, light brown; YF, yellow with dark flecks, and YP, pure yellow without flecks, the adult condition. (From Brown et al., 1983.)

tion, but there is still another important variable that might be causally involved; the ability of the parents to rear young.

Confounding Variables: Parental Age. Parental ability cannot be easily measured, but young breeders are likely to average lower reproductive success than older ones, especially in species with pronounced delays in breeding. Also, younger breeders may be less successful in getting or defending the best combinations of territory quality and labor force. A clear relationship between age of breeder and group size was found in the Grey-crowned Babbler, as shown in Figure 11.2. In this species iris color is indicative of age. Younger birds are smaller (Fig. 5.4) and are less likely to breed (Fig. 5.3). When they do succeed in obtaining breeding status, it is in groups that are smaller than those of older breeders (Fig. 11.2).

Are older breeders better breeders? In some species, such as the Mexican Jay (Fig. 11.3), there is a clear increase in breeding success with age. A large fraction of this increase may be due to factors other than parental ability; nevertheless, some improvement with age in ability to feed young is suggested by data summarized in Chapter 5, such as that on Brown Jays (Lawton and Guindon, 1981).

In communal species that commonly breed at age one year, effects of parental age may be negligible (Rowley, 1965a, 1981a).

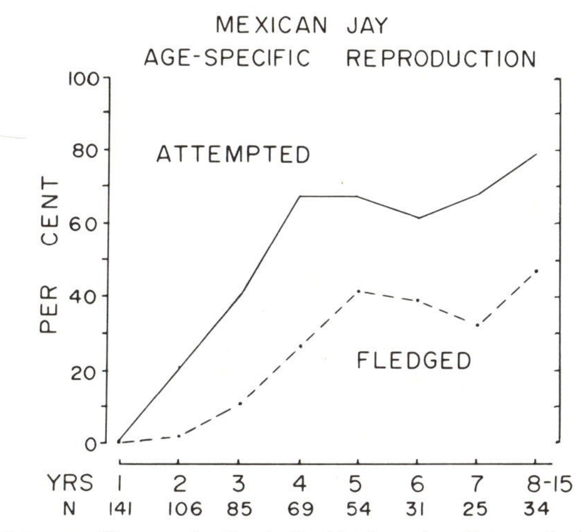

Figure 11.3 Age-specific reproduction in the Mexican Jay. Data are for birds of precisely known age in a ten-year study in Arizona. The percentage of banded birds that attempted to breed at least by beginning to build a nest is shown by the upper curve. The percentage that fledged one or more young is shown by the lower curve. *YRS* = years of age. *N* = number of bird-years in sample. (From Brown, 1986.)

A Strong-Inference Experiment

In behavioral ecology most experiments have been "natural experiments," such as the provision by nature of flocks of various sizes, allowing field workers to examine reproductive success as a function of flock size. This method often works well, but a situation may arise, such as the present one, in which more than one plausible hypothesis fits the data. The inference that helpers benefit recipients can be drawn from the natural experiments discussed above, but it is a weak inference (in the sense of Platt, 1964). If we had controlled all the relevant variables in our comparison of group sizes our natural experiment would have allowed a strong inference (in the sense of Platt) to be made. Thus, natural experiments need not necessarily be limited to weak inference, but when they yield little more than simple correlations, they usually are insufficient to resolve a controversy.

Persuasive arguments have been made that science progresses faster when the method of strong inference is used (Platt, 1964). A few behavioral ecologists have urged greater use of the method of strong inference by use of artificial experiments under natural conditions (Pulliam, 1981; Brown, 1981); but resistance, especially among avian sociobiologists, has been surprising. In this section I describe an experiment that illustrates the strong-inference approach. The experiment was performed by Brown et al. (1982b) in 1976 on the Grey-crowned Babbler in Queensland, Australia. On our study area these babblers lived in flocks of 2 to 13 (Fig. 12.5), each consisting of a breeding pair and a number of nonbreeding helpers (with minor exceptions). We began by finding 20 flocks of approximately the same size (6–8 individuals) and dividing them randomly into two groups. Any characteristic of these flocks that was correlated with flock size would be expected not to differ significantly between the two randomly chosen groups of flocks. For example, the two groups did not differ significantly in age of parents or in various measures of the vegetation on each territory. Other variables that were not measured, such as territory size, would be expected not to differ between the two groups of flocks if they were correlated with unit size.

We then removed all the nonbreeders but one (leaving the mated pair and one nonbreeder) in all the flocks of one group, the experimentals, leaving the flocks in the other group as controls. The removals were done near the start of the breeding season. The dependent variable (*RS*) was the number of fledglings produced per flock over the remainder of the breeding season. This was an unusually good breeding season for this population, and some flocks were able to fledge at least three broods before the season ended. Thus the number of fledglings produced in a unit is the sum from all nests that fledged young during the season in that unit.

A comparison of the reproductive success (*RS*) of reduced flocks versus unreduced flocks of the same original size (experimental versus controls) is shown in Figure 11.4. There was a significant reduction in *RS* of the reduced flocks compared to the unreduced groups. Since this reduction occurred under conditions in which factors other than number of helpers were not different between the groups, we concluded that the difference in *RS* was caused by the loss of helpers in some way (see next section).

By comparing *RS* of reduced flocks to natural flocks of similar size we could estimate the combined effect of factors other than number of helpers. In Figure 11.4 the *RS* of such naturally small units is shown to be less than the *RS* of artificially small units; however, the difference is not significant. Therefore, this particular experiment suggests that the number of helpers is a more important variable for the breeders than other factors; however, the effect of helpers was tested experimentally only with relatively good territories and parents (i.e., original unit sizes were of moderate size). Another design would be needed to assess the impact of helpers on poor territories with younger parents or on richer territories. The role of territory quality and parental ability was not the focus of this experiment,

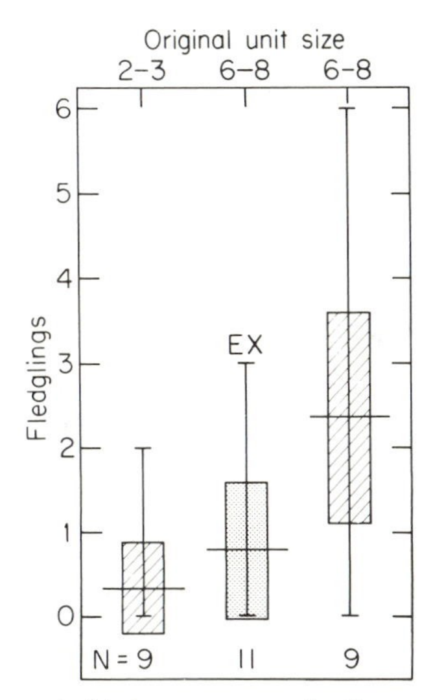

Figure 11.4 Effect of removal of helpers on reproductive success in the Grey-crowned Babbler. Ordinate: Total number of young fledged per social unit from all broods except the first. In experimental units (EX) all helpers but one were removed, leaving three birds, the breeding pair and the nonbreeding helper. The range of unit sizes before removals is given above; the number of units in each group, below. In each diagram the mean is shown by the horizontal line; 95% confidence limits of the mean are shown by the rectangle. Note: The distributions are not normal. (From Brown et al., 1982b.)

and more work appears to be needed before their relative significance can be fairly assessed.

We have demonstrated that controlled experiments in the field are feasible and useful. In my opinion more are needed. For suggestions on further experiments that are needed, see Chapter 14.

Mechanisms of the Helper Effect

Accepting now the proposition that annual *RS* of breeders is commonly augmented in some way by helpers, we may now proceed to ask how and by what mechanism the helpers effect the increase in reproductive success? There are several common theories: (1) by bringing food, starvation of the young is reduced; (2) by bringing food or nesting material, the time and energy drain on the parents is reduced, allowing them to reduce the intervals between broods, raise more broods per year, and to live longer; (3) by antipredator behavior the survival of the young is improved; (4) by sharing costs of territorial defense and antipredator behavior the energy available for breeding is increased, as outlined in Chapter 8, leading to the benefits in (2). These postulated benefits are summarized in Table 11.4.

Of course, combinations of these effects are possible, as well as other effects. Helpers engage in a variety of activities some of which are presumably beneficial and some presumably harmful. Some of these are listed in Table 11.2. The list includes behaviors from a variety of species. Helpers do not do all the listed behaviors in every species. Helpers in many species do not incubate, brood, or share in nest building.

More Food per Baby? Since helpers typically bring food to young, it is natural to begin with the hypothesis that the young are better fed with more feeders (Crook, 1965; Selander, 1964; Snow and Collins, 1962). This hypothesis can be tested by recording rates of delivery of food to nestlings in units with different numbers of nonbreeding helpers but the same number of breeders. Such studies are usually complicated by the predominating influence of other variables, such as number, age, and weight of nestlings, which affect the need for food, and various environmental factors, which affect the availability of food. Controlling these variables by stepwise multiple regression, we demonstrated in the Grey-crowned Babbler that the rate of feeding of young in a nest did *not* increase significantly with number of helpers (Brown et al., 1978). Similar results were obtained with nestling growth (Dow and Gill, 1984). The total rate of delivery did increase as the *need* increased, as reflected in the biomass and metabolic rate of the young (as a function of biomass), and it did respond to tem-

Table 11.2

Behavior of nonbreeding helpers in various species. Helpers may perform some or all of these behaviors, depending on the species. Benefits and costs are to the breeding or dominant pair of a commune.

Potentially Beneficial	*Potentially Harmful*
1. Feed nestlings	1. Consume food needed by breeders or their dependent young
2. Feed fledglings	2. Attract predators to unit or nest
3. Feed alpha female	3. Lay eggs in communal nest
4. Allopreen	4. Cuckold the alpha male
5. Detect predators	5. Displace a breeder
6. Harass predators	6. Kill or injure eggs or young
7. Locate food	
8. Defend territory	
9. Interspecific defense	
10. Incubate and brood	
11. Build nest	
12. Detect brood parasites	
13. Harass brood parasites	

perature and rainfall in a complex fashion. An increase in rate of delivery of food to nestlings in units with more helpers was not found in some other species, e.g., White-winged Chough (Rowley, 1977), Arrowmarked Babbler (Vernon, 1976), Kookaburra (Parry, 1973), Long-tailed Tit (Gaston, 1973), Black Tit (Tarboton, 1981), or Stripe-backed Wren (Rabenold, 1984).

An increase in rate of delivery of food with more helpers has been found in the Chestnut-bellied Starling (Wilkinson and Brown, 1984), the House Sparrow (Sappington, 1977), the Green Woodhoopoe (Ligon and Ligon, 1979), the White-fronted Bee-eater (Emlen, 1984), the Pied Kingfisher (Reyer, 1980a, 1984), and the White-browed Sparrow-weaver (Earle, 1983).

In some species growth rates of nestlings are higher with more helpers, but not in the Grey-crowned Babbler (Dow and Gill, 1984). In the Tody, growth was more rapid at one nest with helpers than at others without (Kepler, 1977). In the Red-throated Bee-eater, Dyer and Fry (1980, pers. comm.) found that there was little or no effect of a helper in broods of two but a significant benefit to the third nestling in broods of three. In the Red-cockaded Woodpecker growth rate of nestlings increased with the number of helpers but only three nests were studied (Ligon, 1970). In the White-browed Sparrow-weaver, growth was more rapid at a nest with four helpers than at one with none (Earle, 1983).

Present evidence then suggests that delivery rates or growth rates are sometimes increased but sometimes not affected by a helper. It is tempting to speculate that the beneficial effects to the nestlings are limited to periods of food shortage or to unusually large broods. So far, this seems not to have been established except for brood size in the Red-throated Bee-eater. It is noteworthy that our failure to detect an increase in delivery rate in the Grey-crowned Babbler occurred in a banner year for reproductive success. It would be interesting to reexamine the relationship in a poor year.

If delivery rates are unresponsive to number of helpers, then we might expect that the number of fledglings per hatched egg would also be uncorrelated with number of helpers, at least when predation at the nest is unimportant. This was true in the Grey-crowned Babbler (Brown and Brown, 1981b).

Lightening the Load. It is clear from Skutch's (1935) pioneering work that helpers share some of the energy costs of reproduction, principally feeding young, but also, in some species, nest building, incubation, and other costs related to reproduction (Table 11.2). The next question is, "What fraction of the work is taken on by the helper?" followed by, "In what ways and how much does such cost-sharing benefit the fitness of the recipients?"

A first step in this direction was simply to quantify the contributions of the helpers and parents. The advent of color-banding revealed that helpers could contribute heavily. In the Mexican Jay deliveries of food by helpers to nestlings were equal in number to those by parents (Fig. 2.2); deliveries to fledglings by helpers approximated 75% of the total (Brown, 1970, 1972). At some nests as much as 92% of the deliveries were by helpers (Brown and Brown, 1980). Not surprisingly, studies on other species have confirmed the generality that a significant fraction of deliveries of food to young comes from helpers (Stallcup and Woolfenden, 1978; Wilkinson and Brown, 1984; Raitt et al., 1984; Rabenold, 1984; and others).

Studies of this sort (Fig. 2.2) do not reveal, however, whether the parents reduce their efforts as helpers are added or whether they maintain the same effort while the helper efforts are added. To find this out one can look at the parental effort with different numbers of helpers. Figure 11.5 reveals the effect of "adding" helpers. We use the study on the Grey-crowned Babbler that is described above, but now we look at feedings by the parents instead of total feedings. On the average, each parent's contribution is reduced from 1/2 to 1/3 by adding the first helper and by smaller decrements with each additional helper, just as described in the model in Chapter 8 for the variable, *Y*. Since the total rate of delivery

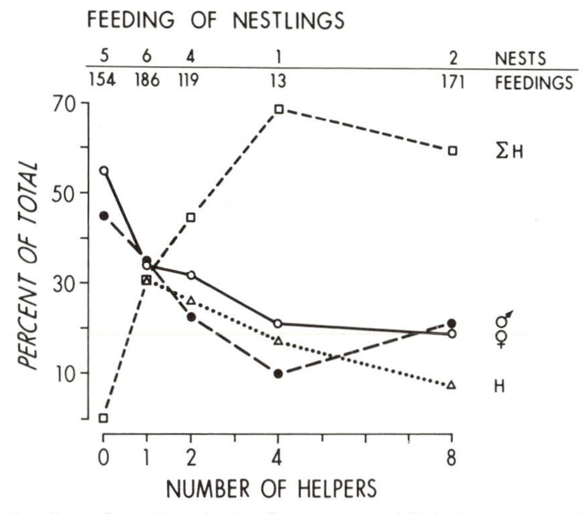

Figure 11.5 Feeding of nestlings in the Grey-crowned Babbler expressed as percentage of the total number of feedings: $\sum H$, total feedings by helpers; ♂, feedings by father; ♀, feedings by mother; H, feedings per helper. The number of nests watched and total number of feedings at those nests are shown above for each number of helpers. (From Brown et al., 1978.)

was unaffected by number of helpers, the reductions illustrated in Figure 11.5 correspond to real reductions in parental effort, not just percentages. With regard to communal nest building by this species, results were similar except that the principal beneficiary was the breeding male (Dow and King, 1984).

There have been few other studies of this kind with reasonable control and analysis of the relevant variables. Lightening of the parental load, Y, by helpers in the manner described above has been shown for the Stripe-backed Wren by Rabenold (1984) and the Beechey Jay by Raitt et al. (1984) with regard to feeding of young.

More dramatic lightening of the workload is evident when we consider the care of fledglings, that is of young recently out of their nest and still dependent on food given to them by others in their unit. In many species care of the young after they leave the nest is taken over by the father and the helpers. This leaves the mother free to restore her energetic losses from the previous brood, to "fatten up" for the next clutch or the non-breeding season, or simply to reduce her risk of predation. Quantitative estimates of the role of mother, father, and helpers in the feeding and care of *fledglings* are almost nonexistent. The first study of this sort was done on the Mexican Jay in Arizona (Brown and Brown, 1980). Females of this population typically do not renest after a successful brood. They usually contribute heavily to the feeding of all the fledglings in the unit, their own and others (see Fig. 2.2 and Chapter 14). The single female known to have nested a second time successfully in the same season abandoned her young to the care of the flock before they fledged and was incubating her next clutch of eggs while her fledglings were being fed by other members of her unit. In the Stripe-backed Wren, helpers at one nest provided over half the feedings to the fledglings (Rabenold, 1984).

In the humid tropics and subtropics, unlike large regions in the temperate zones, multiple broods in a single season are the rule rather than the exception. Certain temperate regions also have a protracted breeding season. For example, in temperate regions of Australia, *Malurus* wrens probably have multiple broods per year with relegation of the care of the fledglings to the father and helpers. This was found for the Superb Blue and Splendid Wrens (Rowley, 1965a, and 1981a) in eastern and western Australia and for the Grey-crowned Babbler in subtropical Queensland. Table 11.3 lists species of communal breeders that typically have two or more broods of young in a year. Since most communally breeding birds inhabit regions with long breeding seasons (see geographic distribution in Table 3.1), many species are multiple brooded. Of the 40 multiple-brooded species listed, 39 are tropical, subtropical, or Australian and only one occurs north of Mexico or in Europe. Of the 30 single-brooded species, 26 are in part tropical or subtropical or Australian, but 8 occur north of Mexico or in Europe.

With respect to time and energy, we have seen that helping may lighten the work load for breeders, but what evidence is there that such savings correspond to effects on direct fitness? The first clue that time-energy savings from helpers affected fitness of breeders came from Rowley's (1965a) work on the Superb Blue Wren. Pairs with helpers had more successful broods in a year than pairs without. The difference was due to the number of clutches laid, not to success per clutch.

In a more detailed analysis of a subtropical population of Grey-crowned Babblers, Brown and Brown (1981b) observed the following evidence for the load-lightening hypothesis: Flocks with more helpers, when compared with smaller ones, (1) began nesting earlier in the year; (2) had a shorter interval between successive clutches; (3) had more successful nesting attempts in the year; and (4) did not have greater success on a per egg or per nest basis. Thus, the observed reduction in parental energy costs in feeding young was associated with increased reproductive success through the mechanism of more frequent breeding during the available time. The data, therefore, are in agreement with the model in Figure 8.5. A more

Table 11.3

Number of successful broods per year in communally breeding species of birds. This list has been compiled from the references in Table 2.2 and from reference books.

One Successful Brood per Year	*Two or More Successful Broods per Year*
Magpie Goose	Galapagos Hawk
Hoatzin	Harris' Hawk
Dusky Moorhen	Native Hen
Lonnberg Skua	Purple Gallinule
Arctic Tern	Pukeko
Puerto Rican Tody	Southern Lapwing
Red-throated Bee-eater	Smooth-billed Ani
Bushy-crested Hornbill	Groove-billed Ani
Ground Hornbill	Speckled Mousebird
Acorn Woodpecker	Pied Kingfisher
Red-cockaded Woodpecker	Kookaburra
Spotted Wren	Green Woodhoopoe
Bicolored Wren	Rufous-fronted Thornbird
Band-backed Wren	Yellow-billed Shrike
Grey-barred Wren	Stripe-back Wren
Black-capped Donacobius	Galapagos Mockingbird
Yellow-eyed Babbler	Grey-crowned Babbler
Striated Thornbill	White-browed Babbler
Buff-rumped Thornbill	Common Babbler
Black Tit	Jungle Babbler
Pygmy Nuthatch	Superb Blue Wren
Brown-headed Nuthatch	Splendid Wren
Australian Magpie	White-winged Fairy Wren
Pinyon Jay	Yellow-rumped Thornbill
Scrub Jay	White-browed Scrub-wren
Mexican Jay	Red-browed Treecreeper
Beechey Jay	Brown Treecreeper
Yucatan Jay	Rufous Treecreeper
Green Jay	Bell Miner
Black-throated Magpie-jay	Noisy Miner
	Blue Tanager
	White-browed Sparrow-weaver
	Grey-capped Social-weaver
	Sociable Weaver
	House Sparrow
	Chestnut-bellied Starling
	White-winged Chough
	Bushy-crested Jay
	Nelson San Blas Jay
	White-throated Magpie-Jay
Total: 30	*Total*: 40

favorable energy budget by the addition of helpers to lighten the parental work load enables a longer breeding season. A similar interpretation has been suggested for the Stripe-backed Wren (Rabenold, 1984).

Lightening the load should also reduce the "cost of reproduction" (Williams, 1966b). In the Splendid Wren females do all the incubation and brooding. Those who had helpers lived 39% longer, and only females with helpers survived to breed again. The difference just missed statistical significance, but I agree with Rowley that the difference is probably meaningful biologically. Woolfenden and Fitzpatrick (1984) have also reported a higher rate of adult survival in larger groups than smaller ones in the Scrub Jay. Contributions to inclusive fitness by means of enhanced survival of breeders may be classified as future indirect (i_H), provided the affected breeders are related to the helpers.

These studies in the field, in conjunction with the models described in Chapter 8, suggest that a closer look at time-energy budgets in communally nesting birds is now justified. Counsilman's (1977b) study is a good beginning, but it was done before the studies just described and so was not addressed to the models in Chapter 8 or to any others.

One reason for doing time-budget research is to determine the relative importance of D, V, and Y in the energy savings from helpers. We have studied savings in care of young (Y); but the savings in territorial defense costs (D) postulated since Brown's (1969a) early discussion of group territoriality have not been well documented. Savings in vigilance costs (V) have been documented for Southern Lapwings (Walters, 1982) and in other group-living birds (Caraco et al., 1980a,b; Bertram, 1980b). Savings in D are likely to affect the dominant male primarily. Do such savings free him to feed the female and young more? Clearly the mother must be affected primarily via Y, but this needs quantitative verification. Presumably the helpers are also benefited by cost-sharing, as shown in Figure 11.5, but what effects do such savings have on helper direct fitness? In general, energy is important for survival, for breeding, and for decisions about dispersal. Helping is, above all, a cost-sharing strategy.

Enhanced Survival. It has often been suggested that survival is likely to be higher for a member of a communal group then for a lone individual, but data on survival as a function of unit size are difficult to obtain in adequate samples and tricky to interpret. The advantages of group-living for survival have been much discussed for noncommunal birds and other animals; for a review see Pulliam and Caraco (1984). We are concerned here with the possibility that the benefit conferred by helpers on breeders is mediated by enhanced protection of the breeders' offspring from predators.

Woolfenden (1978) has reported that in Florida Scrub Jays, which nest in low bushes, there is no apparent benefit from the feeding by helpers and that the benefit arises through greater protection from predators, such as snakes. A similar situation was described by Rabenold (1984, 1985) for the Stripe-backed Wren, which nests in trees on the savannas of central Venezuela.

The impact of predation on wren nests was demonstrated experimentally in the Bicolored Wren (Austad and Rabenold, 1985). Sheet metal was wrapped around the trunks of experimental nest trees but not of controls. The experimental nests experienced significantly fewer losses than the controls. Presumably the sheet metal made it too difficult for ground predators to climb the nest trees. In Stripe-backed Wrens at least two helpers are needed for successful nesting (Fig. 11.1). Presumably at least four wrens are needed to drive off predators from the bare tree trunk beneath the nest. The predators are unidentified, but birds can be excluded since they would not have been deterred by the metal collars.

Paradoxically, the antipredator hypothesis explains how parents benefit from the presence of helpers, but it does not explain how the parents benefit from the actual *feeding* of the young. Since it is quite feasible for nonbreeders to harass predators but not feed young, we must conclude that allofeeding of young in Florida Scrub Jays and Stripe-backed Wrens is not adequately explained by any of the current hypotheses. One possibility mentioned by Rabenold is that helpers, by feeding nestlings, free the parents, who are more experienced, for antipredator behavior. Another is that with more birds visiting the nest, predators are more likely to be sighted in time for effective deterrence. The hypothesis of lightening-the-load should also be investigated; feeding of the young by helpers may extend parental survival.

Parental Facilitation

It is not necessary for the helpers to increase the survival or reproductive success of the parents for the parents to benefit from retaining their offspring in their natal territory. If the parent either reduces the risk of dispersal for its offspring or makes high-quality resources preferentially available to its offspring for breeding by retaining them as helpers, then the parent can profit even without increased lifetime reproductive success of its own. This, of course, is the situation modeled in Figure 7.4, where it is shown that the parent can even experience a small nonaltruistic *reduction* in lifetime reproductive success and still profit in direct fitness. This is an example of kin selection that may apply to species like the Green Woodhoopoe, in which an increased number of helpers was not associated with increased parental reproductive success.

Conclusions

The hypothesis that helpers somehow increase the lifetime reproductive success of breeder-recipients is surprisingly controversial. The idea that the food brought by helpers is responsible for such increases is also controversial. In some cases the young that are fed by the helpers appear not to benefit from the food, the benefit going instead to the parent by lightening the workload. In other cases neither parents nor the fed offspring appear to benefit from the food itself, the benefit arriving instead by antipredator behavior perhaps associated with feeding. Measurements of the augmentation of the lifetime reproductive success of the breeders as a result of helper activities vary from not detectable (usually with small sample sizes) to dramatic. Table 11.4 summarizes the principal mechanisms by which helping might benefit breeders.

Table 11.4

Postulated indirect advantages for nonbreeding helpers. Examples, discussion, and references in text. Recipients may be parents on natal territory or other kin on another territory. I_H, i_H, h_x: see Table 4.1.

Inclusive Fitness Component	*Recipient's Life History Equivalent*	*Mechanisms*	
		General	*Specific*
1. I_H	$h_x > 0$	Clutch size ↑	Energetic savings by lightening load or cost sharing.
2. I_H	$h_x > 0$	Clutch size ↑	Female fed by helpers; energetic gain.
3. I_H	$h_x > 0$	Survival of y/n ↑	Starvation ↓ by mean food ↑.
4. I_H	$h_x > 0$	Survival of y/n ↑	Starvation ↓ by variance in food ↓.
5. I_H	$h_x > 0$	Survival of y/n ↑	Lightened load of parents allows increased antipredator behavior by parents.
6. I_H	$h_x > 0$	Survival of y/n ↑	Helpers add to antipredator behavior.
7. I_H	$h_x > 0$	Nest attempts/yr ↑	Lightening the load and feeding of female enable earlier and later nesting with shorter intervals between nests.
8. i_H	s_x↑	Augmentation of group size	Antipredator behavior ↑.
9. i_H	s_x↑	Augmentation of group size	Lightening of load.
10. i_H	$h_{x+1} > 0$	More helpers for parent	As in 1–7.

The assumption that the lifetime reproductive success of the parents is augmented in some way by the helpers is essential for theories that invoke any of the following concepts: indirect selection, reciprocity, variance utilization, and parental manipulation. This assumption need not be made for parental facilitation. The implications of the data for these theories will be discussed later, after the concepts and theories have been more fully developed.

12 The Genetic Structure of Social Units

The theory of inclusive fitness has in the two decades since its introduction (Hamilton, 1964) added a new dimension to population genetics, namely, the study of genetic structure *within the deme*. D. S. Wilson (1975, 1977) referred to the new view as one of *structured demes*. The building blocks of these structures are the social units that characterize a population—not just its individuals, but its families and flocks.

The earlier view tended to ignore social units and to focus on the frequencies of genes within demes. A *deme* is a local population that is small enough that mating within it is random, meaning that no individual is prevented from mating with another because of geographic isolation. In effect this means that each individual is within normal dispersal distance of all others in its deme. This definition is convenient for mathematical modeling, where the assumption of random mating simplifies matters.

The new view is a consequence of a new question: What is the importance of indirect selection in evolution? To answer this it is first necessary to describe the genetic structure of the building blocks of the deme in social species, its social units.

To know the frequencies of genes in demes would be useful for the question of geographic differentiation, but it is not very helpful for the question about indirect fitness. For the latter we must know the *pattern* of occurrence of genes within the deme. Are alleles distributed evenly or randomly through a deme or are they highly segregated so that like is with like? If they are segregated, then how did they get that way?

From what we have already learned in this book, it is clear that demic structure is created mainly by dispersal and that dispersal patterns are themselves a product of natural selection. A *nuclear family* by definition consists of parents and their offspring. It becomes more important as a structured unit when its existence as a unit is prolonged and its size increased, as happens in some types of communal breeding. When the family includes additional relatives, such as grandparents, grandoffspring, uncles, aunts, and cousins, it becomes, according to anthropological usage, an *extended family*. In this chapter we take a close look at nuclear and extended families and other social units, such as sibships, as elements of demic genetic structure. We consider their origins through the timing, mode, and extent of dispersal and the consequences of such structure for natural selection. In particular, we consider the extent to which observed degrees of genic segregation are consistent with theories that require indirect selection.

Age Structure

In communal parent-offspring systems the young help the old. The value of the total help in a social unit depends upon the ratio of young to old in that unit. Consequently, with respect to the importance of indirect fitness, age structure and genetic structure are interactive.

In many communal species the immature age classes from previous years are recognizable in the breeding season by differences in coloration (see Chapter 6). In such species an approximation of the age structure of the population can be obtained in one year. An application of this method is shown in Figure 12.1. In the Grey-crowned Babbler, breeding

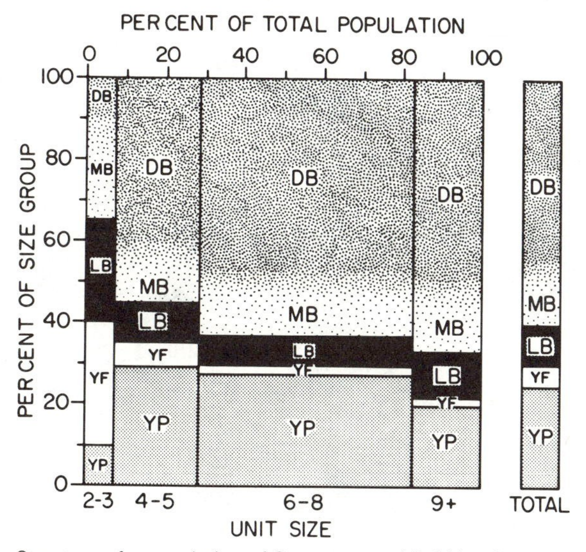

Figure 12.1 Structure of a population of Grey-crowned Babblers by eye-age categories early in the breeding season, excluding birds of the year. The eye-age categories are arranged from youngest at the top to oldest at the bottom. See Figure 5.3 for ages. The percentage of the total population ($n = 265$) in each eye-age category is shown at right. The population is divided into subgroups according to unit size along the abscissa. The width of each unit-size group is proportional to its representation in the population on an individual basis. (From Brown et al., 1983.)

status is correlated with color of iris and with age; the older birds (YP) have yellower eyes and are the breeders (Fig. 5.3). The young birds (DB and MB) were the largest age class and were unusual in the smallest units (Fig. 12.1). The age structure of the breeding birds was strikingly correlated with unit size; the younger breeders were predominantly in the smaller units (Fig. 11.2).

A more detailed view of age structure is available in those species for which a population of color-banded individuals has been followed for many years. This procedure permits an increasingly accurate depiction of age structure in successive years as shown in Figure 12.2. In the Mexican Jay age determination is precise only for birds banded within two years of hatching. Ages of birds three or more years old are known only by keeping track of individuals over a period of years. Only by doing this did we realize how important in the population are birds in the 5–15 year range. Indeed, on average, 5% of the population is ten years old or older.

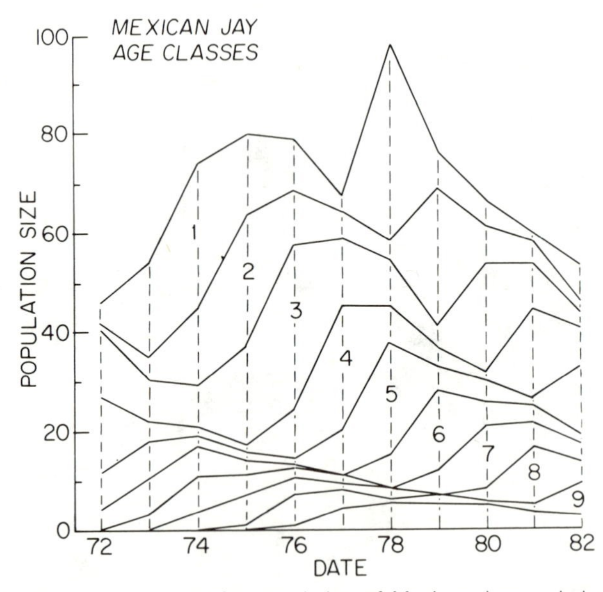

Figure 12.2 The age structure of a population of Mexican Jays varied greatly from year to year. The curve above each of the ages in the figure shows the number of individuals of that age and older. The number in a given age class for a year is given by the vertical difference between the line above and the line below the indicated age. Data are for the entire population. Included are some unbanded birds and banded birds whose age is the stated age or older. The latter category declined steadily from 50% in 1972 to 15% in 1982. (From Brown, 1986.)

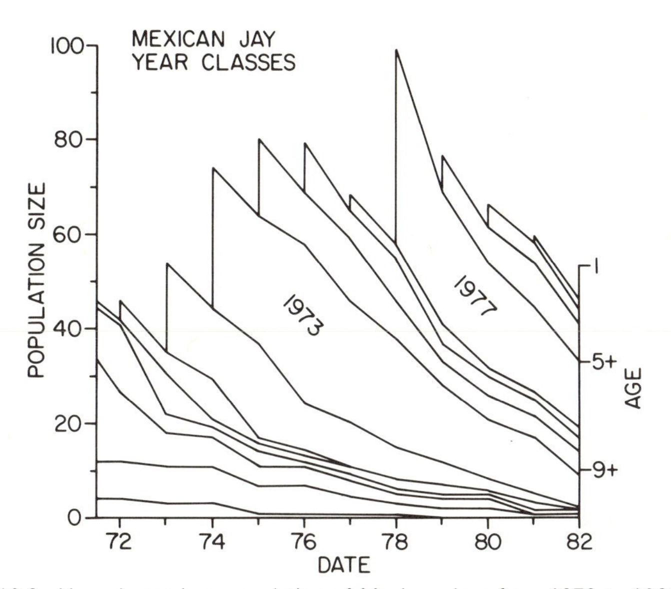

Figure 12.3 Year classes in a population of Mexican Jays from 1972 to 1982. Data are as in Figure 12.2. (From Brown, 1986.)

Although not the most numerous in the population these birds are the most successful at reproduction. Since nonbreeding helpers come primarily from the cohort of one- and two-year-olds, we may use the same data plotted differently (Fig. 12.3) to illustrate dynamic changes in age composition from year to year. The predominance of certain age classes is pronounced. In certain years (e.g., 1974, 1978) many yearlings were present; in other years (e.g., 1977, 1981), very few.

A similar age structure was found in the congeneric Scrub Jay in Florida (Woolfenden and Fitzpatrick, 1984), as shown in Figure 12.4. In this species the helpers are all nonbreeders. The figure shows clearly the importance of yearlings for the helper category, but also reveals a diminishing fraction of helpers at older ages. This type of structure appears to be typical of nuclear-family helper systems.

Of more tangible importance to the recipient of help than the age structure is the number of helpers per breeder. This is a joint function of the age structure and the age at first breeding. Expressed differently, this ratio corresponds in singular-breeding species to the fraction of the population that is surplus, N_F. The dependence of N_F upon age structure and recruitment is indirectly shown in Figure 3.2, where it is expressed as a function of annual survival rate.

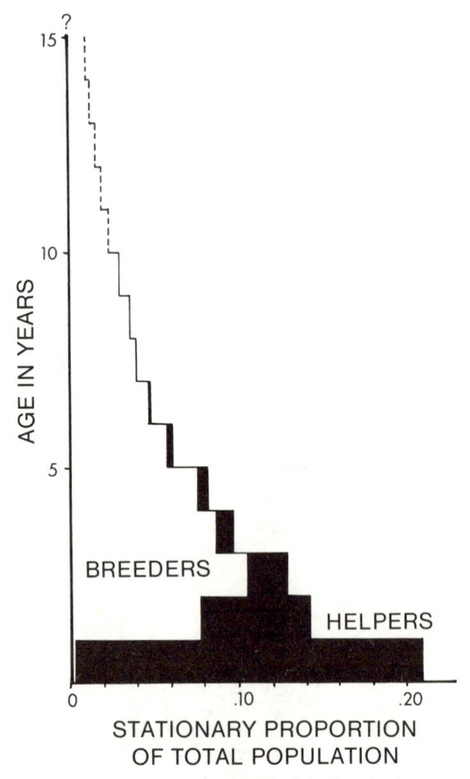

Figure 12.4 Stationary age distribution of a Florida Scrub Jay population at the onset of an average breeding season. Dashed line and question mark signify incomplete knowledge of mortality rates among the oldest age classes. (From Glen E. Woolfenden and John W. Fitzpatrick, *The Florida Scrub Jay: Demography of a Cooperative-breeding Bird*. Copyright © 1984 by Princeton University Press. Figure reprinted with permission of Princeton University Press.)

The number of helpers per breeder in singular breeders varies widely among species. The following examples illustrate the spectrum of variation. Some species, such as the Superb Blue Wren, often have no helpers (67% of 43 unit-years) but sometimes a single helper and rarely more. In the Florida Scrub Jay 20–59% of pairs have no helpers while the rest have one or two and rarely up to six, as shown in Table 12.1. Other species that are primarily singular breeders may have even more helpers per unit. The Grey-crowned Babbler, for example, in our one-year study had up to thirteen members in a unit (Fig. 12.5). All but two of these in each unit were nonbreeding helpers, with possibly a few exceptions. Most units had four to six members (two to four nonbreeding helpers) and only 10% of

Table 12.1

Number of helpers per family in the Florida Scrub Jay, 1969–79 (Table 5.1 in Woolfenden and Fitzpatrick, 1984).

Breeding Season	*Number of Pairs*	*Number of Helpers*							*Helpers/Pair*
		0	1	2	3	4	5	6	
1969	6	33.3	50.0	16.7	—	—	—	—	0.83
1970	16	50.0	25.0	12.5	12.5	—	—	—	0.88
1971	25	20.0	36.0	28.0	8.0	4.0	4.0	—	1.52
1972	30	50.0	20.0	20.0	10.0	—	—	—	0.90
1973	26	57.7	26.9	11.5	—	3.8	—	—	0.65
1974	29	48.3	31.0	13.8	3.4	—	—	3.4	0.90
1975	27	59.3	18.5	18.5	3.7	—	—	—	0.67
1976	27	29.6	25.9	18.5	14.8	11.1	—	—	1.52
1977	28	42.9	21.4	32.1	—	3.6	—	—	1.00
1978	32	50.0	31.3	3.1	9.4	6.3	—	—	0.91
1979	34	29.4	41.2	17.6	8.8	—	2.9	—	1.18
1969–79	280	43.2	28.6	17.5	6.9	2.9	0.7	0.4	1.00
$\bar{X}$ or $\bar{x}s$	—	42.8	29.7	17.5	6.4	2.6	0.6	0.3	1.00

Note: Based on censuses taken early in each breeding season (March—April).

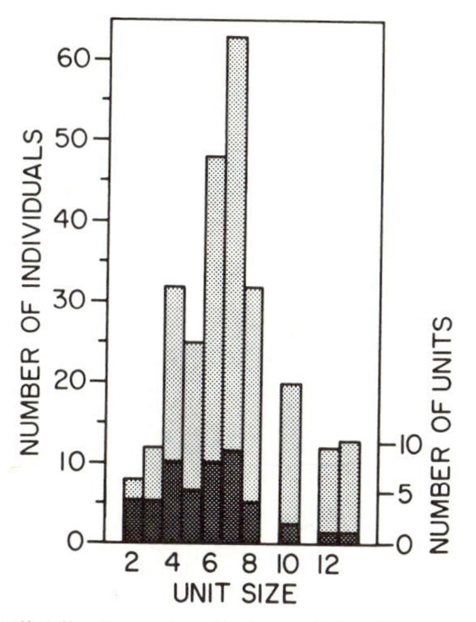

Figure 12.5 Frequency distribution of unit sizes of the Grey-crowned Babbler early in the breeding season, excluding nestlings, fledglings, and other birds of the year. The number of individuals in the population in each of the unit sizes is shown by the upper bar and left ordinate; the number of units is shown by the lower bar (dark shading) and right ordinate. Total units = 46. Total individuals = 265. Mean unit size = 5.76. (From Brown et al., 1983.)

units had no helpers. The extreme with respect to helpers per pair in singular breeders is the Yellow-billed Shrike, which in 1973 had an average of ten helpers per nest and a range of 4–23 (Grimes, 1980). In the largest group, which was unstable, females were competing to breed. One could refer to species whose units sometimes lack helpers as facultative or opportunistic, and those which always have helpers as obligate communal (or cooperative) breeders. Since there is a spectrum on this gradient I prefer not to employ a single dichotomy here because it hides much variation.

Relatedness in Singular-breeding Species: Is it High Enough?

Singular breeding with nonbreeding helpers typically involves a nuclear family. In the purest sense a nuclear family consists of two parents and their offspring. In reality, however, many social units among singular-breeding species are complicated by immigration. A common cause of immigration is the loss of one parent followed by remating of the other. Units in inferior territories may leave them and split up, their members joining others. Offspring may lose both parents, become separated, and join different units. Takeovers may also occur, resulting in much reshuffling. In short, there are good reasons to avoid the naive assumption that social units in communally breeding birds are nuclear families pure and simple. Therefore we need to look closely at detailed, long-term empirical studies.

Nearly every study comments on relatedness, so it would be difficult to review them all. We shall consider some representative examples. In Australian fairly wrens (*Malurus*), relatedness is close but details are lacking. In the Superb Blue Wren males "are frequently allowed to stay in the parental group and this appears to be the commonest way in which supernumerary associations . . . arise" (Rowley, 1965a). In the Splendid Wren, "Most of the helpers were progeny retained in the family after they had reached maturity" (Rowley, 1981a). Similar results were reported by Payne et al. (1985). In the Kookaburra the picture is similar except that the sexes may be more evenly represented among the helpers. Parry (1973) wrote, "Those juveniles that survived their first year functioned as auxilliaries by helping their parents. . . ." Adults too "remained within their parents' territories all year." In the Grey-crowned Babbler, of 45 banded young that survived until the age of one year on the study area, all were still with both parents. In the Beechey Jay, "most helpers definitely were associated with at least one parent" (Raitt et al., 1984). In the Jungle Babbler, the mean relatedness of nonbreeders to the breeding pair was

$r = 0.29$, based upon a complex calculation involving immigration ratios and replacement of breeders (Gaston, 1978b). In the Hoatzin, 90 of 95 banded nonbreeding helpers were offspring retained from the previous year (Strahl, 1985). In Speckled Mousebirds helpers are "in most cases . . . subadult males from the preceding generation helping their own parents" (Decoux, 1982). In Galapagos Mockingbirds "two of the helpers were siblings of the first brood associating with their parents at the second brood . . . in at least three instances" (Grant and Grant, 1979). In the Bicolored Wren, 82% of 17 nonbreeding helpers were rearing full sibs and 18%, half sibs or nephews (Austad and Rabenold, 1985). In the Stripe-backed Wren 66.4% of 110 first-year, nonbreeding helpers were full sibs of the nestlings; 27.3% were half-sibs, and 6.4% were nonsiblings (Rabenold, 1985). Nonbreeding helpers aged two years or older ($n = 43$) were more distantly related; 22.0% were full sibs, 58.5% were half-sibs or quarter-sibs, and 19.5% were nonsibs. Average relatedness of first-year helpers to the nestlings was $r = 0.39$ and of older helpers, $r = 0.23$.

Longer studies have provided more detailed answers with regard to relatedness between helper and helped. Table 12.2 shows how the non-breeders (presumed helpers) were related to the breeders whose offspring

Table 12.2

Relationship of helpers to recipient breeders in the Florida Scrub Jay (Table 5.3 in Woolfenden and Fitzpatrick, 1984).

		Helpers				
Breeders	*r*	*Males*	*Females*	*Sex Unknown*	*Total*[a]	*Percent*
Brother X Mother	0.75	1	—	—	1	0.3
Father X Mother	0.50	69	76	16	161	64.1
Grandfather X Mother	0.38	—	1	—	1	0.3
First Cousin X Mother	0.35	1	1	—	2	0.8
Brother X First Cousin	0.31	—	1	—	1	0.3
Father X Stepmother[b]	0.25	21	9	—	30	12.0
Stepfather X Mother	0.25	10	17	4	31	12.4
Brother X Nonrelated	0.25	2	1	—	3	1.2
"Brother" X Nonrelated[c]	0.25	4	—	—	4	1.6
"Uncle" X Grandmother	0.19	2	2	—	4	1.6
"Uncle" X Half Sister	0.19	—	1	—	1	0.3
Half Brother X Nonrelated	0.13	3	—	—	3	1.2
Nonrelated X Nonrelated	—	2	6	1	9	3.6

[a] Based on 165 individual jays of known age and parentage during 251 helper seasons.
[b] Stepparent designation used for nonrelated breeders that paired with a helper's parent.
[c] Quotes indicate probable family relationships.

they probably fed in the Florida Scrub Jay. In 90% of the sample at least one parent was alive; in 64%, both parents were present. *In only 3.6% of cases in this long-studied species was the helper unrelated to the young being fed.*

In singular breeders we expect average relatedness of helpers to recipient young, $\bar{r}_{HR}$, to be determined primarily by the survival rate of the parents, s, and the age of the helper, t, as follows (Brown, 1978a):

$$\bar{r}_{HR} = s^t/2$$

If survival were 100% in the first year ($t = 1$), $\bar{r}_{HR}$ would be 0.5; if 50%, then $\bar{r}_{HR} = 0.25$.

According to Woolfenden and Fitzpatrick (1984), the data for the Florida Scrub Jay almost exactly fit this equation. With $s = 0.82$, the average relatedness at age 1, when helping is most frequent, is 0.41; at age 2 it is 0.34; at age 3, $\bar{r}_{HR} = 0.28$; at age 4, $\bar{r}_{HR} = 0.23$; at 5 it is 0.19. Comparing this series with the percentage helping at each age (Fig. 12.4), one cannot help but be impressed with the fact that *helping is most frequent when the relatedness values are the highest*. The decline in proportion helping follows the decline in $\bar{r}_{HR}$, as shown also for the Grey-crowned Babbler (Fig. 13.1).

The data may be said to agree well with expectations based upon kinship theory. It is unlikely, however, that this close agreement arises from an assessment of relatedness by the nonbreeder, nor is this required by kinship theory (Hamilton, 1964; Brown and Brown, 1980). Both the average relatedness and the fraction not breeding are dependent on a third variable, the mortality rate of breeders. As breeders die, their places tend to be taken by individuals from the older end of the age spectrum of nonbreeders. Thus, in general, as breeders die, nonbreeding helpers become less closely related to the recipients of their help (i.e., a helper's parent dies and is replaced by non-kin), and they accumulate more opportunities to breed themselves.

Since I suspected when I wrote my 1974 theory that the contribution of a helper to its inclusive fitness would be greater by breeding (if it had a suitable mate and territory) than by helping ($D_B > I_H$; see Fig. 13.1, for example), I did not require in the theory that individuals give up such opportunities to breed in order to help. Instead, I theorized that the bird should help only if it could not breed, and that this contribution to inclusive fitness would be better than no contribution at all ($I_H > 0 = D_B$). Of course, one could have theorized that situations might occur in which a bird with a territory and mate would contribute less to its inclusive fitness than by helping ($I_H > D_B$), as in small units of Stripe-backed Wren (Rabenold, 1984, 1985); however, I did not require this. In the Florida Scrub Jay, therefore, the birds need only follow a simple rule to bring about a good fit to my theory, namely: "Help at home if you cannot breed." As the data convincingly show, this rule allows the helpers to experience a high degree of average relatedness and also to breed as soon as possible.

I regard these data as beautifully consistent with my model for the evolution of helping based upon habitat saturation and indirect selection (Brown, 1974; Appendix). Woolfenden and Fitzpatrick, on the other hand, feel that the data are *not* consistent with "strictly kin-selected helping behavior" (1984:88). Their conclusion is: ". . . within their first three years of life at least, helpers appear to remain regardless of the degree by which they are related to the offspring they assist." I believe that their observation that the helpers were related to their recipient young in 96.4% of the cases (1984:87, Table 5.3) makes it unnecessary for the birds to assess fine gradations of relatedness. It is possible that Woolfenden and Fitzpatrick expected helping under indirect selection only at the higher values of $\bar{r}_{HR}$ (how high they did not say); however, helping by *breeders* at *lower* average values of $\bar{r}_{HR}$ has been described for cases in which parenthood is probabilistic (Fig. 12.6; Chapters 9 and 10; Fig. 10-3.) Thus the observed average values of $\bar{r}_{HR}$ for Florida Scrub Jay helpers fall well within a range that justifies parental behavior by real parents in nest-sharing systems. For further discussion of this case see below under "Is Kin Recognition Necessary for Kin Selection?"

Another species in which the importance of relatedness is in dispute is the Green Woodhoopoe. Ligon and Ligon (1983) reported that in 92% of 168 helper-years the helpers were related to the young being fed. They have interpreted these data as an argument *against* a role of indirect selection in the evolution of helping in this species (Ligon and Ligon, 1979; Ligon, 1983). Their reasoning, which has been adopted implicitly by Woolfenden and Fitzpatrick (1984) also, appears to be as follows: if indirect selection had been important, then helpers should *never* help unrelated birds since natural selection would have provided them with a way to avoid such mistakes. I believe this reasoning to be fallacious. No published theory for the evolution of helping by indirect selection makes such requirements (Brown, 1979b, 1983a, this chapter; see Hamilton, 1964 or Maynard Smith, 1964 or D. S. Wilson, 1975). Therefore, indirect fitness cannot be rejected on this basis.

Surveying the various studies, it is apparent that *the fraction of unrelated helpers is strikingly small in a variety of species.* As shown in Table 12.3, estimates range from 10.1% to none at all. In the species for which the sample size is largest, the Florida Scrub Jay, only 3.6% of the helpers were unrelated to their recipients. In species with smaller sample sizes more variation was found and some relatives were necessarily misclassified

Table 12.3

Fraction of nonbreeding helpers that are related to the young they feed in some singular-breeding species. FSIB = full sibling. HSIB = half sibling. NEF = nephew. NONSIB = neither half sibling nor full sibling. YR = yearling. References in text.

Species	*Percentage*	*Relationship*	*Helper Ages*	*Number of Helpers*
Hoatzin	94.7	in natal territory	1 yr	95
Green Woodhoopoe	92	related	all	168
Stripe-backed Wren	89.9	related	all	153
	53.9	FSIB		
	36.0	HSIB		
	9.9	NONSIB		
Bicolored Wren	100	related	all	17
	82	FSIB		
	18	HSIB, NEF		
Grey-crowned Babbler	100	FSIB or HSIB	1 yr	45
Florida Scrub Jay	96.4	related	all	251

as unrelated for lack of information. The data for Hoatzins and Grey-crowned Babblers are for yearlings, and the proportion of unrelated helpers would probably be higher among older birds.

How low can average values of relatedness between feeder and nestling go and still be significant for kin selection? I cannot answer this question, but I can put the matter into a useful perspective. In Figure 12.6 the observed values of average relatedness between nonbreeding feeders and nestlings in nuclear-family systems are compared to those between breeders and nestlings in nest-sharing systems. The idea that relatedness is always 1/2 between breeders and the young they feed is a myth. Many birds rear brood parasites of other species, and intraspecific brood parasitism affects many nests in some populations (C. Brown, 1984; Andersson, 1984). Cuckoldry may also be more common in monogamous species than is generally realized (Gowaty and Karlin, 1984). Nevertheless, parental care persists in spite of such deviations, so $\bar{r}$ need not be as high as 1/2. More relevant are the cases of nest-sharing discussed in Chapters 9 and 10. The Groove-billed Ani serves as an example in Figure 12.6, but as shown in Figure 10.4 the principle is the same for polyandry as for joint nesting. In all of these "natural experiments," uncertainty of paternity or maternity leads to a rather low degree of average relatedness between the feeder and the pooled young in its nest. Nevertheless, there seems to be no skepticism that kin selection is an important element in nest sharing even though the average relatedness is often lower than for nonbreeding helpers in nuclear-family systems.

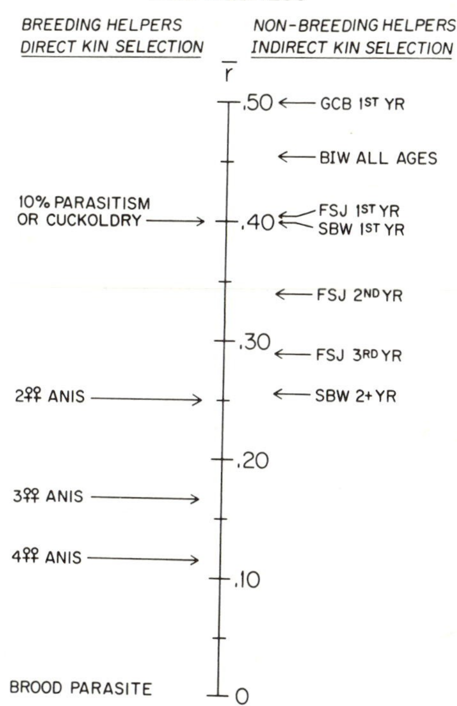

Figure 12.6 Estimates of average relatedness of feeders to recipient nestlings in two kinds of helpers. GCB = Grey-crowned Babbler; BIW = Bicolored Wren; SBW = Stripe-backed Wren; FSJ = Florida Scrub Jay.

Relatedness in Nest-sharing Species: How Low Can It Go?

Relatedness of breeding helpers to the brood of young they feed is consistently significantly above zero because the helpers are parents of some of the young (Chapters 9 and 10). We have not considered the relatedness of nonbreeders in such species. Although breeding may occur at age one in Groove-billed Anis, Pukekos, and Acorn Woodpeckers, some individuals in these species refrain from breeding and stay in their natal uints. In California Acorn Woodpeckers 75% and 42% of one- and two-year-olds respectively are nonbreeding helpers who leave by their third (females) or fourth (males) year (Koenig et al., 1983). Since the units of Acorn Woodpeckers vary in number of breeders of each sex, no breakdown by relatedness has been attempted. Just as with the breeders, however, relatedness of nonbreeders to young in the nest varies widely, being higher on average in smaller units and for younger nonbreeders.

It appears that *average relatedness of breeding helpers to the young they feed (including their own) is typically 0.25 or lower in nest-sharing species. Therefore, as previously noted, average relatedness may be lower in such systems than that of nonbreeding helpers in singular-breeding species* (Figs. 10.3, 12.6)!

Relatedness under Plural Breeding with Monogamy

Plural breeding is found in many communally breeding species on a regular basis (Tables 2.1, 2.2). In some of these species pairs nest separately and usually behave as if they were monogamous. The birds consort together as a pair when building and laying. If they are separate-nesting species only one pair builds the nest, and only one female lays eggs, incubates them, and broods the young. Although for most of the species studied it is mentioned that young stay in their natal unit for at least a year, only in the Mexican Jay have studies been continued long enough to reveal the complex genealogical patterns that exist in such groups.

The details of relatedness in social units of the Mexican Jay have been worked out from the genealogies shown in Figure 12.7 (Brown and Brown, 1981a). The data were gathered by keeping track of individuals banded as nestlings or full-sized birds for ten years beginning in 1969. Immediately apparent is that the units are larger and more complex than a nuclear family. Not only are the expected parent-offspring associations present; but there may be two such lineages in a single unit (as in HI, BY, RC). Furthermore, the offspring may produce young while the parents are still present so that grandparents and grandoffspring may coexist (in CO grandfather C occurs with grandoffspring I, J, and K, although two of the three parents of these were no longer present). This situation also results in the presence in the same unit of aunts, uncles, nephews, and nieces (in HI, D is aunt to L, M, N, O; and J and K are uncles). Cousins may also coexist (in HI, O is a cousin of L, M, N). In the nuclear-family species immigration occurs mainly to replace breeders. In Mexican Jays, however, nonbreeding immigrants were frequent in the less productive units (RC).

In this social system it is normal to find some close relatives in the same unit. Most individuals remain for the first few years on their natal territory, thus keeping them in the same unit with close relatives, such as parents and sibs. At the same time, however, they share the costs and

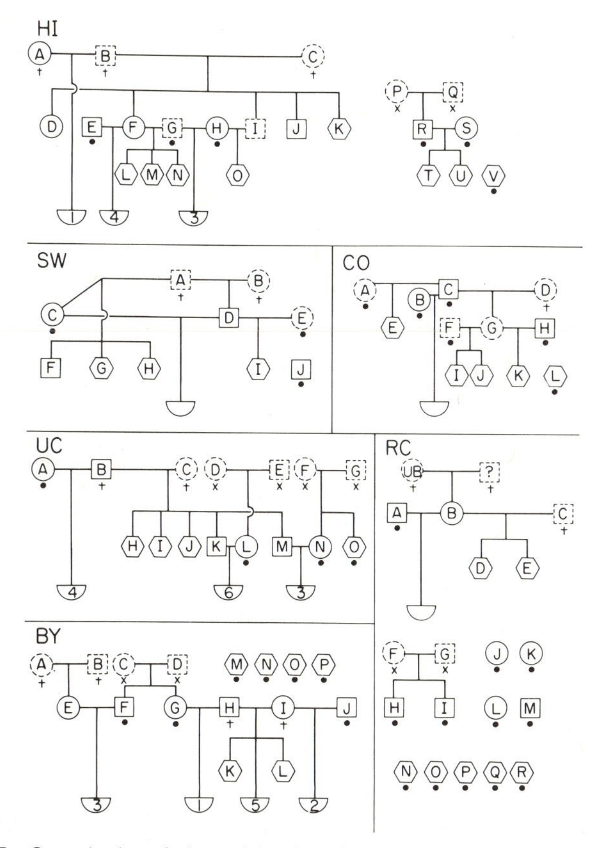

Figure 12.7 Genealogies of six social units of Mexican Jays. All individuals present in May 1979 are indicated by symbols with solid lines. Selected ancestors no longer alive or in another unit are indicated by dashed symbols. Symbols: (square) male; (open circle) female; (hexagon) sex unknown; (semicircle) nest with young; (dot) immigrant; (†) present at start of study, ancestry unknown; and (x) in neighboring unit. (Brown and Brown, 1981a. Copyright 1981 by the AAAS.)

benefits of communal life with genetic strangers in addition to the expected stepparents. The result is that benefits to close kin are retained to some degree but are diluted by immigration. Dilution is minimal in the most productive units (SW). In these, unit size stays relatively large through production of young, who may later be "exported" to neighboring units as shown in Figure 7.3. Average relatedness of helpers to nestlings in the SW unit was 0.28. The units with least production have the most immigration and are in effect "genetic importers" (compare RC to SW in Figs. 7.3 and 12.6). In RC the average relatedness of helpers to nestlings was only

0.06. Thus, the rather low average relatedness within social units when they are all averaged together is somewhat misleading since most of the young who join the breeding population and leave descendants come from the units with the highest relatedness, as shown in Figure 7.3.

Species like the Mexican Jay, in which average relatedness tends to be lower than in nuclear families, show that alloparenting and other aid-giving behaviors can be maintained even with a relatively low average relatedness. It is not clear just how low relatedness can go and still be significant in natural selection; however, the levels of relatedness in the more productive units of Mexican Jays are not very different from the average relatedness between breeders and their collective offspring in mate-sharing communal birds (Chapters 9 and 10; Figs. 10.3, 12.6). It could be argued, therefore, that if relatedness in parent-offspring situations can be as low as 0.125, then it might still be important in extended family situations at this level. This problem and the existence of conditions favorable to reciprocal feeding of fledglings in the Mexican Jay will be considered in Chapters 13 and 14.

Is Kin Recognition Necessary for Kin Selection?

Kin recognition is associated with kin selection in some contexts but not others. Colonial birds, for example, can often single out their own young from a large group, as in the Pinyon Jay (Balda and Balda, 1978). In contrast, many hosts of brood parasites seem unable to distinguish their own young from those of the parasite. Kin selection operates in both of those examples to maintain care of young even though mistakes are not rare. In the dispensing of alloparental care, effort reflects relatedness in some species, such as the Pied Kingfisher (Reyer and Westerterp, 1985; see Chapter 13) but not in others, such as the Stripe-backed Wren (Rabenold, 1985) and Pukeko (Craig and Jamieson, 1985). Is kin selection unimportant when active kin recognition is not shown?

This is a critical issue. As already mentioned, Ligon and Ligon (1978) and Woolfenden and Fitzpatrick (1984) have questioned kin selection in part because a tiny minority of helpers feed young to which they are unrelated. The implicit reasoning seems to be that if kin selection were important, active kin recognition would have prevented such errors; however, active kin recognition is not a necessary cause or effect of kin selection.

The giving of aid to close relatives can be achieved by following simple behavioral rules without the need of active kin recognition. Hamilton

(1964:24) described examples in which even parents lacked ability to discriminate their own young. Thus, at the outset of the study of inclusive fitness, active kin recognition was acknowledged to be unnecessary in some contexts. Making a similar point, Maynard Smith (1976) wrote as follows: "It is not a necessary feature of kin selection that an animal should distinguish different degrees of relationship among its neighbors and behave with greater altruism to the more closely related."

What rules might a bird use in becoming a helper? Remember first of all that the feeding of young by parents is a very basic behavior that must have been already present in the ancestors of nearly all species with helpers. For breeders in many noncommunal territorial species, simple rules such as "Feed any young in your own nest" or "Feed any fledgling of your own species," appear to suffice because any young they encounter are likely to be their own.

No new motor patterns are involved in the evolution of helping. A simple rule for a potential nonbreeding helper would be "Breed if you can; but if not, care for any young in your social unit," This "membership rule" (Brown and Brown, 1980) embodies the assumptions of my 1974 theory (habitat saturation; indirect selection). A similar rule was formulated by Rabenold (1985) to describe the origins of close relatedness between nonbreeding helpers and breeders in the Stripe-backed Wren.

This "rule of thumb" would bring about the situation in the Florida Scrub Jay described above in which (1) the helpers are mainly yearlings who are full or half sibs of the young being fed, and (2) the fraction of birds breeding declines sharply with age along with their average relatedness to the young. Since a very high degree of association between kin is achieved in such species *without* active kin recognition, selection to improve on the result by favoring active recognition must be weak indeed. Furthermore, it is doubtful that such a mechanism would be able to eliminate all mistakes.

In what context then would selection favor active kin recognition? Two situations may profitably be contrasted, the colonial and the communal. In breeding *colonies* pairs or larger families exist within large aggregations of unrelated individuals. In *communes* nuclear or extended families occupy largely exclusive territories and actively chase intruders out and are themselves chased out should they intrude. In a colony, parents often must find their own young in a sea of unrelated young. Orphans try to scrounge from others but have a hard time. In a commune, access of the feeders to young in another unit is largely precluded by territorial behavior. Therefore, the chances that any dependent young encountered are related to the feeder are high. Feeders behave accordingly and feed young indiscriminately.

An illustrative comparison is afforded by the colonial Pinyon Jay and the communal Mexican Jay. In the Pinyon Jay, parents find their own young and feed them preferentially. In the communal and group-territorial Mexican Jay, they do not; parents feed each others' fledglings indiscriminately even though ample opportunity exists for the learning of discrimination (Brown and Brown, 1980).

The lesson from this comparison is that active mechanisms for kin recognition may be needed in colonial species but not in group-territorial, noncolonial species. The available evidence fits this theory. Evidence for active kin recognition exists in the following colonial species: Pinyon Jay (above), Pied Kingfisher (Reyer, 1984; Reyer and Westerterp, 1985; Chapter 13), and possibly the Bell Miner (Clarke, 1984) and White-fronted Bee-eater (Emlen, paper at 1984 AOU meeting). Evidence for nondiscrimination when a choice could be made between the individual's own and other young exists only for the Mexican Jay; however, several studies of group-territorial species have reported feedings by helpers that were not closely related to the helped. (Brown, 1972; Ligon and Ligon, 1978; Rabenold, 1985; Woolfenden and Fitzpatrick, 1984).

To answer the question posed at the beginning of this section, probably active kin recognition of some sort is needed in colonial species. In the vast majority of all-purpose-territorial species, however, active kin recognition is not needed. In this situation simple "rules of thumb" provide a sufficiently high degree of association among kin for indirect and direct kin selection to operate on the giving of care to dependent young. Therefore, we cannot on the bases of Ligon's and Woolfenden's arguments reject indirect selection as being important in the evolution of helping behavior in group-territorial species.

Inbreeding

Inbreeding is a pattern of breeding in which the members of mating pairs are nonrandomly close relatives. A high frequency of inbreeding can result in a relatively high frequency of homozygotes in a population and a high average relatedness.

Does a significant amount of inbreeding occur in communal birds? If so, is it a consequence of the population genetic structure that is brought about by the conspicuously reduced dispersal in communal birds? And does the increased relatedness resulting from inbreeding influence natural selection in a manner that might be expected from Hamilton's rule? These questions will be considered briefly.

In theory the level of inbreeding in a population can be estimated in two ways: (1) from pedigrees, such as that in Figure 12.7, and (2) from observations of the relative frequency of homozygotes and heterozygotes. Unfortunately, the former requires long-term, intensive studies and the latter is subject to several statistical assumptions. For communal birds we have now considerable anecdotal information about instances of inbreeding derived from genealogical fragments, but the only reliable published estimate of the coefficient of inbreeding was done using the second method. More progress can be expected as the genealogies improve with time and, thereby, stimulate analyses. In the meantime, it is illuminating to consider the present evidence.

It is useful to distinguish *close inbreeding* as a special case. Close inbreeding occurs when mates are related by $r = 1/2$ as in brother-sister or parent-offspring mating. Mating between half-sibs may also be considered as close inbreeding. Close inbreeding is relatively easy to detect. It is only necessary to follow one cohort of offspring to maturity and observe its pattern of mating. To detect cousin-cousin inbreeding is more difficult. To identify such a mating it would be necessary to prove that the cousins had a common grandparent, which requires two generations of observation instead of one. Not only is more time required, but a larger initial sample of individuals must be followed to allow for mortality over a longer period. Consequently, cousin-level inbreeding is rarely reported and its frequency is poorly known.

The following cases of inbreeding in communal birds have been reported:

Rowley (1965a:287) reported three cases in the Superb Blue Wren: father-daughter, mother-son, and brother-sister matings. In the Splendid Wren little inbreeding occurred despite the fact that 11 of the 29 banded males attaining breeding status inherited their (presumably natal) territories along with the widow, who was usually not the mother (Rowley, 1981a). A father-daughter and two mother-son matings were observed in 57 unit-years. Later, with a larger sample, 10 of 28 mated females were reported to be related to their mates (Payne et al., 1985). The related males included 4 fathers, 3 sons, 2 brothers, and 1 nephew. Inbreeding is hindered in these species (1) by the wider and earlier dispersal of females than males, (2) by the high likelihood that a male's mother will die before he inherits his natal territory, and (3) an apparent mating preference for nonrelatives, to judge by two sets of observations.

In the Acorn Woodpecker ten cases of inbreeding were reported for the period 1975–82 (Koenig et al., 1984). Three of these involved sib-sib matings; and four were between father and daughter, the remaining cases being more remote and more questionable.

In the Florida Scrub Jay Woolfenden and Fitzpatrick (1984) reported five cases of inbreeding, four of which involved individuals that had not "functioned as members of the same breeding group" (p. 69). The five included one mother-son mating but the other four were rather distantly related.

The coefficient of inbreeding, F, is really an index of genetic fixation rather than of inbreeding *per se*. In theory, it may be estimated by making use of the deficiency of heterozygotes (and excess of homozygotes) that must result from inbreeding (Crow and Kimura, 1970). F may be calculated from the observed number of heterozygotes, H_O, and the expected number based on the Hardy-Weinberg equilibria with various corrections, H_E, by means of the following formula:

$$\text{coefficient of inbreeding} = F = \frac{H_E - H_O}{H_E}.$$

This method has been used for the Grey-crowned Babbler (Johnson and Brown, 1980). Based on 8 polymorphic loci in 80 individuals, estimates of F ranged from 0.015 to 0.031, depending on certain assumptions. These values are not significantly different from zero. Thus, although some inbreeding might go undetected by this method, inbreeding is not an important factor in the population genetics of Grey-crowned Babblers.

Although studies of color-banded populations of communal birds have been going on since 1969 for the Mexican Jay, Florida Scrub Jay, and Acorn Woodpecker, and for almost as long in a few other species, no workers except Payne et al. (1985) have reported more than a few anecdotal cases of inbreeding; nor is there any conclusive evidence of any kind that inbreeding is extensive in any species of bird. This situation suggests that mechanisms exist which tend to prevent close inbreeding. Three are obvious: (1) the greater dispersal distances of females than males tends to prevent sib-sib mating; (2) delayed breeding tends to reduce the possibility of parent-offspring and other intergenerational mating, since more time elapses before breeding by the offspring during which the parent can die or disappear; (3) furthermore, it is usual among temperate-zone species for all the offspring to leave the natal territory before breeding and for the parents to return to or remain on territories where they have bred successfully. This pattern holds for many singular-breeding communal birds, at least for females, and tends to prevent parent-offspring mating.

There seem to be other, more specific mechanisms too, but there have been few published studies. Koenig and Pitelka (1979) observed that nests with more than one adult female Acorn Woodpecker averaged more eggs in the nest than with only one adult female. They then showed that this effect was absent for five units in which the extra females were yearling daughters of the breeding pair, although such yearlings could breed if outside their natal unit. In two cases daughters bred in the presence of their mothers but absence of their fathers. They interpreted these cases as an indication that the inhibition of breeding lies between father and daughter, rather than between mother and daughter. Their result is surprising since prevailing opinion would have ascribed the suppression of breeding to intrasexual dominance (e.g., Carrick, 1972; Borowsky, 1978). Despite such incest-avoidance mechanisms, ten cases of known or probable inbreeding have been recorded in this population (Koenig et al., 1984).

Absence of inbreeding in cases in which it might be expected is known in other species too. In the Stripe-backed Wren, when the breeding male died, the widow left the group rather than breed with her sons, who remained (Rabenold, 1985).

Another mechanism that should reduce the frequency of sib-sib mating is dispersal of members of the same sex together but not members of different sexes together. Such unisexual social dispersal is common in Acorn Woodpeckers (Koenig, 1981a), Green Woodhoopoes (Ligon and Ligon, 1978, 1979, 1983), Native Hens (Ridpath, 1972), White-browed Sparrow-weavers (Lewis, 1982a), Yellow-billed Shrikes (Grimes, 1980), and African Lions (Bygott et al., 1979), and occurs sporadically in other species.

Kin Selection as Group Selection

The following quotation was published by me in 1966, but it serves well to introduce this section since its prediction has proven to be true and its lesson remains important. Group selection has received more attention, and *the distinction between types of groups* remains fundamental.

> The terms group selection and intergroup selection have been used in referring to a variety of phenomena including recently the evolution of behavioural social systems involved in population control. Since the phenomenon of group selection promises to receive greater attention from students of evolution in the future, I feel that a clarification of terms at this time will help to prevent confusion. The "groups" under consideration may be quite different from each other and the action of natural selection on them will depend on the nature of the group.
>
> At least two extreme types of group can be recognized: the family or lineage group and the population. Intermediate types occur, especially in species in which culture is important, but only the extreme types will be discussed here. The lineage group is a sub-group within a population, and is defined on the basis of kinship. A population in the ecological sense, although it may also have a common lineage, is generally defined from other populations spatially, irrespective of lineage. Theoretically the two concepts can overlap, but in practice this is not embarrassing. The term kin selection may be applied to lineage groups. The evolution of

parental care, whether behavioural, morphological or physiological, may be attributed to kin selection.

The simplest lineage group is composed of one or more parents and their offspring. In the typical case the family group disperses before the first generation has reached maturity. In some social insects, a few species of birds, and in some other animals, the young remain with the parents and help to raise subsequent broods, as in some cases of "helpers at the nest". Among birds, in the Mexican Jay, Fairy Wren, Australian magpies, and some other species, territories may be defended by social groups larger than pairs, in which incipient communal behaviour may occur; but the most extreme development of this evolutionary potentiality in birds appears to be the situation found in certain anis in which all phases of breeding may be communal. . . .

Assuming lineage to be important in the latter cases, we can then recognize a continuum among various species in the importance and complexity of the family or lineage group and in the number of generations in which it may stay together as a unit. The existence of such a continuum is provided for in the expression "inclusive fitness" in Hamilton's model for kin selection. It is possible at all positions on this continuum to recognize the existence of competition between lineage groups. This may be done whether the unit is a pair of titmice and their brood and the competition influences the evolution of clutch size or if the unit is a flock of anis and the competition determines the social organization and behaviour of the flock and species. The critical point in all these cases is that the group is characterized (to varying degrees) by a common lineage, which is characteristically maintained by social bonds within a population and with no important ecogeographic isolation from other groups. The evolution of the ability to form such social bonds is also provided for in Hamilton's model. . . .

In my opinion, because these cases may be recognized as lying along a continuum (provided for in Hamilton's model) they should all be interpreted as representing one phenomenon, which may conveniently be called kin selection, or perhaps lineage selection. To use the term "group selection" for any of these cases is confusing, as Wynne-Edwards (1962) has used "group selection" and "intergroup-selection" in explicit reference to intraspecific interpopulation competition, which is quite a different phenomenon. In order to avoid confusion I propose the following definitions: selection between lineage groups within populations will be called kin selection; selection between spatially defined (allopatric) populations of a species will be called interpopulation selection.

Whether the term group selection will eventually prove useful as a generic term to include both kin selection and interpopulation selection remains to be seen. The differences between the two are significant. Kin selection depends on sympatric competition between lineage groups within a population and can be treated theoretically in a manner comparable to individual selection, since the basic time unit is most frequently the generation and the lineage groups are not prevented from interbreeding except partially in the complex cases, as in species with communal territories. Interpopulation selection depends primarily on allopatric competition; the basic time units are the time constants for extinction and colonization, and it may require a different type of mathematical expression, such as game theory.

Of course, the theory of inclusive fitness was not proposed as a type of group-selection theory; it has been rightly claimed to be more akin to individual selection. Nevertheless, the basic principle of kin selection inevitably involves groups. In Hamilton's rule the group consists of only two parties, but the principle is easily generalized to include larger groups. In such groups the essential quality is that the members of a group share the same genes to a greater than random extent. By far the most important origin of shared genes in a group is kinship based on common lineage; but as D. S. Wilson (1975, 1977) has rightly insisted, nonrandom assortment by genotype may have similar effects. Since nonrandom assortment is relatively unimportant for us, we shall neglect it in our treatment, noting only that it is formally included in the concept of indirect fitness.

It does not take a mathematical model to perceive the similarity between two-party kin selection and n-party kin selection, which has been called kin-group selection (Brown, 1974); however, the models do bring out nicely the relationship between the average coefficient of relatedness in a group and the genetic structure of social groups. Perhaps the most influential model has been that of D. S. Wilson (1975), but the mathematical essence of the relationship was, according to Wilson (1983), brought out independently by at least four others (Hamilton, 1975; Charnov and Krebs, 1975; Matessi and Jayakar, 1973, 1976; Price, 1972).

The key factor in these intrademic kin-group selection models is the average relatedness of the unit members to others in the group, relative to the deme, R. For kin-group selection to operate, within-group relatedness must be above average for the population. This has two related consequences. (1) For a segregating locus, animals will share genes more within groups than between groups. (2) Therefore, genetic variance between groups will be greater than expected by chance.

It is useful to portray the relationship between genetic variance and average relatedness mathematically and graphically. Using Grafen's (1984) formulation for a haploid model with constant group size, genetic similarity, S, is a function of the ratio of the actual between-group genetic variance, v, and the expected binomial variance, v_b, scaled to mean group size, n.

$$\text{genetic similarity within groups} = S = \left[\frac{(v/v_b) - 1}{n - 1}\right]$$

When genes are randomly distributed among groups, $v/v_b = 1$ and $S = 0$. When genes are identically distributed among groups, $v = 0$ and $S < 0$. The interesting case occurs when between-group genetic variance exceeds expectation, $v > v_b$. Now $v/v_b > 1$ and $S > 0$, conditions that are conducive to kin-group selection.

Graphically, the effects of the variance ratio, v/v_b, are striking. When there is no genetic between-group variance and $v = 0$, only classical individual or direct selection occurs (unless there is active kin recognition). In this case the dashed line in Figure 4.2B separates traits that can be selected by classical, direct selection from those that cannot. This situation typifies most older selection models, in which possible subdivision of the deme into social units was not considered.

When an intrademic structure is introduced such that individuals are distributed into social groups randomly with respect to genotype, the evolution of cooperation becomes easier. In the classical model, cooperative donor traits that benefited the donor more than the recipient (donor cooperation in Fig. 4.2C) could evolve but those that benefited the recipient more than the donor could not evolve without additional assumptions (*R*-cooperation in Fig. 4.2C). With random variance, they can. As variance among groups increases, therefore, a greater range of relationships benefiting recipients becomes possible.

Finally, when grouping by kinship causes a high between-group genetic variance, such that $v/v_b > 1$, donor traits whose cost-benefit effects lie in the altruism quadrant may be selected by indirect selection. The line that separates donor traits that can be selected for from those selected against now passes through the altruism quadrant, the origin, and the selfishness quadrant, as in Figure 12.8b and c.

An interesting feature of these diagrams is that the slope of the break-even lines, which separate selection for from selection against, varies with the between-group variance, which in turn is a function of the genetic

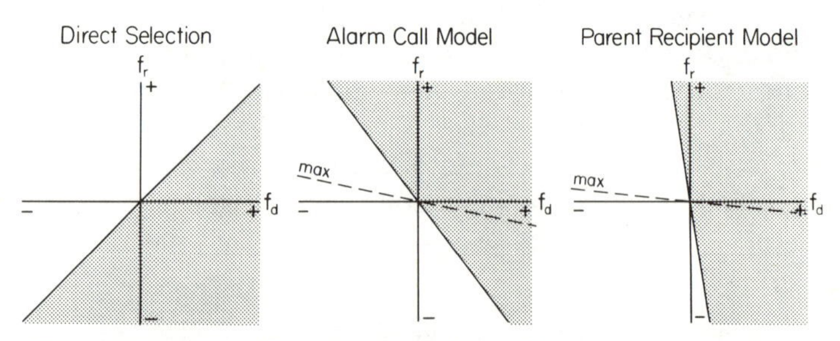

Figure 12.8 Fitness space diagrams for (a) direct selection, (b) alarm call model, and (c) parent-recipient model (D. S. Wilson, 1975, 1977); f_d and f_r are the changes in fitness of donor and recipient respectively. *Stippled areas* indicate traits which could be selected in donors. Thresholds in (b) and (c) are means for all variable allozymes in trait groups of the Grey-crowned Babbler. *Dashed lines* represent the maximum possible slope of these thresholds for $N = 5.8$, the average size of babble trait groups. (From Johnson and Brown, 1980.)

composition of the groups. This fact allows us to test a prediction of indirect-fitness theory using empirically gathered genetic data.

If we sample social units for their proportion of various alleles at polymorphic loci, the prediction is that in highly social species with conspicuous helping behavior, electromorph alleles should have a between-group variance such that its break-even line runs through the altruism quadrant. In other words, alleles should be unusually common in some groups and unusually rare in others. This test has been performed for the Grey-crowned Babbler (Johnson and Brown, 1980). The results (Fig. 12.8) are presented in terms of two models. In the alarm-call model all members benefit from one caller; in the parent-recipient model only the parents and their dependent offspring benefit. In both cases the break-even line passed through the altruism quadrant, as predicted. Similar results have been obtained for the Acorn Woodpecker (Mumme and Zink, pers. comm.).

Conclusions

The expectation that members of social units would prove to be closely related has been confirmed for social systems that are based upon a single breeding pair with a variable number of nonbreeding helpers. Contrary to the claims of some authors, relatedness between nonbreeders and breeders in species such as the Green Woodhoopoe and Florida Scrub Jay is in excellent agreement with predictions based upon an important role for indirect selection. Extended families are less closely related on average.

In nest-sharing species relatedness of breeding helpers to the young being fed is probabilistic but is consistent with kinship theory. Relatedness of the breeders to each other occurs in some species but is probably not characteristic of all such systems. This tentative conclusion needs confirmation using genetic methods.

Close relatedness arises through retention of offspring on the natal territory in the case of nuclear-family systems, and through dispersal in unisexual cohorts from the same social unit in some species. Both phenomena may occur in the same species.

Close genetic relatedness within social units on a population basis has also been established by examination of the distribution among units of electromorph alleles in two species. Both the genealogical and electrophoretic methods have revealed that the genetic structure of populations of singular-breeding communal birds with nonbreeding helpers is suitable for indirect selection.

The expectation of a high frequency of close inbreeding has not been confirmed in any species despite genealogical and electrophoretic studies on a variety of species. Therefore, inbreeding is probably unimportant in the evolution of helping in birds.

This chapter has shown that the genetic structure of nuclear-family species with regular helping is consistent with the requirements of indirect kin selection. Whether or not indirect selection is important depends on its magnitude, which is estimated for a variety of species in the following chapter, and especially its *relative* magnitude compared to direct selection, which is considered in Chapter 14.

13 Indirect Selection for Helping

In this chapter and the next we evaluate various hypotheses for the evolution of helping by *nonbreeders.* Here we consider theories that invoke indirect selection. In the following chapter we consider theories that exclude an important role for indirect selection.

There exist two simple ways with which to reject the hypothesis that indirect selection is important in the decision of a nonbreeder to become a helper. The hypothesis must be rejected if (1) donor and recipient are typically *unrelated* or no more than randomly related, or (2) recipients *do not benefit* from the decision to help. Chapters 12 and 11 respectively have presented the evidence on these points. The data do not allow us to reject the hypothesis, though the evidence is not strong for all cases. Nonbreeding helpers in mate-sharing systems are either rare or difficult to interpret because of the complexity of the social structure. Therefore, we restrict our present discussion of helping by nonbreeders mainly to singular-breeding species.

The evidence considered in previous chapters allows us to conclude that indirect components of fitness may be involved, but it does not give a clear picture of how important they are or how they vary with age and other factors. In this chapter, therefore, we consider the few cases that quantitatively place indirect fitness in the wider perspective of the animals' life history. The data fall into two categories: (1) detailed accounts of age-specific ecological costs and benefits analyzed as described in Chapter 4, and (2) comparisons of helper effort toward recipients as a function of kinship.

Estimates of Indirect Fitness by Age

Grey-crowned Babbler. The first attempt to examine quantitatively changes in the direct and indirect components of inclusive fitness with age was made for a population of Grey-crowned Babblers in Queensland, Australia (Brown and Brown, 1981b). The research was done by two parents, their helper offspring, and some unrelated helpers. The related helper was probably the hardest worker of the entire crew.

The first step in this analysis was to determine the effect of helpers on reproductive success of breeders. Through stepwise regression we determined that an average helper was responsible for an increase of about 0.5 fledglings per breeding pair per year. This is $h_{x,R}$ in Table 4.2. This did not vary significantly with the age of helpers and was, therefore, kept constant with age in the calculations, even though a small increase might be expected.

The next step was the calculation of r_{HR} for each age. Since we had observed that offspring stayed with their parents in their first year, we used the formula from Chapter 12, $\bar{r}_{HR} = s^t/2$.

To obtain I_H we took the product of these two factors. In this example $I_H = h_{x,R}r_{HR}$. Note that I_H (given as Gr_{HR} in Fig. 13.1) declined slowly with age. This decline is caused by r_{HR}, not by $h_{x,R}$.

The direct component of fitness was estimated for the breeding option, D_B, by observing actual numbers of fledglings per breeding pair per year in babblers of known age (estimated from color of iris as in Fig. 5.2). To make these estimates comparable with those for I_H we weighted them by multiplying by r_{HO}, the relatedness of parents to offspring, which we assumed to be 1/2. In terms of Table 4.2, $D_{B,x} = m_x r_B$. In Figure 13.1,

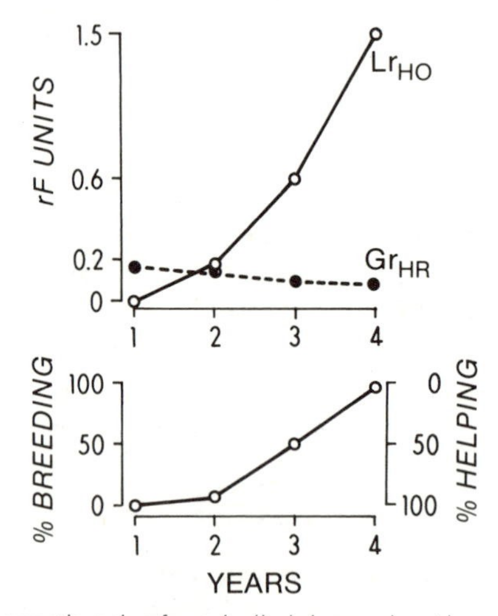

Figure 13.1 Annual genetic gains from I_H (helping and not breeding) vs. D_B (breeding without helping) as a function of age in the Grey-crowned Babbler. *Above*: $I_H = Gr_{HR} =$ average genetic gain by helping in *rF* units (product of expected relatedness of helper to young times marginal gain in number of fledglings per unit per helper). $D_B = Lr_{HO} =$ average reproductive success for all individuals (including breeders and nonbreeders) in *rF* units. *Below*: Percentages of babblers breeding and helping, by age. After four years of age all birds are breeders. (From Brown and Brown, 1981b.) See also Figure 5.2.

$D_{B,x} = Lr_{HO}$. Therefore, $D_{B,x} = m_x r_B = Lr_{HO}$. Expressions of this sort, in which a number of fledglings (F) is weighted by relatedness to them (r) are referred to as rF units in Figure 13.1. They may be thought of as gene-equivalents or genome-equivalents.

Having estimated both I_H and D_B for a series of ages, we are now able to compare them at each age. We wish to know if the criterion, $I_H > d_B$, is satisfied at each age for an average babbler. Figure 13.1 shows that it is satisfied at age one, equivocal at age two, and not satisfied from age three onward. The change in relative profitability of the two options is determined mainly by D_B, which increases rapidly while I_H declines slowly.

D_B is relatively high for two reasons. First, when breeding, a bird has the help of its mate (Charnov, 1981). Second, it is often aided additionally by helpers. For a babbler, therefore, it is better to receive than to give. The advantage of receiving aid from helpers was also estimated in this study by making use of the regression of reproductive success per unit (RS) on number of helpers (H). When this is plotted, the reproductive success without helpers can be read as the y-intercept, i.e., RS when $H = 0$ was 1.74 fledglings. Summing RS unaided over all 36 units ($36 \times 1.74 = 63$) and subtracting from total RS for all units (122) gives the number of fledglings attributable to helpers (59). This was 49% of the total. Helpers, therefore, are very important to breeders.

Calculations of this sort have been misinterpreted.. They do not imply that a one-year old should necessarily delay breeding in order to be a helper (*contra* Woolfenden and Fitzpatrick, 1984). Although it is true that an *average* yearling benefits its inclusive fitness more by helping than by breeding, this does not mean that a yearling should delay breeding *if it gets a chance*. The average Lr_{HO} is low at young ages because chances for such birds to breed are few, as shown in the lower part of Figure 13.1. Although no yearlings bred in this study, some two-year-olds did; and their success was comparable to that of adults. As I stressed earlier (Brown, 1978a:139), the most important factor determining the rise in D_B (or Lr_{HO}) is probably F, the probability of obtaining breeding status (a territory and a mate). The reproductive success if breeding status is obtained is L_{max}; therefore, $L = FL_{max}$. The low value of Lr_{HO} in Figure 13.1 at the early ages is likely to be due mainly to low F.

The onset of breeding in these babblers varied considerably among individuals. Individual variation is not illustrated for values of I_H and D_B. If it were, the agreement of observations with inclusive fitness theory would be more impressive; the D_B of those individuals that were able to attain breeding status at a young age (e.g., two or three would be well above and I_H curve, while the unsuccessful individuals would lie below I_H at their age.

Future effects have not been explicitly considered in these calculations. For simplicity we have assumed that $d_H = d_B$ and $i_H = i_B$. The goal here has been to see what we can learn from present effects (I_H and D_B). To some extent the future is revealed by these present effects. Clearly the probability of achieving breeding status increases with age. This is a logical deduction from the lower part of Figure 13.1. Anything a helper could do to hasten this increase would be to its advantage; however, it is debatable whether helping *per se* has this effect.

The genetic contribution by helping is relatively small compared to that from breeding, but its importance is magnified for some individuals. Half of the yearlings are expected to die before breeding age, based on unpublished data from the same study. These individuals will make no other genetic contribution to the next generation. In brief, when examined age by age these data are consistent with theories that invoke indirect selection. Future effects on direct fitness (d_H and d_B) are considered in the next chapter.

The Stripe-backed Wren. A more detailed analysis with more striking effects of nonbreeding helpers on the indirect components of inclusive fitness has been carried out on the Stripe-backed Wren in open woodland on the llanos of Venezuela (Rabenold, 1984, 1985). This singular-breeding species lives in communal groups of 2 to 14, but averages 5 full-grown birds in a unit. Like the Grey-crowned Babbler these wrens use nests as dormitories (Skutch, 1961b) for roosting at night. In general, their social structure and behavior closely resemble those of the babbler.

In a preliminary analysis Rabenold (1985) performed calculations similar to those just described for the Grey-crowned Babbler. Combining the sexes, as in the babbler, he obtained the following estimates:

Criterion:	$h_{x,R}$	$r_{HR} = I_H$	$>$	$D_B = m_x$	r_B
Results:					
Age: $x = 1$	0.65	$0.39 = 0.35$		$0.11 = 0.22$	1/2
Age: $x = 2+$	0.65	$0.23 = 0.15$		$0.29 = 0.58$	1/2

Comparing I_H with D_B the results show that birds should refrain from breeding at age one but should start at age two. This is what they do.

These calculations are based upon averages for the population. Some individuals may through ability or circumstances do better or worse. It is also possible that individuals might take unit size into account. As shown in Figure 11.1, helpers significantly increase reproductive success but not equally at all group sizes. The second, third, and fourth helpers add more than the first and fifth. Forsaking helping at age one to breed in a group of two would not pay coming from a group of four to six, but it might from groups of three or seven. Since groups of four to six are the most common, the distribution of unit sizes is appropriate for maximum benefits from helping and is consistent with the hypothesis that dispersing birds take group size into account when changing groups.

This situation offers a *test of the importance of the indirect components* for delaying breeding and dispersal, as well as for helping. If the wrens do not take indirect components into account, they should take the first opportunity to breed even if the inclusive fitness gain would be less than by helping ($I_H + i_H > D_B > 0$ and $d_H = d_B$). The data, however, show the opposite. In spite of the fact that most helpers in the population reside in groups of four and five, just above the size threshold for successful breeding, very few dispersers, come from groups of these sizes. Dispersers of both sexes are less likely to originate from groups of four and five than if they were drawn at random; they are disproportionately likely to come from trios or groups of six or more, where their aid would be superfluous or futile (K. Rabenold, paper at Midwest Population Biology Meeting, 1985). Since males and females are reluctant to leave groups where their departure would lower their indirect fitness, they seem to be taking this factor into account and weighing it as expected under inclusive fitness theory. This interpretation is especially appropriate for females, who never breed on the natal territory (males have a direct fitness interest in building the natal group—they could breed there). This study, therefore, strongly suggests that indirect selection plays an important role in delaying breeding and dispersal as well as in helping behavior.

Another factor deserving emphasis is the value of F. As noted, F varies systematically with age; however, it may also vary with population density, which should reflect reproductive success in preceding years, as well as the suitability of the habitat for wrens. Therefore, we may expect variation in F and, consequently, in the decision to help from place to place and year to year.

The future indirect component of fitness, i_H, in Stripe-backed Wrens depends mainly on the fact that the juveniles produced become helpers in the following year for their parents or for an older brother of the helper. Rabenold's calculations have shown that this component is not trivial. Even by itself i_H for a yearling can be as great as average reproductive success.

Other Species. Earlier reviews compared indirect fitness from helping and direct fitness from breeding in several species (Brown, 1975a, 1978a; Emlen, 1978). These studies usually revealed a significant effect on I_H, which was usually less than D_B. Brown (1978a:140), therefore, suggested "that the most important determinant of the tactics of a potential helper is F, not G/L_{max}." F reflects mortality during dispersal, availability of territories, and availability of mates. Since F is frequently low in noncommunal species too, a low F is not a sufficient reason for helping. Biologically significant values of r_R and G are also needed.

Feeding Preference and Effort: A Test of Indirect Selection

A powerful and distinctly different approach to the problem of evaluating the importance of indirect selection has been used only in recent years. It depends upon comparisons of the behavior of helpers toward kin and non-kin or toward individuals differing in relatedness, such as full-sibs and half-sibs. This approach has been used with striking results by researchers investigating kin recognition and its mechanisms (reviewed by Holmes and Sherman, 1983). With this approach the comparison is of the type $D^-B^-H^-$ versus $D^-B^-H^+$. We consider cases in which helping is either present or absent, as well as cases in which helping is graded according to kinship.

Pied Kingfisher. The most revealing discoveries using this approach with communal birds have come from studies of the Pied Kingfisher in Africa (Reyer, 1984). We first provide some background. These birds differ from the vast majority of communally breeding birds by nesting in genuine colonies with clumped nests and without group territories or defended foraging areas, although they may also nest in solitary pairs (Douthwaite, 1978). Consequently, since members of a colony share the same feeding and roosting areas, offspring who stay with their parents in a colony probably experience the same energy environment and risk of mortality as offspring not with their parents but in the same colony. At one stroke these facts control variables based on habitat, colony size, costs of territorial defense, and antipredator behavior. In this species all helpers are males. Females and some males breed at the age of one year. An excess of males over females forces some males into nonbreeding status. Helpers were not seen to copulate, incubate, or brood young.

The key feature of the helper system in the Pied Kingfisher that permits the present approach is the presence of two modes of helping by nonbreeding first-year males (Reyer, 1980a).

Primary helpers accompany breeding pairs as they arrive at the colony to breed. They feed the breeding male, support him in feeding the female, and help defend the nest site. All ten primary helpers of known history were sons of at least one of the breeders, and were one or two years of age. Their average relatedness to the young was $r_{HR} = 0.32$ (Reyer, 1984). At two colonies 28% and 33% of pairs had primary helpers.

Secondary helpers did not appear at the nests until three to seven days after hatching. They would fly through the colony with fish in beak, land near some nests, and wait. They approached various nest holes repeatedly but were initially repelled by the owners. At the food-rich colony secondary helpers were consistently repelled throughout nesting. At the food-poor colony, however, they were eventually accepted. Most secondary helpers were thought to be unrelated to their recipient young, resulting in a low average relatedness ($r_{HR} = 0.06$).

Helpers contributed significantly to reproductive success, especially in the energy-poor habitat. Therefore, the primary helpers received a benefit to their indirect fitness (I_H, 0.15 to 0.90 genetic equivalents), but the secondaries did not (Reyer, 1980a:225). Therefore, these results are consistent with the hypothesis that primary helping is based on indirect selection and secondary helping is not.

Having considered the background, we can proceed to compare the feeding preference and efforts of related (primary) and unrelated (secondary) helpers. The point of this comparison is to determine whether helping based on kinship (I_H) differs in ways not already mentioned from helping based presumably on future direct benefits (d_H). Reyer (1984, 1986; Reyer and Westerterp, 1985) observed that *related helpers worked as hard as the breeders and that unrelated helpers worked less hard.* This was true for quantitative estimates of nest guarding and fishing. Unrelated helpers tended to bring fewer and smaller fishes; and their daily contribution in terms of kcal/adult was less than a third of that of breeders and related helpers. Apparently, when the young were unrelated to the helper, the helper looked out more for himself and less for the young. This difference is difficult to explain without invoking indirect kin selection.

What then are the consequences of helping for the helper's future direct fitness (d_H)? These may be divided into two categories: getting a mate the next year, and survival to the next year. Unrelated (secondary) helpers were more likely to be paired in the following year than were related helpers. Indeed, in six of seven cases the helper bred at the same place where it had helped; in three of these the helper was mated to the mother of the preceding year (Reyer, 1982). Related helpers were never observed to mate with their mother. Secondary helpers appear to be helping mainly as a strategy that could increase their chances of getting a female for the next nesting attempt by becoming familiar with at least one female. Therefore, they should prefer unrelated females and should not work hard enough to endanger their survival.

Unrelated helpers were also more likely to survive to the next year (70%) than related helpers (50%). Survival in males was related to feeding effort. Males working hardest survived least well (Spearman rank test,

$r_S = 0.9$, $p = 0.05$, one-tailed). Survival in females was influenced by their helpers; females with two or more helpers survived at a higher rate (92%) than females with fewer helpers (41%), and at a higher rate than nonbreeding males (68%).

These data provide the strongest case for altruism of any study on communal birds. Related helpers receive lower direct fitness effects than unrelated helpers ($D_s + d_s > D_p + d_p$), yet they still choose to help their parent(s). Why? Reyer's (1984) calculations show that the gain in indirect fitness when feeding relatives ($I_p > I_s$) more than compensates for the loss in future direct fitness ($d_s > d_p$). The greater inclusive fitness of primary helpers compared to secondary helpers may be part of the explanation why related helpers "try harder."

Other Species. In few other species has the possibility been examined of a correlation between relatedness and alloparental effort under conditions in which a choice is available to a nonbreeding helper. Clarke (1984) reported a significant correlation between the coefficient of genetic relatedness (r) and mean adjusted feeding rate based on five Bell Miners feeding two broods in Australia. For four of these birds there was a distinct preference for feeding the more closely related young. However, two of the five birds were parents of one of the broods. Of the three birds that were nonbreeders all had a choice of feeding half-sibs ($r_{HR} = 1/4$) or broods in which $r_{HR} = 1/8$. In two of these nonbreeders there was a clear preference for the half-sibs; but two out of three cases is not enough to be convincing.

In European Bee-eaters, a colonial species without feeding territories, many pairs whose nesting attempt failed did not attempt to renest (M. Avery, paper presented at International Ethological Conference, Toulouse, 1985). Among these failed nesters, those who had kin in the same colony tended to help at their nest; those lacking kin did not help at any nests. Nests with helpers achieved greater success than those without. Consequently, failed breeders with kin were often able to increase their indirect fitness while failed breeders without kin were unable to take advantage of this opportunity. In this case the helpers were mostly male, but this bias did not arise from the sex ratio since the birds had been mated. Instead, Avery reported that the preponderance of male helpers was caused by the fact that males tend to breed in their natal colony while females tend to breed elsewhere. Thus, males, unlike females, are likely to have kin breeding in the same colony. In the few cases in which failed females were observed to help, they were—as expected from consideration of indirect fitness—helping kin.

White-fronted Bee-eaters have also been reported to favor close kin in their alloparenting (S. Emlen, paper presented at A.O.U. meeting, 1983).

These unpublished data provide strong evidence for indirect selection in bee-eaters.

Conclusions

Although no study, except on Pied Kingfishers (secondary helpers), has been able to reject indirect selection as being important in the evolution of regular helping by nonbreeders in singular-breeding species, this is only one kind of evidence. In this chapter, therefore, two types of studies are described whose results more strongly support hypotheses that invoke indirect selection as an important force behind helping behavior in the species studied. In the first type, the magnitudes of the components of inclusive fitness at each age have been measured for alternative strategies and found to agree with inclusive fitness theory. In the second and more convincing type, positive correlations between alloparental effort and relatedness (r_{HR}) have been reported. These are consistent with hypotheses invoking indirect selection and difficult to explain otherwise.

Just as in mate-sharing species, average relatedness between feeder and recipient is important. Before accepting kinship as an important factor, however, the controversial issues described in the next two chapters must be considered.

14 Direct Fitness, Mutualism, and Reciprocity

In discussions about the evolutionary causes of helping, one view is that classical individual selection, or direct selection, is primarily responsible, and that the importance of indirect selection is negligible. In this chapter we explore some conceptual bases for this view and consider what facts are needed to test it. As we shall see, there is a paucity of information on critical points.

All theories based exclusively upon direct selection require that helping be classed as mutualism rather than altruism. It does not follow, however, that theories invoking indirect selection require helping to be altruistic; they do not (e.g., Brown, 1974). In theory, this difference offers a way to distinguish between the two approaches. In practice, this distinction has been of little use, although as we saw in the preceding chapter there is compelling evidence for altruism by helpers in one case, the Pied Kingfisher.

For helping in a particular case to qualify as mutualism, a net profit in terms of direct fitness must be demonstrated for both helper and recipient. For the individual choosing a helping option (H) as opposed to a nonhelping option (N) (as well as for the recipient), mutualism requires the following:

$$\text{Mutualism:} \quad D_H + d_H > D_N + d_N$$

Note that the indirect components for the helper are irrelevant for the definition. They may or may not be present.

Especially in this chapter, it is important to dissect the behavior of helpers into its main components. For a nonbreeding helper these are three, as discussed in Chapter 5: (1) nonbreeding, (2) nondispersal, and (3) helping *per se* or alloparenting. When asking why birds help, it is important to specify whether the choices of nonbreeding and nondispersal are included in the question or not. To reiterate the question from Chapter 5, is the question $B^-D^-H^+$ vs. $B^-D^-H^-$ or is it $B^-D^-H^+$ vs. $B^+D^+H^-$? I have attempted to use "helping" in the strict sense of alloparenting but some authors incorrectly lump these three components of the behavior of helpers together and refer to the combination as "helping" (e.g., Woolfenden and Fitzpatrick, 1984; as pointed out by Ligon, 1985). Failure to keep these questions separate has led to serious misunderstandings. The "behavior of helpers" may also include nonbreeding and nondispersal. "Helping" does *not* include nonbreeding and nondispersal; helping is alloparenting.

A Helping Game between Breeders

The direct benefits to be derived from communal living depend upon the behavior of others in the group. Therefore, an approach using game theory may be useful. In order to illustrate the different kinds of mutualism and their ecological determination in communal birds I examine a case of mutual aid-giving in the Mexican Jay that has been modeled as a two-player game by Caraco and Brown (1986). The comparison here is B^+H^+ vs. B^+H^-. Delayed breeding is not involved, nor is dispersal relevant.

Two Types of Mutualism. This game illustrates the distinction between byproduct mutualism and score-keeping mutualism (Brown, 1983a). In *byproduct mutualism* (Type 4 mutualism of West-Eberhard, 1975), aid is given to others because such aid has a net and often immediate self-benefit regardless of whether or not it is reciprocated; such aid would, therefore, be expected even without future payback from the recipient. Aid to others is a byproduct of immediate self-benefit in this case.

In *score-keeping mutualism* the giving of aid is contingent upon future payback by the recipient. Examples are the tit-for-tat strategy (Axelrod and Hamilton, 1981) and the judge strategy of Pulliam et al. (1982), which would result in Trivers's (1971) "reciprocal altruism" (in Trivers's "narrow sense"; Waltz, 1981). Caraco and Brown explore how ecological factors might determine whether byproduct mutualism or score-keeping mutualism obtains in a given situation.

Mutual Helping in the Mexican Jay. Although most helpers in groups of Mexican Jays are nonbreeders, helping by breeders is also quite typical. Shortly before the young leave their nest, the breeders in a social unit begin to feed each other's nestlings (Brown and Brown, 1980). Some of these mutual feedings between young of different mothers are shown in Figure 2.1. After the young fledge, the parents feed the young from all the nests in their territory (often two or three with no apparent preference for their own young (Brown, 1970). This is not simply a case of mistaken identity because feeding of the young of others by breeders begins while all the young are still in the nest. Neither is it based simply on kinship since the parents often are probably unrelated to some young and, in any case, are typically *more closely related to their own young* than to others (Fig. 12.6; Brown and Brown, 1980, 1981a).

A Risk-sensitive Helping Game. The heart of the helping game of Caraco and Brown is a 2 × 2 matrix, which in this case is a *penalty matrix.* The matrix specifies the probability of a penalty to a player under specified ecological conditions. I consider the general nature of the penalty first and then the ecological conditions that affect the payoffs.

The penalty is a threat of death to the fledglings of the parent-player caused nutritionally, in the first case, or, in the second, by the begging calls of the young attracting a hunting Cooper's Hawk (*Accipiter cooperi*). Each parent attempts to feed its young before it suffers nutritionally or its begging attracts a hawk. The parent itself may also be at risk. We designate the time from inception of play to the satiation of the young as T. This depends upon the foraging success of the parent, which varies stochastically. The time allowed before the young are nutritionally endangered or beg loudly enough to attract a hawk is τ. The probability of a penalty, therefore, is $Pr[T > \tau]$. The probability of a safe T is $Pr[T \leq \tau]$.

An interesting feature of this game is that the optimal strategy depends upon the *variance* of T as well as the mean of the rate of food capture, λ of each parent. In other words, using the concepts developed for foraging behavior by Caraco et al. (1980c) and Real (1981), the jay's strategy of feeding its young may be *risk-sensitive*. In the model the mean rate of capture of food, λ, is identical for each player. Under this assumption the probability of a strategy winning or losing depends mainly on the variance of T and not on its mean. If the penalty is incurred only with unusually high values of T, the probability of a penalty increases with the variance of T since the condition, $T > \tau$, will be more frequent with a higher variance.

The players can reduce their variances in the long run by feeding each others' young. On a single play of the game, however, there is no such advantage and sharing is under many circumstances a disadvantage. We now consider two strategies. In *sharing*, S, a jay feeds its own and other young equally. Twice as many young must be fed, but there are twice as many feeders. In *not sharing*, N, a jay feeds only its own young. The mean and variance for each strategy are shown in Table 14.1. Although the means are the same for the two strategies, the variance is reduced when sharing.

Some properties of mutual sharing (SS) and not-sharing (NN) are compared in Figure 14.1 for an example in which the mean foraging rate is $\lambda = 0.5$ units of food per unit of time. The average time until capture, T, of a unit of food sufficient for one's fledglings will be $1/\lambda = 2$ time units under each strategy. The probabilities for an interval centered on any particular value of T under each strategy pair (i.e., both players adopt the same strategy) can be inferred from the probability density functions in the upper part of Figure 14.1. Note that intervals centered narrowly

Table 14.1

Strategies in a helping game.
T = time from inception of play to satiation of own young.
λ = rate of food capture.

Name	*Expected Mean T*	*Expected Variance of T*
S = Sharing	$1/\lambda$	$1/2\lambda^2$
N = Not sharing	$1/\lambda$	$1/\lambda^2$

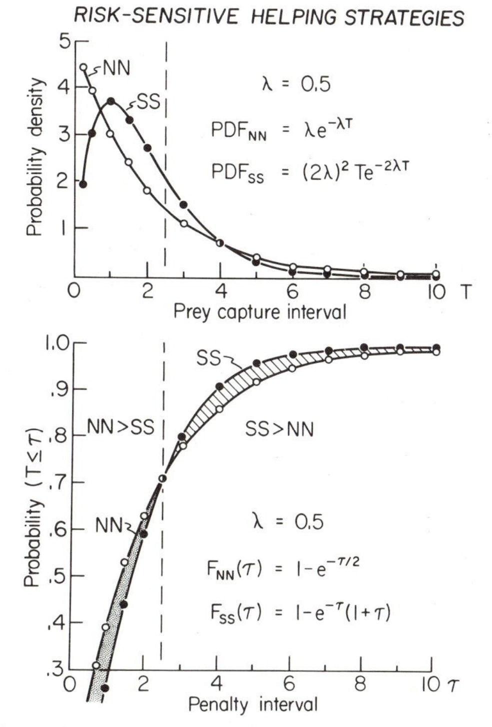

Figure 14.1 Risk-sensitive helping in a cooperative foraging game. The *upper figure* shows the probability density of various values of T for two players sharing (*SS*) or not sharing (*NN*) in the feeding of each other's fledglings for an example in which the mean rate of foraging success (λ) is equal for N and S (0.5). *PDF* = Probability density function. The *lower figure* shows for the above *PDF*s the probability of avoiding a penalty for a range of penalty intervals (τ). The vertical dashed line separates the regions in which N and S are ESSs. The higher variance of N (above) causes S to be the ESS at values of τ somewhat above the mean value of T (2.0).

around relatively high values of T, such as 5.5–6.5, are more likely for NN than SS.

Suppose now that a hawk is attracted if one's young are not fed until 6 time units have elapsed, i.e., $\tau = 6$. What is the safest strategy pair, NN or SS? We need to know for each strategy pair the probability of obtaining a value of T greater than 6. The strategy with the highest probability that $T \leq \tau$ is the safest. To determine these probabilities we integrate the area under each probability density curve above. The results are shown in the lower part of Figure 14.1. As the time allowed before penalty increases, the probability of a safe feeding time increases. For $\tau = 6$ we see that SS has a higher probability of safety than does NN. Therefore, if other conditions for an ESS were met, sharing would be selected with these values of λ and τ. The other conditions may be critically important.

BOX 14.1

EVOLUTIONARILY STABLE STRATEGIES

The concept of an evolutionarily stable strategy (ESS) is valuable when we deal with models of selection for social behavior. An ESS is a strategy that when established in a population cannot lose in natural selection to a mutant strategy. The expected fitness of a strategy, I, when playing against another strategy, J, may be expressed as $E(I, J)$. If I is a stable strategy, then no mutant, J, can do better against I than I itself:

$$E(I, I) > E(J, I).$$

This is a sufficient criterion for I to be an ESS, but I may also be an ESS if

$$E(I, I) = E(J, I)$$

and

$$E(I, J) > E(J, J).$$

These are the "standard conditions" for an ESS in pairwise encounters given by Maynard Smith and Price (1973; Maynard Smith, 1982).

The Prisoner's Dilemma. One determines the winning strategy or ESS using the criteria explained in Box 14.1. To determine precisely the best strategy in a particular environment we need to know the expected penalty for each type of behavior when matched against each possible strategy of the other player in that environment. To know only the equilibrium payoffs (for NN and SS) is insufficient. Consequently, in this game we need mathematical expressions for $Pr[T > \tau]$ for each cell of a 2×2 matrix in an environment that is described by the mean rate of success in foraging, λ, and the time-to-penalty, τ. These expressions for the game with nutritional penalties are shown in Table 14.2. By influencing λ and τ the environment determines the payoff probabilities, the nature of the game, and the ESS.

In reasonably good environments (high λ and high τ), the helping game may become a *prisoner's dilemma.* Under these circumstances it is only the occasional high value of T that incurs a penalty. Therefore, when starvation is an unusual but realistic possibility, the strategy with the lowest variance will minimize the chance of penalty. The prisoner's dilemma is a game between two players in which the payoff matrix is the same for each player (a symmetrical game). It is additionally defined by two rules using the cell designations M, R, P, and Z in Table 14.2. We use these letters to designate the expected fitnesses for the indicated cells. Note that the lowest probability of a penalty provides the highest fitness. Fitness is inversely related to $Pr[T > \tau]$.

Formally, the environment provides the birds with a prisoner's dilemma only when the cells of the matrix in Table 14.2 satisfy the following

Table 14.2

A helping game with nutritional penalties to the young. The payoff to player *A* is shown. In this game the payoff is a penalty, and the winner has the smallest payoff but the highest fitness. The terms *R, Z, M*, and *P* refer to the fitness of player *A* when playing against the indicated strategy of *B*. Thus the mathematical expressions give the probability of penalty but the letters refer to the corresponding fitnesses, which are inversely related. The traditional verbal description of *R, Z, M*, and *P* is given for each cell. (From Caraco and Brown, 1986).

		Player B	
		Share	*Not Share*
Player A	*Share*	*R* Reward for mutual sharing $e^{-2\lambda\tau}(1 + 2\lambda\tau)$	*Z* Sucker's payoff $e^{-\lambda\tau}(\lambda\tau e^{-\lambda\tau} + 1)$
	Not Share	*M* Temptation to not share $e^{-2\lambda\tau}(1 + \lambda\tau)$	*P* Punishment for not sharing $e^{-\lambda\tau}$

relationships:

(1) $M > R > P > Z$, which may be written as
$E(N, S) > E(S, S) > E(N, N) > E(S, N)$; and
(2) $R > (Z + M)/2$, which may be rewritten in a similar way.

These relationships are shown by the sizes of the circles in the center diagram of Figure 14.2. The largest circle or fitness benefit comes with *N* against *S*; however, the player who plays *S* against *N* receives the lowest score and would be expected to shift to playing *N*. Clearly then, *S* can never win on a single play against *N* in a prisoner's dilemma; and *N* can never lose. Also, in the long run an average of *E*(*N*, *S*) and *E*(*S*, *N*) does not pay as well as *E*(*S*, *S*). In single play *N* is an ESS. How then can *S* become a stable strategy if *N* is an unbeatable strategy on single plays?

In order for *S* to be a stable strategy for player *A*, the game must be repeated between the same players indefinitely and *A* needs some knowledge of player *B*'s probable behavior. *A* can acquire this knowledge by keeping track of *B*'s plays. Axelrod and Hamilton (1981) have shown that under these circumstances a cooperative relationship may develop based upon *conditional* strategies in which each player "keeps score" on the other's behavior as a protection against cheating, hence score-keeping mutualism or reciprocity.

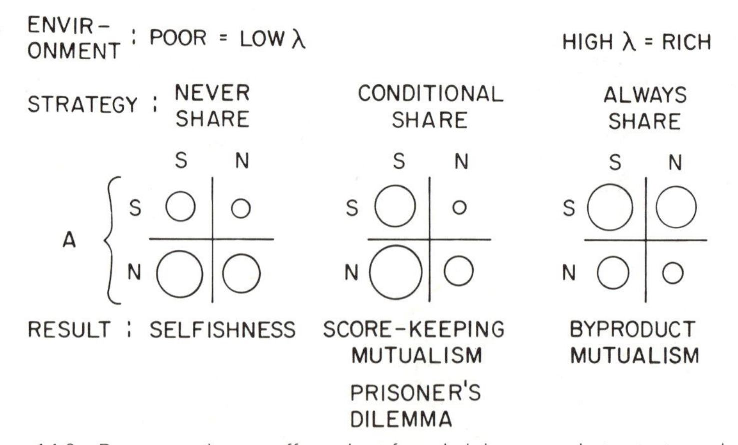

Figure 14.2 Representative payoff matrices for a helping game between two players (*A* and *B*) who may share (*S*) or not share (*N*) in the feeding of each other's young. The options of *A* are represented along the rows; the options of *B*, along the columns. The relative sizes of the circles represent benefits to the direct fitness of the players. The three matrices represent regions I, II, and III in Figure 14.3.

One of many ways to base one's strategy on the "score" is the *tit-for-tat* strategy. If A is playing tit-for-tat, A plays share first and continues to play share as long as the opponent also shares; but as soon as the opponent plays N, A also plays N and continues to do so until the opponent plays S.

Biologists now appreciate that the existence of a score-keeping mutualism as an ESS depends upon the closeness and duration of the association between the players. Supplementing this approach, Caraco and Brown (1986) have shown that simple "asocial" parameters of the environment, such as rate of foraging success, and predation pressure on the young, can by the mechanism of risk-sensitivity create conditions that correspond to a prisoner's dilemma, hence to the situation discussed by Axelrod and Hamilton.

Byproduct Benefits. The transition from score-keeping mutualism to byproduct mutualism may depend additionally upon a socio-ecological parameter. Recall that loud begging by the young of pair A can attract predators to the young of pair B as well. We designate θ as the probability that the young of one pair are placed at risk of predation by proximity to the young of the other pair. In practice θ will be high if the fledglings from different nests tend to perch together and low if they avoid each other. In the Mexican Jay θ is probably often high because fledglings from different nests often perch together and because their risk of capture by hawks is high.

As θ increases, the penalty for playing N worsens (Table 14.3), causing $E(S, S) > E(N, S)$ and $E(S, N) > E(N, N)$, as shown in diagram to the right in Figure 14.2. Consequently, a transition occurs, from a prisoner's dilemma with score-keeping reciprocity as the long-term ESS, to a payoff matrix that no longer qualifies as a prisoner's dilemma—one in which sharing, the ESS, is a byproduct mutualism. This transition is marked by the line between regions II and III in Figure 14.3.

The influence of θ reflects the importance of byproduct effects in the behavior of each party. With sufficiently high values of θ, self interests lead to mutualistic sharing (regions II and III in Fig. 14.3) as the ESS; with low θ, self interests lead to nonsharing or conditional sharing as the ESSs (regions I and II in Fig. 14.3).

There may be a class of factors, such as θ, whose operation favors byproduct mutualisms. Another example of a *θ-factor* would be close relatedness. The indirect benefit from helping a close relative is also not contingent upon reciprocation. Therefore, it may make helping more likely. Nest sharing, whether by polyandry or joint nesting, also provides a high θ-factor, since the young of each player are clumped (in the same nest) with those of other players. Other examples of "θ-factors" may be found in

Table 14.3

A helping game with penalty of death by predation for offspring of the players. Each cell specifies the probability of a penalty ($T > \tau$). The winner has the smallest value. The letters *R*, *Z*, *M*, and *P* refer to the fitness of player *A* when playing against the indicated strategy of *B*. The traditional verbal description of *R*, Z, *M*, and *P* is given for each cell. (From Caraco and Brown, 1986).

		Player B	
		Share	*Not Share*
Player A	*Share*	*R* Reward for mutual sharing $e^{-2\lambda\tau}(1 + 2\lambda\tau)$	*Z* Sucker's payoff $e^{-\lambda\tau}(\lambda\tau e^{-\lambda\tau} + 1)$
	Not Share	*M* Temptation to not share $e^{-2\lambda\tau}(1 + \lambda\tau) + \theta(e^{-\lambda\tau} - e^{-2\lambda\tau})$	*P* Punishment for not sharing $e^{-\lambda\tau} + \theta(e^{-\lambda\tau} - e^{-2\lambda\tau})$

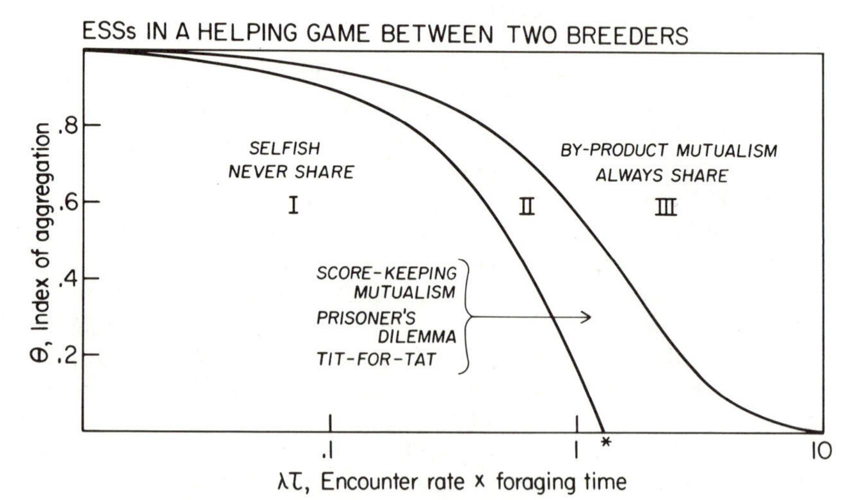

Figure 14.3 A helping game between parents of different broods in the Mexican Jay. In region I, Not Sharing is the ESS. In region II, a prisoner's dilemma obtains and Sharing is the longterm ESS. The result is an example of a score-keeping mutualism or true reciprocity. In region III, Sharing is the ESS, but it represents a case of byproduct mutualism and is not true reciprocity. The asterisk denotes the values of $\lambda\tau$ when $\theta = 0$. (After Caraco and Brown, 1986.)

protection of the nest from predators by colonial seabirds and the seeking of cover by flocking sparrows in response to danger.

The Three Regions of Socio-ecological Space. Each of the three matrices in Figure 14.2 has a diffferent ESS. The combinations of socio-ecological parameters that lead to each ESS are portrayed in Figure 14.3. Each region in this socio-ecological space corresponds to one of the three payoff matrices in Figure 14.2. The figure (14.3) shows the rather complex dependence of the ESSs on the socio-ecological parameters.

The transitions between regions (and matrices) are governed by two factors, θ, the probability of hazardous aggregation, and the product $\lambda\tau$. Theta reflects the quality of the *predator environment*, increasing with the risk of predation. The product $\lambda\tau$ reflects the quality of the *nutritional environment* in relation to the penalty interval; it increases as foraging success improves and as the risk of failure diminishes (greater τ). The transitions between regions are not caused by qualitative or obvious changes in the environment. Instead they are thresholds determined by combinations of continuously varying quantitative parameters. As such the transitions between regions are not intuitively predictable.

The transition between mutualism based upon score-keeping (reciprocity) and mutualism based upon self-interest is shown by our analysis to the rather subtle. It is influenced by a variety of ecological factors. In the Mexican Jay model the transitions were influenced by λ, τ, and θ; however, λ and τ alone were insufficient, and θ was required to be above zero.

In general, θ-factors enable sharing to become an ESS under conditions when sharing would not otherwise be stable. This can be appreciated from Figure 14.3 by noting that without the operation of θ ($\theta = 0$), sharing requires a relatively high value of the other specifying conditions (marked by $*$ on the abscissa). Firstly, when θ increases, score-keeping mutualisms are stable at lower values of the other specifying conditions ($\lambda\tau$). Secondly, if θ is sufficiently high, byproduct mutualism becomes stable at relatively low values of the other parameters (below the $*$ in the figure). Finally, our model, by showing how sharing can be favored by a variety of *combinations of factors*, suggests that the particular combination in a given case in nature may be difficult to identify. In particular, in concentrating on tit-for-tat and other types of score-keeping mutualisms in which $\theta = 0$, recent theoretical studies have neglected the importance of θ-factors.

This analysis has used an observed case of mutual aid-giving between breeding helpers as an example of how various environmental parameters may affect the payoff matrix of a 2×2 helping game leading to three different ESSs. The degree to which this model can be applied to other helping situations and other species remains to be determined. A special case of the model in which $\theta = 1$ appears to represent the situation in nest-sharing species such as anis. Here all young of different pairs are aggregated in the same nest, and unconditional sharing of feedings occurs as expected from the model. Obviously, the model does not apply to the large category of helping by nonbreeders. Even in the Mexican Jay to suggest that this model is a sufficient explanation of helping between breeders would be misleading. The origin of plural breeding, on which θ mainly depends, is tied to other factors (see, for example, the discussions of parental facilitation in Chapter 7 and joint nesting in Chapter 10). Still other uncertainties are discussed in Brown and Brown (1980). Perhaps the principal value of this model is to put the prisoner's dilemma in a broader and more richly detailed ecological perspective with respect to the different kinds of mutualisms and the various environmental influences upon them.

Direct Benefits of Nonbreeding and Nondispersal

We shift now to the case of nonbreeding helpers. Perhaps the most difficult to measure of the components of inclusive fitness and the most misunderstood are those that affect reproductive success and survival in the future, d_H and d_B. Some authors including myself have felt that these are critically important for selection to favor helping (nos. 1–3 in Table 14.4).

Table 14.4

Postulated direct advantages ($d_H > d_B$) for nonbreeding helpers. Examples, discussion, and references in text. F = probability of getting a mate and territory. s_x = annual survival at age x. m'_{x+1} = fecundity at age $x + 1$ if a territory and mate are obtained.

From Delayed Breeding Coupled with Delayed Dispersal	
1. F	↑ due to increased *ability to obtain a territory* and mate stemming from age
2. s_x	↑ due to avoidance of *risks of dispersal*
3. m'_{x+1}	↑ due to increased skill in *foraging* and other skills needed for favorable energy balance
From Alloparental Behavior (Helping) per se	
4. F	↑ due to *association with recipient female breeder*, if not the mother or sister
5. F	↑ due to *augmentation* of unit size favoring budding or social dispersal
6. F	↑ due to formation of *social alliances* with recipient nestlings
7. s_x	↑ due to *augmentation* of unit size favoring vigilance and other antipredator behavior
8. m'_{x+1}	↑ due to *practice in feeding* young (as opposed to no. 3) and *building nests*
9. m'_{x+1}	↑ due to *more helpers*, caused by augmentation and alliances

They are the basis for phase 1 of my three-phase theory (i.e., the "*K*-selection Phase," which was based on habitat saturation; see Appendix). These components were omitted from the earlier calculations using the offspring rule (Brown, 1975a, 1978a; Emlen, 1978) for two reasons. First, no estimates of them existed then. Second, the early inclusive fitness calculations were concerned mainly with the consequences of alloparenting *per se*, not the effects on survival or the probability of getting a territory that stemmed from not breeding and not leaving the natal territory. These are separate issues. In the language of Chapter 5, the early calculations were concerned with the question of $B^-D^-H^+$ vs. $B^-D^-H^-$, as opposed to $B^-D^-H^+$ vs. $B^+D^+H^-$. Both approaches have merit, but they must be kept straight. It is not true that all future effects were overlooked, as some authors have suggested; for they were mentioned in much of the literature on helping, including papers by the authors that did the calculations. Of course, they were not referred to as future direct fitness or "lifetime reproductive success"; it was simply mentioned that helpers probably obtained useful experience, survived better, and had a better chance to obtain a territory in the future if they stayed with their parents for a year.

Indices of Indirect Selection. Using the extended version of the offspring rule and associated symbolism introduced in Chapter 4 we may obtain an expression that indicates the relative importance of the future direct components of fitness in the decision to adopt a strategy of helping while not breeding and not dispersing ($B^-D^-H^+$ vs. $B^+D^+H^-$). The expression will actually be for indirect fitness, but by specifying the relative importance of indirect fitness we automatically specify the relative importance of direct fitness with the same number. Two rather similar indices have been proposed. I use that of Brown (1985) because an essential term was omitted in that of Vehrencamp (1979). The index is simply the proportion of the difference in expected inclusive fitness (E) between the two strategies that is due to indirect components for an average individual at a given age. For nonbreeding helpers we assume $I_B = i_B = 0$, and the expression is:

$$\text{index of indirect selection} = Q_I = (I_H + i_H)/(E_H - E_B).$$

This index resembles Vehrencamp's index of "kin selection," but differs from it by the inclusion of i_H, which is required for reasons given in Chapter 4. In slightly longer form, since $I_B = i_B = D_H = 0$ in this case,

$$Q_I = (I_H + i_H)/[I_H + i_H + d_H) - (D_B + d_B)].$$

If the choice of helping (plus nonbreeding and nondispersal) is entirely due to *indirect selection without altruism*, the numerator equals the denominator, $Q_I = 1$ and $d_H = D_B + d_B$. If the choice is entirely due to future direct selection, the numerator is zero, $Q_I = 0$ and $d_H > (D_B + d_B)$. Values of Q_I between zero and one indicate that both direct and indirect effects contribute to superiority of E_H over E_B. Higher fractions indicate a greater role for indirect effects relative to direct effects. If the choice is due to *indirect selection with altruism*, $Q_I > 1$ and $d_H < (D_B + d_B)$.

Vehrencamp's calculations made a valuable contribution by extending analyses based on the offspring rule to include d_H and d_B. Her estimates did not, however, include all future direct effects. She included only some noncontroversial ones and omitted the controversial ones—although even this is not entirely clear because so little detail was given. It had been widely agreed that birds in helper status probably 1) had a higher rate of survival (s) to the next breeding season than those that left the natal territory early, and 2) had a higher probability of success in competing for breeding status, (F), because of their greater age, experience, and presumbably dominance (e.g., Selander, 1964; Brown, 1974, 1978a; Emlen, 1978). Therefore, $d_H > d_B$ because nonbreeders would be more likely to survive and achieve breeding status.

By making plausible assumptions based on incomplete data involving higher s and F for those birds that stay with their parents than for those that leave at age one, it is indeed possible to arrive at a Q_I of less than 1.00. These estimates of Q_I range down to 0.55 for Scrub Jays (Vehrencamp, 1979) and 0.43 for Splendid Wrens (Rowley, 1981a). Such estimates are completely consistent with the K-selection phase of my three-phase theory (Brown, 1974), but do not provide an indication of the degree to which phases 2 and 3, which invoked "kin selection," are correct. Vehrencamp's and Rowley's estimates deal with $B^-D^-H^+$ vs. $B^+D^+H^-$. Therefore, they do not provide a value of Q_I for alloparenting *per se*, only for alloparenting and nonbreeding and nondispersal combined. The demographic facts which cause Q_I to be less than 1.00 in these cases are those that cause delayed breeding and reduced dispersal.

Scrub Jay. A much more detailed examination of the factors at work in the K-selection period (phase 1 of my theory) has been provided for the Scrub Jay by Woolfenden and Fitzpatrick (1984, their Chapter 10). The K-selection phase explains nonbreeding and nondispersal; it does not explain alloparenting. Their analysis is based mainly upon two factors: (1) the probability of successful *dispersal* by a potential helper at or before the helping age, and (2) the probability of *surviving* and becoming a breeder at a later age if breeding and dispersal are delayed. Their conclusion that nonbreeding and nondispersal (called "helping" by them) are favored in the Scrub Jay by a small value for the first factor and a large value for the second is very consistent with the explanation I advanced a decade earlier. Ironically, they differ from me by invoking indirect selection for this process, at least possibly in some situations. I carefully avoided doing so in 1974; and in 1978 I made my position explicit with the following statements: "The existence of delayed breeding [in communal birds] is also easily explained *without kin selection*" (p. 135, emphasis added). Obviously, I cannot agree with them (p. 324) that their 1984 model is a "major departure" from my views; rather, it is a vindication of phase 1 of my 1974 theory and agrees with my review of the data in 1978 (pp. 134–140).

In the past there has been some confusion over terms. I had treated delayed breeding and dispersal as separate phenomena from alloparenting, and had attempted to single out factors conducive to each (see the Appendix, where my 1974 theory is summarized). Woolfenden and Fitzpatrick have lumped together in their model and under their use of "helping" the three factors that I have attempted to tease apart (*B*, *D*, *H*), even though elsewhere they agree on the importance of treating them separately. This may have been a reason for their misunderstanding of my position in 1978 (see Brown, 1978b). In my opinion we are and always have been in full agreement on the likely bases for nonbreeding and nondispersal in jays and probably in communal birds generally.

To clarify the issues, it may help to think in terms of the three theories outlined in Table 14.5. Theory no. 1, pure direct selection, does not require

Table 14.5

Three theories of communal breeding in nuclear-family systems. These theories differ in the phenomena for which direct and indirect selection were involved as explanations. The entries in the boxes refer to the type of selection that is *required* by the theory; the other type may also occur but is not regarded as critical.

Phenomena to be Explained	1. *Pure Direct Selection*	2. *The "Straw Man," Pure Indirect Selection*	3. *The Combined Theory*
Delayed breeding	Direct	Indirect	Direct
Delayed dispersal	Direct	Indirect	Direct
Helping	Direct	Indirect	Indirect
References	Woolfenden & Fitzpatrick 1978, 1984; Koenig & Pitelka 1981	No published source	Brown 1969a, 1974, 1978

indirect selection to explain any of the three phenomena listed. It has been favored by Zahavi (1974, 1976, 1981), Woolfenden (1976; Woolfenden and Fitzpatrick, 1978, 1984), and Koenig and Pitelka (1981), among others. These authors tend to view the problem as a dichotomous choice between theories no. 1 and no. 2, neglecting to discuss theory no. 3.

Theory no. 2, or pure indirect selection, is basically a straw man. It has been the principal point of attack of the authors favoring theory no. 1. No author known to me has seriously proposed that we accept theory no. 2. It differs from both theories no. 1 and no. 3 by invoking indirect selection to explain delayed breeding and reduced dispersal. In 1978 I employed the offspring rule to evaluate the evidence relevant to theory no. 2. My conclusion was that direct selection (caused by habitat saturation) was the principal cause of delayed breeding and reduced dispersal and that the gains in indirect fitness were relatively unimportant (p. 140, ". . . the most important determinant of the tactics of a potential helper is F, not G/L_{max}"). Despite my clear rejection of theory no. 2 both Koenig and Pitelka (1981) and Woolfenden and Fitzpatrick 1984:324) have continued to represent the issue as theory no. 2 vs. no. 1. In my view it is no. 1 vs. no. 3.

Theory no. 3 combines direct selection to explain delayed breeding with indirect selection to explain care of young by helpers. This theory was not considered at all by Woolfenden and Fitzpatrick (1978, 1984) or Koenig and Pitelka (1981), who directed their criticisms at theory no. 2. This is unfortunate because it is a viable theory supported by much evidence, whereas rival versions of theory no. 1 are tenuous with respect to empirical support. In short, *confusion of the basic issues has caused much effort to be devoted to rejecting a theory that had never been seriously proposed and that had already been rejected by me in 1978.*

The reader may well ask at this point why, if there is so much agreement with me on the importance of direct selection, have Woolfenden and Fitzpatrick persisted in describing my position simply as "kin selection underlies the evolution of helping behavior," not mentioning my theory based on habitat saturation (Woolfenden and Fitzpatrick, 1978) and later, though without naming the source, as "helping behavior is *caused by* or *evolves through* kin selection" (1984:313) when that has *never* been my position (using the term helping in the manner of Woolfenden and Fitzpatrick). The answer seems to lie partly in a different use of the word, helping, partly in their misunderstanding of my use of mathematical models, such as the offspring rule, and partly in their inexplicable reluctance to refer to my theory as I stated it in 1974 and 1969a. With regard to the definition of helping, I have already pointed out that we use the term helping in different ways. As explained above, Woolfenden and Fitzpatrick have not used the standard definition.

My use of models requires some comment. Woolfenden and Fitzpatrick (1984:314) have expressed concern over a statement of mine (Brown, 1978; 140) that Hamilton's inequality is satisfied. They agreed that it is satisfied, but they still objected as follows: "The concern we have over this statement, and with most other references to Hamilton's inequality, is the implication that helping would not otherwise occur. . . ." Such reasoning "ignores a host of *other* gains that might accrue to an individual by helping. . . . Gains pertaining to survival, dispersal, territorial ownership, and later reproduction are not included. . . ." (Note that "helping" in this quotation includes nonbreeding and nondispersal). Their discussion distorts my conclusions, implying that I neglected survival, dispersal, territorial ownership, and later reproduction. Actually these were part of the K-selection phase of my theory. I used the mathematical model which is the offspring rule to generate predictions against which to compare the very simple data that were available prior to 1975 and 1978. Data with which to judge more elaborate models were not available then. The data were consistent with a role for indirect selection to explain helping (not delayed breeding) as long as F (the chance of acquiring a territory and mate) was low; so theory no. 3, involving relatedness, could not be rejected by that means. The data did not allow the rejection of various other hypotheses, including their augmentation hypothesis; *nor did I argue or imply that they did.* Consideration of the data did, however, stimulate my revision of the 1974 model in 1978 to include F explicitly and in effect to reject theory no. 2. Emlen (1982a) and Woolfenden and Fitzpatrick (1984) have subsequently adopted F as the heart of their theories (using ψ and D as synonyms for F). The offspring rule continues to be useful, but it is not my theory for the evolution of helping. It can be used to express a variety of theories, including those of Woolfenden and Fitzpatrick.

Direct Benefits of Alloparental Care: Learning

No discussion of the future direct consequences of helping while not breeding would be complete without mention of the learning of skills during this period. Virtually every author mentions learning as a potential future direct factor, but a critical role for it has not been clearly established. We have already discussed the possible role of *learning to forage* in the treatment of delayed breeding (Chapter 5). It is clear that some learning of foraging skills should occur and that, at least in some waterbirds, the process requires more than one year. It appears, however, that such learning would occur whether the bird was helping or not (no. 3 in Table 14.4).

Building a nest is a complicated and time-consuming process where birds also show improvement with age. It too might be a reason for delaying breeding (Brown, 1984; no. 8 in Table 14.4), but it too can be learned by trial and error whether helping or not. Indeed, in many species helpers are discouraged from participation in nest building.

Dominance is another skill that is often correlated with age, at least for the first few years. Just how the ability to be dominant improves with age or by helping is unclear, but its correlation with age is quite precise for Stripe-backed Wrens (Wiley and Rabenold, 1984; Rabenold, 1985) and could well be a large part of the reason for delayed breeding in Scrub Jays, Splendid Wrens, and possibly other species (no. 1 in Table 14.4). In the Mexican Jay some individuals spend their first winter at the top of the dominance hierarchy in their social unit, above even the breeders (Barkan et al., 1986); so at least one species does not always show such a simple relationship between age and dominance. Although improvement with age in the chances of obtaining a breeding territory is commonly an implicit assumption of habitat saturation models, a role for "learning to be dominant" remains undocumented.

The feeding of young is a skill that appears to improve with age in some species (see Fig. 2.1 for example; no. 8 in Table 14.4). In the Brown Jay, naive, nonbreeding helpers improved visibly with experience even during the brief period in which the young were in the nest (Lawton and Guindon, 1981).

Generational Mutualism

The concept of reciprocal altruism advanced by Trivers (1971) did not at first find favor among students of communal birds. There were at least two reasons. Firstly, it was not at all clear whether helping was altruistic or mutualistic. Secondly, and more important, the evidence for a score-keeping kind of reciprocity was nonexistent, since attention had centered on nonbreeding birds that fed nestlings. No signs of repayment by the nestling or parent were obvious. It was clear, however, that a mutualistic relationship existed *between the generations* (Fig. 14.4). The old (breeders) depended on the young (nonbreeders) for help. The nonbreeders then matured and became recipients rather than donors. The young nonbreeders depended on the old for the privilege of remaining on the natal territory for prolonged periods after independence. Still, it seemed too much to

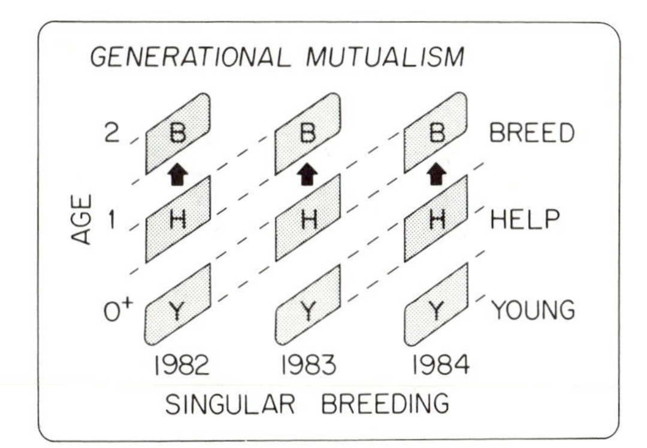

Figure 14.4 Generational mutualism in a simple case, a singular-breeding species. The young (*Y*) develop into helpers (*H*) and then breeders (*B*). As helpers they give aid to the older generation, but as breeders they receive aid from the younger generation.

believe that the young fed by a helper remembered who fed and who did not and then consciously repaid the debt selectively in future years only to those who had earned repayment.

Nevertheless, it was recognized in 1975 that the young produced by helpers might later directly benefit the helpers (nos. 5, 7, 9 in Table 14.4).

> The young so raised . . . contribute to keeping the size of the unit large enough to reap the hypothetical survival benefits of group living. Furthermore, the young produced may in turn help the present helper in acquiring a breeding territory (supposed by Woolfenden, 1975, in *Aphelocoma c.*) and might become helpers to the present helper should the latter accede to breeding status in the same territory, which in *A. ultramarina* happens not uncommonly. . . . These advantages are of particular interest because they constitute a previously unrecognized form of reciprocation, which may be termed unconscious. This reciprocation is achieved without the necessity of remembering social obligations. . . . It would be maintained by selection pressure against nonreciprocators (Brown, 1978a: 136; see also a similar statement in Brown, 1975a:208–209).

Later I preferred to call this relationship generational mutualism (Fig. 14.4) rather than reciprocity (Brown, 1975a) or unconscious reciprocation. It seems likely, in my opinion (Brown, 1983a), that most of these cases are byproduct mutualisms arising out of self-interest and not from the special circumstances of a prisoner's dilemma (see Ligon, 1983, for an opposing opinion). More work is needed in this area.

In the Mexican Jay generational mutualism as just described is found in the same units as mutualistic reciprocal feeding of each others' young by members of the same generation (Fig. 14.5). This appears to be another case in which two or more "origins" of helping may occur in a single species.

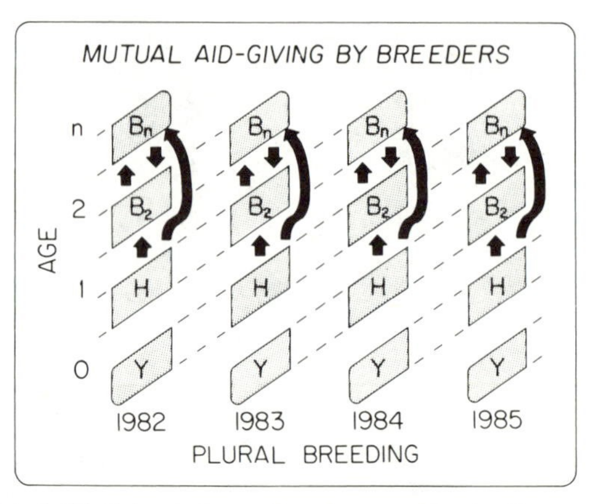

Figure 14.5 Mutual aid-giving between breeders combined with generational mutualism, a complex case involving two kinds of potentially mutualistic relationship. Helping relationships of these two types are both found routinely in the Mexican Jay. *B, H, Y* as in Figure 14.4.

The concept of reciprocation has been developed by several students of communal birds. It was first invoked as a substitute for indirect selection by Woolfenden (1975, 1976; Woolfenden and Fitzpatrick, 1978). These papers were not couched in terms of reciprocity or future direct fitness, but they clearly emphasized the future self-benefit to the helper resulting from helping. The concept of reciprocation was first explicitly invoked by Brown (1975a, 1978a). Later, Ligon and Ligon (1978) advocated it, based on a few cases of unrelated helpers. Emlen (1981) described a few cases of apparent return of favors. Wiley and Rabenold (1984) used "reciprocity" as an explanation of delayed breeding. Authors have varied in the weight they have accorded reciprocity as a unifying explanation of helping behavior. Probably the greatest use of this theory has been made by the students of Scrub Jays and Green Woodhoopoes. Therefore, I use these species for a critical examination of the importance of the future direct components of inclusive fitness.

The Augmentation Hypothesis in the Scrub Jay

Woolfenden (1975) presented evidence that helpers benefit recipients in the Florida Scrub Jay (see Chapter 11). Although he recognized that indirect selection was involved, he minimized its importance and concluded with the following statement: "By increasing the size of the group to which they belong, older helpers may be improving their own chances of obtaining space in which to breed." This was an example of a more general hypothesis, based on future direct fitness that was later termed the augmentation hypothesis (Brown, 1980) because it depends entirely on the additional young reared due to the helper *beyond those due to the unaided parents* (no. 5 in Table 14.4).

It is important to delineate the point at issue. To do this we must establish how Woolfenden's theory (further elaborated in Woolfenden and Fitzpatrick, 1978, 1984) differs from that of Brown (1969a, 1974, 1978a), which also invoked future direct fitness but not by augmentation. As we have seen, my theory was based on the idea that remaining on the natal territory with the parents is a strategy of young birds for attaining breeding status, often by inheritance of the natal territory (Nos. 1 and 2 in Table 14.4). Later Woolfenden and Fitzpatrick (1978) arrived at exactly the same conclusion.

Going further, Woolfenden and Fitzpatrick (1978) hypothesized (interpreted here using my terms) that the augmentation of group size, G, caused by the additional reproductive success of the breeders due to alloparental care by the helper, a, would increase the probability, F, that a male helper would be able to obtain a territory and mate in a future breeding season. The hypothetical increase in *F caused by a* is the essence and the principal new part of their theory (no. 5 in Table 14.4). Woolfenden and Fitzpatrick have hypothesized that this effect on F is the principal selection pressure leading to helping in the Florida Scrub Jay. Effects of a on survival of the parents and other unit members are also conceivable.

This is the original and controversial part of their theory. It is the germ of the interest in "reciprocity" that followed. Their increment to F resulting from a (no. 5) is *not* the same increment to F that results simply from remaining on the natal territory in nonbreeding status (Nos. 1 and 2), which could be achieved *without helping* (reviewed above). The theory of Woolfenden and Fitzpatrick differs from mine by strong emphasis on the augmentation hypothesis and by de-emphasis of indirect selection. In other important respects we are in agreement, although this is not readily apparent in their papers.

In order to evaluate the augmentation hypothesis for the evolution of alloparenting in the Scrub Jay we must understand how Scrub Jays of each sex acquire breeding status. In Chapter 7 we saw that some *male* Scrub Jays acquire breeding status by budding a new territory off the territory of their group into and partly at the expense of the territory of a neighboring group. Other males simply fill a vacancy created by the death of a breeder in a contiguous or noncontiguous territory. Call the former method *budding* and the latter, *replacement*. For budding, but not replacement, it is natural to expect that a male coming from a large source-group would have an advantage since his co-group members presumably facilitate territorial gains from a neighbor group. The group would *not* influence success by the replacement method since replacement is done by the male *alone* on the new territory, while his group remains on its original territory.

Consistent with the augmentation hypothesis is the observation that territories tend to enlarge as group size grows, and this is about the extent of the supporting evidence. Such territorial expansion was thought *a priori* to facilitate budding but not replacement. We should, therefore, expect budding more from large groups and replacement more from small groups. This was *not* the case. Woolfenden and Fitzpatrick (1984) found no significant difference in group size between units from which a male attained breeding status by budding ($G = 5.00 \pm 1.51$) and units from which a male reached breeding status by the replacement method ($G = 4.76 \pm 1.52$).

Perhaps more important than the augmentation achieved by helping is the dominance status of the helpers. We would *a priori* expect a more dominant male to have a higher F than would a subordinate. Woolfenden and Fitzpatrick (1984:171) have reported that the annual probability of achieving breeding status by male helpers was 0.58 for a dominant older helper, 0.50 for a dominant yearling helper, and 0.31 for a subordinate helper—all with sibs. For a lone helper the estimate was 0.29. Thus there seems to be an advantage to being in a group with sibs as opposed to being without sibs, but *only for the dominant helper*, not for the subordinates. If this difference really depends entirely upon the presence of the source-group when budding, then it should be restricted to cases of budding. Unfortunately, these data are not available. There is reason to expect, however, that the difference is not restricted to budding. A dominant male should also have an advantage over a subordinate even in a replacement. If this were true, the argument that these data support the augmentation hypothesis would be weakened.

The evidence that F for males is a function of group size is restricted to dominant males and even for them is not strong. It should also be remembered that most of the increase in group size in the breeding season is due to the parents. The influence of a helper is weaker, especially if there is more than one. For example, group size would normally increase *without* the aid of a helper by about 0.6 independent young for an experienced pair; with the helper's aid, the increase would be 1.0. Thus the group on average would go from 3.0 to 3.6 without the helper's aid and from 3.0 to 4.0 with it. The augmentation attributed to the helper is $a = 4.0 - 3.6 = 0.4$. Before the augmentation hypothesis can be accepted it is necessary to show that F of a male helper is significantly raised by $a = 0.4$. No doubt F is raised somewhat by the augmentation due to the helper, but it is probably raised more by the parents—a benefit the helper could obtain without helping by remaining with the parents but not caring for young. Is the increase in F due to a really important? This question has never been directly addressed by students of Florida Scrub Jays. It should be and could be, using field experiments.

Considering all the available evidence, I have to conclude by *rejecting the augmentation hypothesis* as a major influence on F *for females*, since they do not acquire breeding status by budding (Fig. 7.3). For males the hypothesis is plausible, but there are major difficulties. First, half the males who achieve reproductive status use the replacement method. Second, even for those who are successful in promoting a large augmentation, it may turn out that if they are subordinate they cannot profit from this augmentation attributable to them. Third, whatever explains helping by females could probably also explain helping by males without invoking Woolfenden and Fitzpatrick's scenario for males. Theoretically, the augmentation hypothesis has been stimulating. Empirically, it has not been supported by the necessary critical data in the Scrub Jay or in any other species.

The augmentation hypothesis received little discussion in the monograph on Scrub Jays by Woolfenden and Fitzpatrick (1984). The only remaining hypothesis that is completely consistent with their data is phase 2 of my 1974 model, namely, indirect selection. Since they provided no valid reasons to reject this hypothesis, it appears that helping in the Florida Scrub Jay is most parsimoniously explained by my 1974 theory.

The Augmentation Hypothesis of Reciprocity in the Green Woodhoopoe

A slightly different version of the augmentation hypothesis was proposed by Ligon (1981a) for the Green Woodhoopoe. As in Woolfenden and Fitzpatrick's version, it invoked the idea that young reared by a helper, above and beyond what the unaided parents could rear, would increase F as well as benefit the helper in other ways (no. 5 in Table 14.4).

> First, a helper may gain directly by the production of younger flockmates, especially those of its own sex. Helpers of both sexes use younger birds of the same sex, often sibs, to gain entry to and establishment in a new territory (Ligon and Ligon, 1978b). These younger birds subsequently become provisioners for the oldest former helper's own young. Thus young woodhoopoes in the nest can be viewed as an essential resource that can be utilized by the current helpers for their own personal gain (Ligon and Ligon, 1978a,b). (From Ligon, 1981a:240)

Since social dispersal is prominent in Green Woodhoopoes this approach has considerable appeal. However, no augmentation attributable to helpers was found: "Our data over three years do not convincingly demonstrate that more helpers per flock yield more surviving young woodhoopoes" (Ligon, 1981a:242). In other words, the nonbreeding woodhoopoes could apparently achieve the same increase in group size without helping at all. The failure of helpers to benefit the parents or their dependent young in any easily measurable way also forces us to reject for the moment, the Ligons' version of the hypothesis of reciprocity. This hypothesis requires that the helpers benefit the young to whom they give alloparental care. Without the giving of benefit by the first party (helpers) there is no debt to pay back by the second party (the young), hence no reciprocity.

Faced with this dilemma, one could assume that failure to augment the number of surviving young does not have to mean that benefit is absent altogether. It could be that the helpers somehow benefit the young, even if that benefit cannot be readily identified or measured. This idea seems implicit in the Ligons' work discussed above (see quotation and discussion in Chapter 11). As soon as such an assumption is made, however, the hypothesis of indirect fitness must also be resurrected, since most helpers in this species are closely related to the nestlings (Table 12.3). If benefit is assumed, then a plausible hypothesis is that indirect fitness is of primary importance and that the reciprocal aspects are secondary.

"Reciprocity" in Other Species

To what extent do young fed by helpers aid the helpers later in life? Two mechanisms have been suggested. In the first the return of benefits occurs through *aid in dispersal* (no. 5 in Table 14.4). In social dispersal, as found in the Green Woodhoopoe, dispersing two-year-olds were frequently accompanied by the one-year-old young they had helped to feed. This pattern of dispersal (in unisexual sibling groups) is, however, not very common in other communally breeding birds. This hypothesis of selection for helping is clearly untenable for most species, including Scrub Jays, Stripe-backed Wrens, and Mexican Jays.

The second mechanism postulates that former helpers when they begin to breed are helped by young they helped to rear (no. 9 in Table 14.4). This does occur in Mexican Jays, especially in those that breed in their natal area; however, it is much less likely in jays that emigrate before breeding. In fact, only a small fraction of feedings at nests of Mexican Jays falls in the category of "reciprocation" (unpublished data). In the Stripe-backed Wren, this kind of return of favors does not occur in females and occurs in only a minority of cases for males (Rabenold, 1985). In general, the parents of the helpers are the principal recipients of yearling helpers, not their siblings. The evidence that this kind of "reciprocity" is important in communal birds appears to be very spotty and weak.

Social Bonding as Mutualism

It is clear that a corps of potential helpers forms a social resource for the breeders in a social unit (Brown, 1975a:208; Brown, 1978a; Brown and Brown, 1980; Ligon, 1981a; Ligon and Ligon, 1978) and it has been suggested that helpers might be feeding young as a form of social bonding (no. 6 in Table 14.4). Behaviors whose effects seem to include social bonding are conspicuous and widespread among communal species. The rallies, choruses, corroborees, and greeting ceremonies of the Noisy Miner (Dow, 1971, 1975), Grey-crowned Babbler (King, 1980), Acorn Woodpecker (MacRoberts and MacRoberts, 1976), and other communal birds are well known, but allopreening and allofeeding have also been considered to have social functions. Therefore, it is quite possible that *social bonding may come about in many ways* that do not require giving up food by the helper, who may not be efficient at finding food.

An objection to the social bonding hypothesis in Green Woodhoopoes (Ligon and Ligon, 1983) is that it does not explain why a young bird chooses to leave or stay in its unit. The nestlings are fed not only by helpers who later leave, but also by the parents and other helpers who do not leave. It is difficult to explain why a social bond formed by alloparental feeding would take precedence over a social bond formed by parental feeding.

It has also been verified (Ligon and Ligon, 1983:483) that effective bonds for social dispersal could be formed in Woodhoopoes *without* allofeeding of nestlings. Unrelated females from different flocks were observed thirteen times to merge into the same social unit. Perhaps more important than allofeeding in such cases are two factors, the energetic profitability of the merger (as in Fig. 8.7) and a stable dominance relationship, which the Ligons emphasized. Allofeeding might be a good way to establish such a relationship, but it is not the only way and we have no idea of its importance relative to others.

Not only is there little evidence of a favor being given by the helper to the nestlings it feeds, but there is good evidence that the nestlings would perform the "repayment" even if it were not owed. Subordinate young often accompany a dominant former helper when it assumes breeding status in another unit (or even in the same one) as a tactic for later gaining reproductive status themselves. Ligon and Ligon (1983:487) reported: "... younger birds, by assisting older ones, are not simply paying back the latter for their previous beneficence. Because breeding or alpha males usually die before beta males ... a subordinate male will live to inherit breeding status." It would appear that allofeeding is not needed to secure such "repayment" since it is probably to the advantage of the young to aid the former helper in dispersal anyway. Therefore, a tit-for-tat strategy need not be invoked. I suggest instead a byproduct mutualism.

In short, Green Woodhoopoes live in social units whose members are often but not invariably closely related (details in Chapter 12). There is no good evidence as yet that alloparental care by helpers benefits the fitness of the young or parents in this species in any way, although clearly the parents achieve a savings in energy. It has been suggested that allofeeding benefits the allofeeder by recruiting future subordinates for dispersal attempts; however, it appears that the subordinates would willingly aid the dominants whether they had been reared by the dominant or not. Thus Green Woodhoopes, in my opinion, remain enigmatic. The evidence for indirect selection is just as weak as the evidence for score-keeping reciprocity, since both depend upon a measurable benefit to the parents or offspring, which has not been demonstrated. If such benefit is assumed, then the arguments for both indirect selection and for byproduct mutualism would be strengthened. A tit-for-tat kind of reciprocity, as suggested by Ligon (1983), would not, even with this assumption, be sustained by the data, since failing to "reciprocate" provides no immediate benefit to the "defaulter" and causes no "punishment" or change in behavior by the "victim."

Other Direct Advantages for Nonbreeding Helpers

Two other advantages of being a nonbreeder and a helper were discussed in earlier chapters. In Chapter 5, I developed a model showing how age at first breeding might be related to learned skills at foraging in various environments. Foraging skill improves with age. It could be argued that being a helper intensifies this effect, as did Rowley (1981a) for the Splendid Wren; but the nonbreeder must in any case forage at least for itself whether it helps or not, so I have listed foraging in Table 14.4 as being independent of alloparenting (no. 3).

In the Pied Kingfisher, male secondary helpers had a higher chance of mating with a female in the following year than did male primary helpers

(Reyer, 1984; Chapter 13). This suggests that being a helper for an unrelated female can be part of a strategy for bonding with that female or possibly others (no. 4 in Table 14.4). Support for this hypothesis in other species is lacking, but more consideration of it is needed.

Conclusions

If helping with delayed breeding is mutualistic, then the helper should enjoy better lifetime reproductive success as a result of the experience. This chapter considers some of the ways in which helping might enhance the helpers' future direct fitness, causing helping to qualify as mutualism, provided the recipient also benefits. This area has suffered from three types of confusion.

(1) Not all mutualisms should be termed reciprocity. *Score-keeping mutualisms* are based on strategies like tit-for-tat and can be considered as true reciprocity (in the "narrow sense" of Trivers, 1971; see Waltz, 1981). *Byproduct mutualisms* are not based upon genuine reciprocation but upon self-interest that incidently benefits others. The available evidence suggests that byproduct mutualisms are common. No case of score-keeping mutualism has been firmly established for communal birds, although the case of mutual feeding of each others' young by breeding Mexican Jays is attractive.

(2) The most obvious direct benefits of nonbreeding ("helper") status in singular-breeding species stem not from alloparental care but from nonbreeding and nondispersal (Table 14.4). This point never has been controversial even though some authors have made it appear to be. The choice of nonbreeding and nondispersal may benefit helpers in some species by raising their chances of survival and improving their chances of ultimately attaining breeding status. This is discussed further in Chapters 5 and 6.

(3) Lack of close relatedness to the young in a small minority of helpers in a species is not proof that indirect selection is unimportant (see Chapter 13), and certainly does not indicate true reciprocity. In three species for which future direct benefit or reciprocity has been suggested (Scrub Jay, Green Woodhoopoe, Stripe-backed Wren), relatedness of helpers to recipient young is unusually close on average. Of the reported cases of reciprocal aid-giving in birds, only in breeding Mexican Jays does indirect relatedness seem unlikely to be important.

The hypotheses of "reciprocity" based on augmentation or on social bonding are not strongly supported by available data, in my opinion; but many questions remain in this area, and more work is needed, particularly in the form of field experiments.

15 Parent-Offspring Relationships

Perhaps surprisingly, the pioneers in the evolutionary study of parent-offspring relationships have been entomologists. Students of the social insects have long been impressed with the fact that in eusocial species the queens and male reproductives are given "royal" treatment during their development while the nonreproductive workers are confined to smaller cells and a diet that does not allow them to become reproductives. What is the significance of such observations for social birds and mammals? Although birds and most mammals do not show caste systems like those of eusocial insects, the theories developed for social insects have stimulated biologists to speculate about their application to the vertebrates.

Historically, modern interest in possible conflict of genetic interest between parent and offspring began with Trivers (1974). A heated controversy occurred among students of haplodiploid social insects over this issue (Trivers and Hare, 1976; Alexander and Sherman, 1977), but it shed little light on the situation in diploid vertebrate animals. The discussion of parent-offspring conflict in relation to helping in communal birds was introduced by Brown (1978a), with the concepts of variance-utilization and variance-enhancement, and by Vehrencamp's (1979, 1980, 1983) models of dominant-subordinate relationships among relatives. The theoretical work of Emlen (1982b) has renewed interest in the possibility of parent-offspring conflict in communal birds.

In this chapter we begin with a general consideration of donor-recipient conflict, and then look at the particular kinds of conflict that might occur between helpers and breeders. The key feature of this chapter is the development of the concept that *variance* in the reproductive potential of offspring is a critical parameter in models that optimize parental fitness in helper systems. In other words, the *quality* of the offspring must be considered as well as their quantity. For this purpose an approach to modeling the parents' interests is presented that maximizes the number of *grand-offspring*, as in Chapter 7. This is not a new approach in sociobiology, having been used by Oster and Wilson (1978) for social insects and Parker and MacNair (1978) for parent-offspring conflict; however, it is new for communal birds. A similar approach has been used to model the optimal tradeoff between offspring size and clutch size for parents (Smith and

Fretwell, 1974; Brockelman, 1975). Lack (1947, 1948a,b) employed similar thinking in his treatment of family size in birds.

Donor-Recipient Conflict

To illustrate the idea of conflict it is sufficient to ignore for the moment the quality of the offspring. Therefore, we now express gains and losses to donor, D, and recipient, R, in terms of expected numbers of young produced as a result of the donor's actions in present and subsequent breeding seasons combined: Y_D for the donor and Y_R for the recipient. Since these represent individuals rather than genes they must be weighted by the appropriate values of genetic relatedness, r.

r_O = relatedness to one's own young (typically 1/2),
r_{RD} = relatedness of the recipient to the donor's young,
r_{DR} = relatedness of the donor to the recipient's young.

The criterion for giving to be selected in the donor is simply that genes gained exceed genes lost. The genes gained by a donor are counted in the recipient's offspring and are calculated as $r_{DR}Y_R$. The genes lost by a donor are in the future offspring of the donor that do not materialize as a result of aid to the recipient; they amount to $r_O Y_D$.

For the recipient, the offspring gained are identical to those for the donor, namely, Y_R; but the recipient is related to them by r_O instead of r_{DR}. Similarly, the offspring lost by the recipient are identical to the donor's loss, namely, Y_D; but the recipient is related to them by r_{RD} instead of the donor's r_O. In summary, the criteria are as follows, where $K = Y_R/Y_D$:

	Weighted Gains		*Weighted Losses*	*Threshold*
Donor's Criterion:	$r_{DR}Y_R$	−	$r_O Y_D > 0$	$K > r_O/r_{DR}$
Recipient's Criterion:	$r_O Y_R$	−	$r_{RD}Y_D > 0$	$K > r_{RD}/r_O$

The result of these differing criteria for donor and recipient is a possible conflict of genetic interest, depending on r_{DR} and r_{RD} as shown in Figure 15.1. When the donor is less related to the recipient's offspring than to its own young (and the recipient is less related to the donor's offspring than to its own) a conflict of genetic interest can exist if Y_R/Y_D falls in the stippled region. The donor's requirements become rapidly more stringent as its relatedness to the recipient's offspring (r_{DR}) diminishes. The recipient, however, becomes less and less considerate of the donor as r_{RD} diminishes.

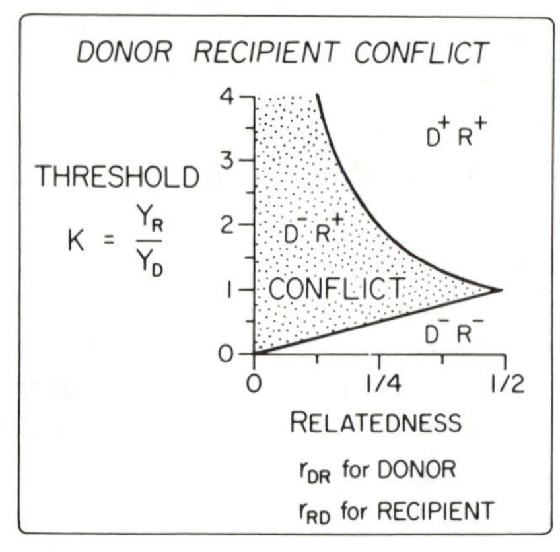

Figure 15.1 Donor-recipient conflict. In the region marked D^+R^+ both donor and recipient benefit in terms of inclusive fitness if the donor aids the recipient. In the region marked D^-R^- both parties benefit more when the donor does not aid the recipient. In the region marked D^-R^+ the recipient gains while the donor loses. For explanations of other terms see text.

The requirements of the donor are more sensitive to r than are those of the recipient.

In the zone above $r = 1/2$ the relatedness to one's own young is less than to the other's young. This is not realistic in diploid organisms for most genes (but see Whitney, 1976). Therefore, this case is not considered further here; however, it is a common occurrence in haplodiploid organisms and has been considered extensively for the Hymenoptera (Hamilton, 1964; Oster and Wilson, 1978).

The relatedness of the donor to a recipient's young, r_{DR}, varies with the mating system in a potentially critical manner. For the case in which the donor is an offspring of the recipient the donor is related to the recipient's young, Y_R, via two pathways, through the mother and through the father. If the father of future young of the recipient is not the same as the father of the potential donor, then the only common parent is the mother and future young of the recipient would be half-sibs ($r_{DR} = 1/4$). On the other hand, if the father of the potential donor is the father of the future young too, then the young produced, Y_R, would be full sibs ($r_{DR} = 1/2$). As shown in Figure 15.1, this difference in relatedness is expected to increase donor-recipient conflict considerably (Parker and MacNair, 1978).

If this theoretical consideration is important in nature, then we should expect less parent-offspring conflict in monogamous species than in non-monogamous ones. Although a convenient empirical measure of donor-recipient or parent-offspring conflict is not available, it is noteworthy that monogamy with nonbreeding offspring as helpers is common in singular-breeding communal birds and mammals, and that nonmonogamy is limited to mate-sharing, in which case nonbreeding helper-offspring are not common. Polyandry, polygyny, and polygynandry do occur in communal birds, but not outright promiscuity within a deme ("promiscuity" may occur *within* a social unit in some species, such as Acorn Woodpeckers, but this is not true promiscuity because mating is strongly hindered between units).

Assuming that both parents try to stay together with the offspring, the expected average relatedness of offspring-donors to future offspring of recipients (r_{DR}) in their unit depends upon the annual survival rate of mothers, s_M and of fathers, s_F. In a singular-breeding species,

$$r_{DR} = s_M^t/4 + s_F^t/4,$$

where t is the age of the offspring in years. Since communal breeders have high survival rates relative to noncommunal breeders, these demographic parameters should tend to reduce parent-offspring conflict.

Implicit in Trivers's approach is the realization that either offspring or parent can in theory behave altruistically. Likewise, either parent or offspring can in theory suppress or parasitize the direct fitness of the other. Thus, we can view either parent or offspring as "donor" or initiator in diagrams such as Figure 4.2 or 12.8. Therefore, in Figure 15.1 the region marked D^-R^+ could be either P^-O^+ or O^-P^+ (P = direct fitness of parent, O = direct fitness of offspring), depending on whether the parent or offspring is the donor.

The parent-donor situation differs from the offspring-donor situation in the value of r_{DR} when both parents remain together. When the offspring is the donor, $r_{DR} = 1/2$; but when the parent is the donor, $r_{DR} = 1/4$. This is because the offspring is related to the parents' offspring (donor's full sibs) by $r = 1/2$; yet the parents are related to the offsprings' offspring by only $r = 1/4$. Thus, the offspring as donor values its full sibs and its own offspring equally when helping both parents, but the parents would experience genetic conflict were they to sacrifice their own offspring for an equal number of grandoffspring.

For this reason and, as more commonly realized, because the parent is usually dominant in its own territory, it has seemed unlikely to most authors (except Zahavi, 1974, and Emlen, 1982b:49) that parents would allow their future reproduction to be unduly jeopardized by present offspring. In my view, it is in the parents' interest to aid the survival of their young and facilitate their acquisition of breeding status, as described in Chapter 7 (see Fig. 7.4), but these activities are expected to remain mutualistic in general. From the offspring's viewpoint conflict is minimized because of its close relatedness to its full sibs, because the high survival rate of breeders in communal birds tends to keep r_{DR} high, and because offspring are commonly in a situation in which $Y_R > Y_D$.

The above generalizations should not be interpreted to mean that conflict between parent and offspring never occurs. Extreme cases do sometimes occur. When offspring starve or parents become so weakened that they might die, conflict must occur. Normally, however, these extremes are avoided along with their accompanying conflicts. The *potential* for conflict may be more important than its observed frequency suggests.

In contrast to social birds, the social insects illustrate vividly that the reproductive potential of some of the offspring can be greatly reduced by the manner of rearing. This has suggested to some the possibility that parent birds may at times influence their offspring to remain as altruistic helpers when the offspring could actually achieve some success at breeding on their own. "When breeders are clearly dominant over auxiliaries, we might expect them to somehow manipulate the latter to induce them to remain as helpers. . . . Breeders might achieve the same effect . . . by influencing the outcomes of breeding by auxiliaries . . . taking the form of harassment of would-be-breeders . . . or destruction of the nest or eggs" (Emlen, 1982b).

Harassment and destruction of nests and even eggs have long been known in detail for Mexican Jays (Brown, 1963a), and more recently destruction of eggs and young have been discovered among female anis (Vehrencamp, 1978) and Acorn Woodpeckers (Mumme et al., 1983a). Similar destructiveness is strongly suspected in Dunnocks (see Chapter 9). In these plural-breeding or mate-sharing species, however, the destruction is directed not at offspring by their own parents, but at offspring of unrelated or sibling rivals. In singular-breeding systems the offspring are typically not allowed even the preliminaries to breeding within the natal territory while the parents remain.

This should not be interpreted as "manipulation" or suppression, in my opinion, since allowing the offspring to remain as nonbreeders constitutes a *relaxation* of the aggressiveness commonly shown by parents toward independent offspring in noncommunal species (as discussed in Chapter 7). The transition between the situation in which young depart from their parents immediately upon reaching foraging-independence to the situation in which young remain with their parents long after foraging-independence is most parsimoniously attributed to a change in the environment (as reflected in F, see Chapter 5) coupled with an increase in tolerance by the parents rather than a change in "manipulativeness" of the parents. In the language of the mathematical arguments given later in this chapter, variance utilization provides a cheaper route to communal breeding for the parents than does variance enhancement by parental suppression or "manipulation."

Emlen (1982b:50) has stated that conflicts of the parental-suppression variety (his type II) "will be most prevalent during benign years in fluctuating, unpredictable environments." He reasoned that offspring will be most likely to leave the parents to breed on their own in such years. The counterargument, however, also deserves consideration: such conflict should be *least* frequent in benign years because these years provide the conditions under which parents are least in need of help and in which parents are likely to achieve most benefit in terms of grandoffspring by encouraging their offspring to breed.

Vehrencamp's Suppression Models

In a series of papers Vehrencamp (1979, 1980, 1983) has elaborated the basic model depicted in Figure 5.8. In a group whose members average higher fitness together ($\bar{W}$) than apart (W_i), the dominant is assumed to be able to raise its fitness (W_a) at the expense of subordinates, depressing the fitness of subordinates (W_w) to the point at which subordinates would do better to leave (W_i). Reasons why the subordinates might tolerate such exploitation include limited options elsewhere for reasons concerned with habitat saturation, sex ratio, and skill-environment interactions as elaborated in Chapter 5. In Vehrencamp's models the gain by the dominant is equal to the sum of the losses by its subordinates. The maximum loss by an individual subordinate is determined by the difference between its fitness without dominance ($\bar{W}$) and its fitness if it left the group (W_i).

If relatedness between subordinate and dominant, r_{SD}, is zero, then subordinates should leave if their direct fitness elsewhere, W_i, exceeds their direct fitness as subordinates in the group (W_w). When subordinates are related to dominants, then to an extent determined by r_{SD}, the indirect gains to the subordinate allow it to stay even though its direct fitness would be higher if it left than if it stayed. This is an act of altruism by the subordinate, which is justified only because its inclusive fitness if staying exceeds its inclusive fitness if leaving.

Vehrencamp's model thus resembles my own (Brown, 1969a, 1974) with the exception that I did not invoke altruism with respect to the decision to stay or leave. My own position, as will become clearer below, invoked nonsuppressive constraints on the subordinates outside their natal territory, followed by indirect benefits from alloparenting. Implicit in my theory was the assumption of parental suppression, which is the common condition in birds.

Vehrencamp (1983) elaborated on her basic model by exploring the effects of additional variables. Her goal was to provide a model that would explain the differences between "despotic" and "egalitarian" societies in terms of selection pressures generated by ecological factors. Vehrencamp (1983:667) wished to recognize "a continuum in terms of the degree to which fitnesses of individuals within social groups are biased" or "skewed" by the dominant. Her examples of "egalitarian" societies mainly involved nest-sharing or joint-nesting species. Her examples of "despotic" societies were probably all singular-breeding species whose social systems are based on the nuclear family. Therefore, *although such a continuum might be imagined mathematically, her examples suggest instead a biological dichotomy between nuclear-family systems and systems based on probabilistic parentage* (joint-nesting and mate-sharing systems). As we have seen (Chapters 9 and 10), in order to understand probabilistic-parenthood systems, it is necessary to examine sex-specific options and adopt an approach that differs much from that used for nuclear-family systems. Some mate-sharing systems, such as that of the Dunnock, are characterized by simple inability of the dominant to exclude the subordinate. There is no evidence of egalitarian restraint by the dominant in his efforts to evict the subordinate. Thus, Vehrencamp's model does not obviously fit the behavior of the dominant in the Dunnock. Furthermore, it suggests a mathematical continuum where a biological dichotomy seems more heuristic.

According to Vehrencamp (1983), the helping of a parent by its offspring "is a special case of the related group member model." She wrote, "Clearly, parents will not want to bias all of their offspring to this extreme degree all of the time" (p. 671). Therefore, another approach is needed. This is provided below in the discussion of variance in quality of offspring.

Species Differences in Patterns of Fitness Variability

As background for considering possible influences of parents on variability of fitness in their young it may be helpful to take a brief overview of the range of patterns of variability found in a gamut of species from asocial to eusocial. For simplicity the comparison is limited to nuclear- and extended family species.

The individuals in a population at any one time may be arrayed as a frequency distribution of E_B (expected inclusive fitness if breeding), of which reproductive potential is the main component. The population may be divided into a fraction above a threshold for breeding ($E_B > E_H$) and a fraction below threshold ($E_H > E_B$), who are assumed here to be helpers.

In order to describe in simple terms the range of variability among species, four "levels" of sociality are contrasted in Figure 15.2 and the corresponding Table 15.1. The strategies of individuals are hypothesized to vary according to their chances of becoming a successful breeder.

At level (a) of Figure 15.2 and Table 15.1, we have species that usually do not have helpers and in which chances to reproduce are not curtailed by habitat saturation or other factors. Interference competition, such as territoriality, is minimal, and the typical individual strategy is "bullish" or *r*-selected, with the emphasis on early breeding and higher fecundity.

At level (b), some individuals are prevented from breeding because of foraging inefficiency or because they lost out in interference competition, causing an increase in the variance of E_B. In some cases these individuals may become helpers, though still with the primary goal of breeding. This intensifies interference competition.

At level (c) are many avian species with regular helping. Conditions for reaching breeding status are sufficiently constrained that individual compromises and default strategies, such as delayed breeding, mate sharing, and helping, become adaptive. These differences characterize the stage transitional between nonbreeding and full-breeding conditions. These adaptations become profitable for some individuals because of an increased variance in E_B that has been brought on primarily by various environmental conditions. The general nature of these adaptations is flexibility

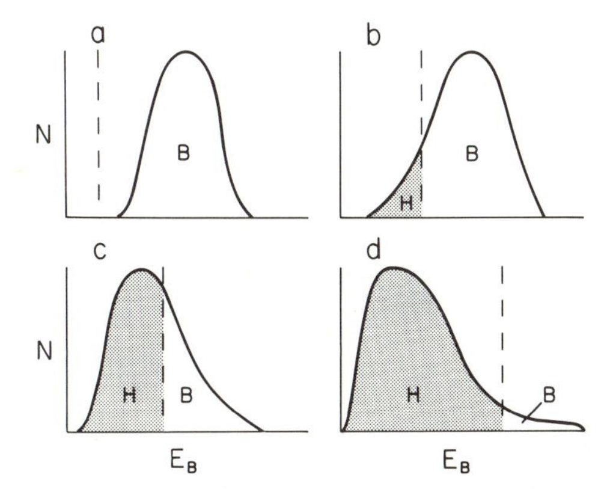

Figure 15.2 Variance in expected inclusive fitness resulting from choosing to breed ($E_B = D_B + d_B + I_B + i_B$) at four levels of breeding sociality, as described in the text. (a) Asocial or colonial without helpers; no surplus. (b) Helpers regular but not a large fraction of the population. (c) Helpers regular and numerous. (d) Eusocial. The vertical dashed line marks the point in the distribution of E_B at which $E_B/E_H = 1$. This is the decision criterion for an individual faced with only two options, breeding and not helping, *B*, or helping and not breeding, *H*. (From Brown, 1983a.)

Table 15.1

Spectrum of individual strategies in ontogenetic and evolutionary perspective. (From Brown, 1983a.)

Level	*Name*	d_H*	I_H*	*Habitat Saturation*	*Dispersal Benefit Ratio*	*Variance in* E_B*	*Developmental Stage Affected*	*Adaptations*
a	Bullish	±	0	–	Highest	Small	None	Exploitation competition
b	Cautious	High	Slight	±	Intermediate	Moderate	Mature adult	Interference competition; opportunistic helping
c	Compromise	Medium	Moderate	+	Low	Large; environmentally enhanced	Subadult	Delayed breeding; mate sharing; regular helping
d	Resignation	Low	Large	+	Lowest	Very large; developmentally enhanced	Immature	Extreme role specialization

* See text and Chapter 4 for explanation of d_H, I_H, and E_B.

and the ability to utilize a variety of compromises to achieve reproductive status.

At level (d), the eusocial level, determination of alternative developmental paths is typically environmental, but the switching mechanism has been shifted (genetically) to an earlier stage of the life history, and changes from helper to breeder after this switching stage have become more difficult. Consequently, there is little that an individual can do to alter its social role. Such individuals are "resigned to their fates," and they accept the resulting specialization in role—except for a small minority of special and borderline cases, such as potential replacement breeders, laying worker Hymenoptera, and sneaky kleptogamists.

There are many sources of variance in E_B, and I can do little more than enumerate a few examples. A frequent source of variance is dominance and the conditions that force individuals to tolerate it. Social rank is a good predictor of reproductive success in some species, though not in all (Gauthreaux, 1978; Brown, 1978a). Chapter 5 discusses other sources.

Variance Utilization and Variance Enhancement

Although the causes of variance in E_B are many and diverse, it is convenient to divide them into two types: (1) *natural*, imposed by genetics and the environment (excluding the parent), and (2) *socially imposed* by a parent. The higher, narrower curves in Figure 15.3 show variance in E_B that arises independently of parent-offspring relationships, from such causes as age, genotype, experience and behavior of other members of the population. Such a situation allows the parents to utilize the aid of helpers (to the left of the threshold line) without parental manipulation, hence the term *variance utilization.*

On the other hand, the parents may in theory increase or *enhance the variance* among their offspring or others under their influence by suppressing the reproductive potential of some, and/or facilitating the reproductive potential of others. By such processes they may be responsible for increasing the variance in E_B, as shown in Figure 15.3.

The lower and wider curves in Figure 15.3 reflect enhanced variance. The effect of variance enhancement (with symmetrical distributions) depends upon the location of the threshold for breeding (the vertical line in Fig. 15.3) with respect to the mean. When the threshold lies above the mean, as in the upper diagram, an increase in variance increases the number of individuals that exceed the threshold. When the threshold lies

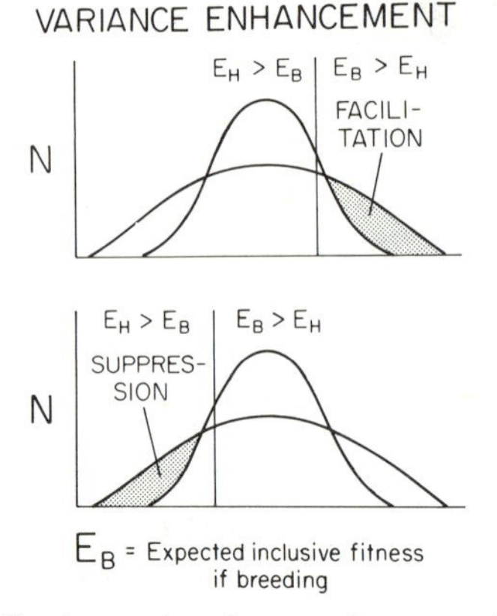

Figure 15.3 Variance utilization and variance enhancement. The distribution of expected inclusive fitness when breeding is shown without parental effects by the narrower curves (above and below). The distributions with parental influence are shown by the wider curves. With symmetrical distributions, increasing the variance increases the number of breeders if the threshold for breeding is above the mean, resulting in parental facilitation (upper diagram). If the threshold is less than the mean, increasing the variance increases the number of nonbreeders, resulting in suppression (lower diagram).

below the mean, as in the lower diagram, increasing the variance decreases the number of breeders and increases the number of helpers.

Obviously, if a parent suppressed breeding in all its offspring it would have no grandoffspring. In suppressive systems, therefore, mechanisms to facilitate breeding by other offspring are required in the model. Although it is obvious that this must be true, parental facilitation has received much less attention than parental suppression.

Using simple mathematical models Brown and Pimm (1985) have shown that *helping is more likely to have originated via variance utilization than variance enhancement* because the latter is more expensive to the parent. Their simplest models dealt with discrete, nonoverlapping generations in which the animals breed only once and then die. In the variance enhancement model a parent has the opportunity to divide its offspring into two classes according to "quality," namely, nonhelping breeders and nonbreeding helpers. It is assumed that nonbreeders help their siblings, thus increasing the number of parents' grandoffspring. The number of grandoffspring, g, of a parent depends upon the number of offspring breeding, p, their reproduction if unaided, k, and the collective effectiveness of the helpers, which is represented by a function, $h(p)$. Thus,

$$\text{grandoffspring} = g = pk[h(p)].$$

A simple form for the helping function, $h(p)$, is shown in Figure 15.4 (upper). The collective influence of helping increases linearly with the frequency of helpers, $(1 - p)$, and a factor, a, that represents the effectiveness of helpers:

$$\text{helping function} = h(p) = 1 + (1 - p)a.$$

When $a = 0$ or if there are no helpers, then the reproductive success of the breeders is not augmented (Fig. 15.4, lower), $h(p) = 1$, and $g = pk$. In general the number of grandoffspring is:

$$\text{grandoffspring} = g = pk[1 + (1 - p)a]$$

$$g = \underset{\text{reproduction as if unaided}}{pk} + \underset{\text{augmentation due to help}}{pka(1 - p)}$$

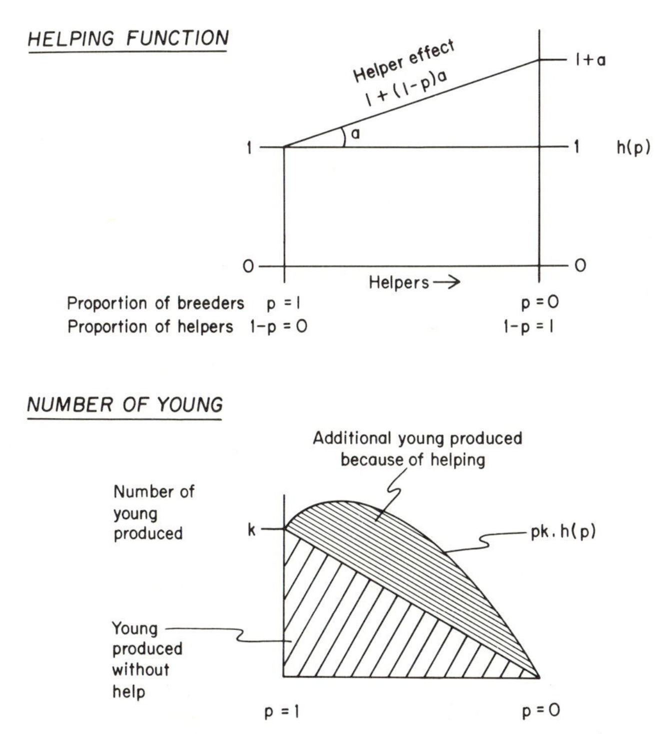

Figure 15.4 Above: Form of the helping function, $h(p)$. When all individuals breed ($p = 1$), no additional young are produced and $h(p) = 1$. When p individuals breed, the number of young produced is increased by a factor of $h(p)$. Below: Number of grandoffspring as a function of p. a = helping value, k = reproductive rate if unaided. (From Brown and Pimm, 1985.)

The threshold value of p at which it becomes more profitable for a parent to have some helpers than to have none occurs when the benefit from augmentation exceeds the loss of grandoffspring the helpers would have had if they bred:

$$\text{benefit} = pka(1 - p) > k(1 - p) = \text{cost},$$

which simplifies to

$$a > 1/p.$$

We obtain p^*, the value of p that maximizes g for the parent by setting the derivative,

$$dg/dp = k + ak - 2akp,$$

equal to zero:

$$p^* = (1 + a)/2a.$$

This is the optimal brood composition for the parents, namely, p^* breeders and $(1 - p^*)$ helpers. Note that helpers must contribute more grandoffspring *per capita* than breeders to justify to the parent the production of nonbreeding helpers.

The model above depends upon parentally imposed variance. Next we develop a model in which the variance arises naturally and is not imposed by the parent. We then compare the two models.

Assume two qualities of potential breeders. Good breeders achieve a breeding success of k, where $k > 0$; bad ones produce no young ($k = 0$) for a variety of reasons, such as territory quality, that do not involve parental manipulation. Assume further that bad breeders find that their best option is to aid their sibs. If there are q good breeders and $1 - q$ bad ones, then by the methods used above the number of grandoffspring per parent with help, G_h, is:

$$G_h = qk(h(q)) = qk + qka(1 - q).$$

If the bad breeders do not help, then the number of grandoffspring, G_{nh} is:

$$G_{nh} = qk(h(q)) = qk(1).$$

It follows that $G_h > G_{nh}$ for all $a > 0$. Potential breeders in bad territories would do better from the parents' point of view and their own to aid their sibs in good situations. Parents do not need to force or manipulate their young to help in such situations since each young acts in agreement with its own inclusive fitness. Note that neither altruism nor parental manipulation is part of this model.

It is much easier to evolve helping by this variance utilization model than by the preceding variance enhancement model, since $a > 0$ is a less stringent criterion than $a > 1/p$.

Similar conclusions were reached in the case of overlapping generations.

Conclusions

Although it is important to appreciate that conflict between parent and offspring is theoretically possible and even expected under certain circumstances, even Trivers characterized the parent-offspring relationship as mutualistic for the most part. Of what use then is the study of parent-offspring conflict? The answer may lie in our understanding of the *outer bounds of the normal mutualistic relationship.*

Trivers expected conflict when the period of parental care draws to a close and the profitability of such care to the parent diminishes. In communal birds, however, the period of parent-offspring interaction has been dramatically *lengthened.* As shown in our model of parental facilitation (Fig. 7.4), the optimal behavior for the parent is not to extract every possible fitness concession from its offspring but instead to facilitate breeding among at least some of its offspring. Much evidence is summarized in this book which shows that this prolonged relationship benefits both the parent (Chapter 11) and some offspring (Chapter 5, 6, 13, and 14). Consequently, the evidence may be described as favoring the conclusion that the parent-offspring relationship in communal birds is characterised by net mutual benefits. This is not to say that offspring may not behave altruistically when the occasion warrants. Likewise we expect parents to accept such altruism on certain occasions.

Conflict has also been postulated based on the assumption that parents benefit most by treating all their offspring equally, while individual offspring prefer better-than-equal treatment. The possibility that the parent might "manipulate" its offspring by forcing less-than-equal treatment upon them has also been examined. These possibilities lead us to consider the general problem of inequality or variance in reproductive potential.

Although one can imagine a situation in which parents suppress breeding in all offspring, or even only in some, such suppression is not necessary to create the variance in reproductive potential that is conducive to helping behavior. As we have seen in Chapter 5, delayed breeding is fostered by a variety of social and environmental factors other than parental supression. These natural sources of inequality among individuals provide circumstances in which some individuals fall below their own criteria for breeding and consequently find helping to be their next best option. Of course, the parents can encourage even more helpers by suppressing reproduction in even more of their offspring; however, this means of creating helpers is more costly to the parent than simply utilizing the helpers that are made available to them by natural, nonsuppressive means. This has been shown mathematically for simple models based on overlapping and nonoverlapping generations.

Conflict between parent and offspring is greatly ameliorated by high rates of survival of the parents. In nuclear families in communal species the survival rates of both parents tend to be remarkably high and often create a situation in which an offspring is equally related to the parent's offspring and its own. This greatly reduces the chance that an offspring would balk at rearing its parents' young and, consequently, tends to minimize parent-offspring conflict. From the parents' point of view, therefore, relatedness is a critical factor. Of course, the parent should benefit even from unrelated helpers; and in general (Fig. 15.1) recipients become less discriminating about costs to donors as relatedness to them diminishes. Donors, however, should be more discriminating (Fig. 15.1). So long as indirect fitness benefits are important to the helper, we should expect it to prefer to help its parents or sibs than to help genetic strangers. The demonstration of such preferences in Pied Kingfishers suggests that relatedness is important for helpers (Reyer and Westerterp, 1985). The lower effort of helpers in Pied Kingfishers when helping genetic strangers is also in good agreement with our treatment of parent-offspring conflict.

16 Infanticide, Dominance, and Destructive Behavior

Communally breeding birds are not always cooperative. That they may act destructively toward the nest and eggs of others in their social unit has been known since nest robbing and egg destruction were described for the Mexican Jay (Brown, 1963a) and Australian Magpie (Carrick, 1963, 1972). More recently Vehrencamp (1977) discovered that female Groove-billed Anis in the same "cooperative" unit routinely destroy each other's eggs. Infanticide, or the destruction of eggs or young, is now known to be common also in the Mexican Jay (Trail et al., 1981) and Acorn Woodpecker (Mumme et al., 1983a; Stacey and Edwards, 1983), and is suspected in Dunnocks (Houston and Davies, 1985), Greater Anis (Strahl and Withiam, pers. comm.), and other species. It is natural to hypothesize that such behavior is related to dominance and rivalry within units. Therefore, we review in this chapter the meager knowledge of all these phenomena in communal birds.

The most important common denominator of the cases of infanticide and nest robbery that have been reported for communally breeding birds is that all occur in species in which two or more members of the same sex often breed in the same social unit. These violent and highly selfish acts have not been reported for singular-breeding species with nonbreeding helpers. In order to have such selfish behavior within an otherwise mutualistic group, reproductive rivalry must be allowed within the group. In singular-breeding species it is not allowed, and the helpers do not behave so selfishly toward the breeders—who are frequently the parents of the helpers.

Nest Robbery by Jays

In a brief study of two flocks of the Mexican Jay in the Santa Rita Mountains of Arizona, Brown (1963a) observed jays robbing nest lining from active nests of their own social unit 33 times. During the same period only 23 visits of a neutral or constructive nature by nonowners were observed. Two of the four active nests in each unit were seen being robbed.

Five of the eight jays in one unit and one jay in another unit participated in robberies. In both flocks pairs that were behind in the breeding cycle plundered the nests of pairs that were more advanced. When the owner resisted by begging, the raider on at least four occasions pecked it vigorously on the crown or in its gape. Defense by the owners was surprisingly slight and consisted principally of sitting on the nest. Some owners were stubborn about not leaving. One was found dead sitting on eggs with a beak mark in the top of its skull.

These observations clearly established the competitive nature of the individuals within a social unit in a communal, group-territorial species. Yet because of the strong interest in helping as a potential case of altruism, the competitive nature of life in communal groups was ignored by other authors for many years.

Subsequent work on Mexican Jays in the Chiricahua Mountains has confirmed the earlier findings and documented the rather frequent occurrence of the destruction of eggs by other members of the same unit (Trail et al., 1981). These cases, including a few observations of the killing of nestlings, have been referred to as infanticide to emphasize the point of central importance and to draw attention to similar cases in mammals.

Such destructive behavior is not characteristic of all Mexican Jays, nor is it found in all social units of the species. Much of the helping is done by nonbreeding jays, while robbery is confined almost entirely to breeders. Although later in the nesting cycle even the breeders may feed each other's fledglings, competition between breeders is conspicuous when they are building nests and courting. Disruption of the otherwise peaceful relations in such groups is, therefore, transient, being limited mainly to periods in which copulation occurs. It also tends to be most frequent in flocks composed mainly of unrelated jays; it was not seen in units with many related jays. Despite this transient destructiveness, it is common for two or more pairs in a unit to rear young successfully.

Infanticide by Anis and Woodpeckers

Female Groove-billed Anis live in groups of one to four monogamous pairs whose eggs are all laid in the same nest. Vehrencamp (1977) discovered that the females at joint nests "were deliberately rolling each other's eggs out of the nest." A given female, however, ejected eggs "only before she laid her first egg" suggesting that females do not recognize their own eggs in a mixed clutch. "As a result, early-laying females lose more eggs than females laying later, and the last-laying female loses no eggs." Early-laying females reduce their disadvantage in part by laying more eggs, but they still have fewer eggs in the final clutch than the last female.

"The females in a group can . . . be ranked in order of decreasing number of incubated eggs, with the alpha female laying last and incubating the most eggs." Last-laying females were consistently older than early-laying females. The size of males was correlated with the order of initiation of laying by their females, the smallest male being mated to the earliest female. Males behave as if they are aware of the inequalities in representation of the eggs of their females in the joint clutch. Their efforts in incubation and feeding the young are greater if their share of the incubated eggs is larger.

In the population of Acorn Woodpeckers studied by Mumme et al. (1983a), laying females sharing the same nest also removed each other's eggs until all had initiated laying. This behavior is necessarily less frequent in these woodpeckers than in the anis since 63–74% of woodpecker units have only one female in the California population and the figure is nearly 100% in New Mexico and Arizona populations (Chapter 10).

In the six social units of woodpeckers studied, the rival females were full siblings, maternal half-sibs, or at least probably full sibs. In 20% of the cases mothers removed their own egg, though it must be added that these were "runt" eggs, which lack a yolk and never hatch. As in the anis, the jointly layed clutch ended up having more of the remover's eggs than the victim's.

Behavioral Dominance

When individuals live together socially we may expect some manifestations of behavioral dominance in various contexts, particularly in contests over food, mates, and use of a shared nest. However, to achieve the benefits of sociality, individuals in communal species must restrain aggressive behavior that might disrupt their social unit. In contrast to species that form groups only in the winter or nonbreeding season, the requirement of nondisruption persists in communal birds even in the breeding season. Evidence of both aggression and its control among communal birds is the ubiquity of dominance relationships or hierarchies. These relationships have been noted in Superb Blue Wrens, Australian Magpies, Long-tailed Tits, Kookaburras, Scrub Jays, Native Hens, Acorn Woodpeckers, Pukekos (references in Brown, 1978a), Mexican Jays (Barkan et al., 1986), Dunnocks (Davies, 1985), Galapagos Mockingbirds (Kinnaird and Grant, 1982), White-browed Sparrow-weavers (Collias and Collias, 1978c), Stripe-backed Wrens (Rabenold, 1985), captive Sociable Weavers (Collias and Collias, 1980b), and Grey-crowned Babblers (King, 1980). Compared to noncommunal group-living birds, aggression in communal species is relatively restrained. A conspicuously low frequency and intensity of aggression within units has been casually noted in nunbirds (Skutch, 1972), anis (Davis, 1942), the Native Hen (Ridpath, 1972), White-winged Chough (Rowley, 1977), Hall's Babbler (Balda and Brown, 1977), Apostlebird (Mack, 1967), Yucatan Jay (Raitt and Hardy, 1976), Acorn Woodpecker (Joste et al., 1982), and Jungle Babbler (Gaston, 1977a). In the last-named species intragroup aggression was very rarely seen, and even at artificial feeding sites "no evidence of conflict over food was ever observed, except among first-year birds." Restraint of aggression within communal species relative to noncommunal ones is especially noticeable in the jay genera *Aphelocoma* (Brown, 1963a) and *Cyanocorax* (Hardy, 1974). The Noisy Miner, which, because of its extreme aggressiveness can be regarded as an exception to the generality of restraint, does not actually live in close-knit, territorial units (see Chapter 9, specifically Figs. 9.3 and 9.4) as do the other species discussed here.

Two detailed studies of dominance have been done on the genus *Aphelocoma*. In the Scrub Jay studied in summer in Florida the order of dominance within social units was as follows: male breeder > male nonbreeder > female breeder > female nonbreeder > juveniles (Woolfenden and Fitzpatrick, 1977). In the Mexican Jay studied in winter in the mountains of Arizona the hierarchies were more complicated (Barkan et al., 1986). Some of the nonbreeding Mexican Jays of both sexes were at the top of their hierarchies even though less than a year old in some cases. There was also considerable overlap between the sexes among adults, with at least one example of a female ranking at the top. The significance of these findings on the Mexican Jay is not well understood.

It is easy to hypothesize advantages to being dominant in communal birds. A clear advantage confirmed by systematic field observations is that dominant Mexican Jays do not need to wait for access to food found first by others while subordinates must wait (Craig et al., 1982). Carrick (1972) reported that subordinate females in the plural-breeding Australian Magpie suffer interference with attempts to breed and may have "retarded ovaries." The dominant hen typically began nesting first. Testes in Australian Magpies developed more slowly in subordinates, but yearling males "have fertilized the eggs where the adult males were caponized."

Some authors have supposed that Vehrencamp (1977) found a relationship between dominance and order of initiation of laying by anis in a joint nest. Vehrencamp did not say this. She ranked her birds by order of laying, not by behavioral dominance. Her ranks are not dominance ranks; they are egg-laying ranks. She did suggest the above relationship as a hypothesis. Craig (1976), in contrast, emphasized an advantage for subordinates in the Pukeko. Subordinate Pukekos were able to extend their activities a little farther into neighboring territories in the presence of a dominant from their unit than they could if unaccompanied. For the Acorn Woodpecker it is tempting to hypothesize that dominant males have a higher confidence of paternity than subordinates, but this has not been established. Unfortunately, Joste et al. (1982) were only able to see four dominance interactions in the one unit they studied and were unable to establish paternity. The relationship between dominance and likelihood of paternity has not yet been tested systematically in any species of communal bird, so I offer the following observations.

For males, dominance is commonly assumed to ensure access to females. For example, in 1983 in one unit (UC) of Mexican Jays there were six adult males and one female (a rare event). Dominance ranks had been established prior to the breeding season on the basis of hundreds of interactions. During the period just before laying, when the eggs should have been fertilized, the dominant male guarded the female and kept all the other males away at all times when they were under observation. Dominance in this case provided easy access to the female and, so far as we could see, completely precluded access to her by the other males. One such case is not enough, and the matter is under study using electrophoresis to assess paternity.

Conclusions

Although manifestations of established dominance relationships are ubiquitous among communally breeding birds, their intensity is often so subtle that they can be difficult to study. In those types of social organization in which two or more breeding members of the same sex coexist in the same unit, however, such highly selfish acts as competition for females, destruction of eggs, and robbing of nest material have been observed. Although behavioral dominance seems likely to be a good predictor of the outcome in such cases, systematic studies remain to be done.

17 Diet and Group Territoriality

Social defense of food resources occurs in a wide variety of orders and families of communal birds (Table 2.2) and mammals. According to a recent model (Brown, 1982a and Chapter 8), cooperative defense of food resources is profitable for such species when the benefits of cooperation in sharable tasks exceed the costs stemming from defense, vigilance, care of young, and the more rapid consumption of resources in a group than alone. Previous papers on group territoriality and associated communal breeding have focused almost entirely on their possible benefits (Brown, 1969a, 1974; Gaston, 1978c). In order to develop a predictive theory of the occurrence of social defense in various environments, the *costs* of group formation in these environments must also be considered. This chapter suggests some relationships between diet and costs of group living. A reconsideration of the prevalence of communal breeding in the tropics ensues.

For a social group that forages in a fixed area shared within the group, the resource yield per member may be expected to diminish as group size increases, in the absence of cooperative hunting. The food-cost function is the effect of such changes in resource yield on individual direct fitness as a function of group size ($L_{i,j}$ in Fig. 8.4). With respect to a given benefit curve (individual benefit in fitness as a function of group size; $A_{i,j}$ in Fig. 8.4), food cost functions may range from permissive (rising slowly with group size, as in Fig. 8.4b) to prohibitive (steeply rising, as in Fig. 8.4a) with corresponding consequences to the profitability of group exploitation of shared resources. Therefore, to understand why some species are group territorial and others not, it should help in some cases to understand the determinants of food-cost functions.

Why Are Nectar Feeders Not Cooperative?

The relations between fitness and foraging are complex but their importance is widely appreciated. A high rate of individual foraging success is probably beneficial because it provides energy and time for other behaviors in the time-energy budget that benefit fitness more directly, such as territorial defense, vigilance, antipredator behavior, reproduction, and parental care. We shall, therefore, examine factors that affect foraging success rate.

Foraging success requires a standing crop of utilizable food resources. This is determined jointly by the renewal rate of the resources, the rate of harvest by the territorial group, and the rate of depletion by other sources, such as competing consumers and weather. The relevance of renewal rates for group size was probably first modeled theoretically by Waser (1981). Data showing a positive correlation between renewal rate and group size have been presented for territorial wagtails by Davies and Houston (1981). Clearly, the renewal rate must be adequate to support the group and may partially determine group size, but under what circumstances is this likely to be so? To answer this question we must consider additional factors.

Generalizing from these considerations of food-cost functions, I predict that species heavily dependent upon foods that are adapted to being eaten should have steeply rising (prohibitive) food-cost functions and, consequently, should not defend resources cooperatively. The following discussion illustrates the argument with nectar feeders, but with minor restrictions it may apply to some fruit eaters too.

Flower patches are defended by nectarivores when the patches are rich enough to support the defender, including defense costs, and when such defense raises the defender's energy profits above the level achievable without defense (Gill and Wolf, 1975; Carpenter and McMillen, 1976; Carpenter et al., 1983). The period of defense is relatively brief, since the period of high nectar yield tends to be short and seasonal for each species of plant. The defender has little need to defend a larger patch than it needs for the short run. Carpenter et al. observed changes in patch size for hummingbirds that they interpreted as adjustments to immediate needs. If the patch fails, the nectar feeder can relatively easily find others—if they exist locally—because flowers advertise the presence of nectar. Territorial nectar feeders are often forced by the seasonal or generally transient nature of flower patches to change patches.

Accessibility and Vulnerability. There are several conditions under which the standing crop of a food supply may be large and adequately renewed yet mostly inaccessible. Many insect prey are cryptic or hide in inaccessible places, being detectable mainly when not hiding. Fish beneath the water surface and fossorial animals below the land surface may become accessible to aerial or terrestrial predators only for brief periods.

Most foods are, not surprisingly, adapated to avoid being captured; but there are exceptions. Nectar is produced by plants in conspicuous structures that appear to be adapted to attract consumers. Similarly, many fleshy fruits hang where their seed dispersers can find them, and are colored so as to attract attention. These plants produce *food that is adapted to being eaten.*

Nectar and some fruits are not only more accessible to their consumers, but also are more vulnerable. Once found, they are defenseless and may be cheaply and totally consumed without harm to the consumer. Many insects, other invertebrates, and other parts of the plant are protected by toxic chemicals, warning signals, spines, or defensive behavior. The nectar in a small patch of flowers may, therefore, be quickly and totally harvested because it is readily accessible and totally vulnerable.

Consequently, nectar eaters (1) can find rich flower patches relatively easily and quickly with minimal sampling, (2) can find food within a patch quickly and easily, (3) can, with experience, relatively accurately estimate the standing crop of food in a patch, and (4) can rapidly and severely deplete a food patch with a low renewal rate. A nectar-feeding bird, therefore, can defend a patch whose standing crop is just above threshold for its own energy needs and one that would be quickly lowered below this threshold if other birds were admitted. Two or three birds could in theory defend a larger patch cooperatively, but this is likely to be more expensive in terms of travel time within the territory (C. C. Smith, 1968). Nectar feeders can relatively easily locate new patches, should the present one fail. Therefore, they have little need to defend a larger patch than needed in the immediate future. For these reasons nectarivores may be said to have prohibitive food-cost functions: they rise steeply.

Why Are Omnivores More Likely To Have Permissive Food-Cost Functions?

Omnivores, on the other hand, such as many jays and babblers, can more easily shift foods as the seasons change than can nectar specialists. Therefore, they can often subsist on the same territory through all or most of the year. Mexican Jays, for example, feed heavily on caterpillars in May, cicadas in June, grasshoppers in July, acorns in August and September, pinyon nuts in October, and stored foods through the winter. In such species the *best territories are spatially predictable,* although their food supply may vary yearly and seasonally. Consequently, since their diet allows them to overwinter on the breeding territories, the best strategy of individuals in such species may frequently be to maintain control of the best areas in order to ensure breeding success when the environment permits (Brown, 1978a:146) and to have a longer breeding season.

In a group-territorial species that retains its young for one or more years, as in the Mexican Jay, yearly variation in average group size is determined mainly by variation in population size, since the number of territories is relatively stable (Fig. 17.1). Population size in the Mexican Jay is influenced predominantly by the number of young surviving from the preceding breeding season (Brown, 1986).

In years when the average group size is large, average territory size should be sufficient to accomodate a large group. At times of less than peak density—most of the time—group size is smaller but space per group tends to remain nearly constant. In the Mexican Jay from 1972 to 1982 the population varied between 49 and 99 birds while the number of groups varied from six to seven. The territories appear to be larger than needed for the number of birds in the group except at peak densities, i.e., in most years. Population size seems to be determined mainly by the production and survival of young. Consequently, the permissive food-cost function in this omnivorous species enables family groups to maintain control of territories even during hard times by retention of their young. Thus, a permissive food-cost function stemming from a generalist diet is also conducive to parental facilitation (Chapter 7).

Production and survival of young in Mexican Jays and many other communal breeders are erratic and probably depend on a variety of factors, such as predation and the effects of climatic variables on food supplies

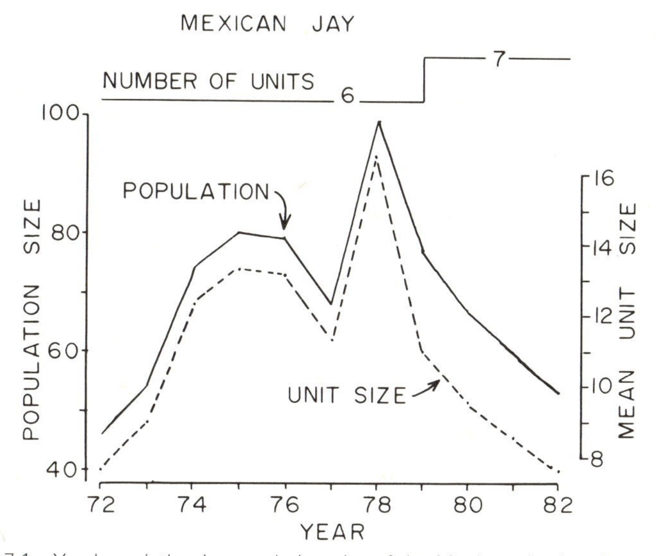

Figure 17.1 Yearly variation in population size of the Mexican Jay is reflected closely by variation in average unit size. The number of units on the study area varied only from six to seven. (From Brown, 1986.)

(Brown, 1986). Because adults and independent young in such species can survive under conditions when reproduction and juvenile survival are jeopardized, it is possible for the group to maintain control of resources even when conditions are insufficient for reproduction.

Extending this reasoning it seems that generalist diets predispose animals toward long-term control of resources. Animals specializing on foods with transient patches are predisposed away from long-term control of resources. Enhanced control of resources is the essence of group territoriality and enables inheritance of territories.

Why Is Helping More Common in the Tropics and Australia?

With the above perspective it is possible to improve the explanation offered previously (Chapter 3) for the prevalence of communal breeding in the tropics and Australia. Previous authors have observed that the number of communally breeding species of birds in a region is correlated with the number of nonmigrant species and with a demographic syndrome that de-emphasizes r_{max} (maximum rate of increase of a population) and emphasizes bet-hedging and survival (Brown, 1974, 1978a; Dow, 1980a; and later authors).

An additional consideration by these authors is that the generally more benign climates of the tropics enable offspring and parents to live together all year on or near permanent territories even during relatively harsh periods. For example, Grey-crowned Babblers in Queensland maintain group territories for feeding at all times of the year whether the food supply is good enough for breeding or not. Such largely insectivorous species have a wide range of foods and can survive nonbreeding periods on their territories provided the nonbreeding season is not too severe. This situation would be more likely and more frequent in warm climates than in temperate regions.

Conclusions

Group-territorial defense in birds is unknown for nectar feeders, rare in frugivores, and occurs most frequently among omnivores and food generalists. This correlation between diet and sociality may be related to the food-cost function of an energy model for cooperative behavior. A basic difference exists between foods whose producers are adapted to advertise their presence, such as nectar-bearing flowers, and foods that are adapted to conceal their presence, such as cryptic insects and most prey of food generalists. Self-revealing foods are highly visible and accessible to consumers. Because individuals can readily estimate the food resources needed to fill their requirements, they defend only as much territory as they need for this purpose. Such territories are vulnerable to rapid and near-total depletion. Consequently, their consumers have a steep food-cost function and rarely encounter conditions that favor group defense. Illustrating this point, offspring of nectar-specialist species do not find it profitable to stay with their parents. Because of their diet, offspring of food generalists are more likely to be able to remain with their parents in the natal territory. Data are presented that demonstrate fluctuations in average flock size in parallel with population fluctuations. These data agree with the hypothesis that territorial groups are usually smaller than the carrying capacity of the territory.

18 Synthesis

We have looked in some detail at a variety of topics and problems. Now it is time to draw together the main points, especially those that span several chapters. With a controversial problem, such as the ecological evolution of helping behavior, it is desirable to employ the scientific method rather than the method of advocacy. In the following overview I shall try to adhere to scientific formalism.

Helping behavior poses an evolutionary paradox. Helpers are "parental" in their behavior although they are not the genetic parents of the young for which they care. Although the definition is simple, the social systems in which helping is found are amazingly diverse and complicated (Chapter 2). We wish to identify the various factors that may have influenced the evolution of helping behavior in a wide variety of contexts. We seek models of natural selection for helping that have a useful degree of generality, knowing that details must supplement principles. The usual models for the evolution of parental behavior do not apply because helpers are not caring for their own offspring.

Natural-History Correlates

Before erecting more formal models it is useful to explore the natural-history correlates of helping. These may allow the formation of useful hypotheses. From our taxonomic survey (Chapter 3) we have seen that helping is widely distributed among many taxa. It is apparent that helping is in most cases produced by convergent evolution and has had many independent origins. Phylogenetic constraints, therefore, explain very little.

One common denominator of species having helpers is that the aid of at least two parents is normally involved in rearing the young. Therefore, helping probably originated from biparental care systems, such as are characteristic of monogamy.

Geographically, an increase in the number of species having helpers occurs toward the equator, at least in the Northern Hemisphere. This pattern leads by itself to no obvious solution, but combined with other information it might be useful.

Habitat also provides few cues. Helping species occur in a variety of aquatic and terrestrial habitats; in aerial, arboreal, and cursorial foragers;

grassland and forest species; as well as uniform and patchy environments. Nonhelping species show similarly diverse patterns. Habitat and foraging requirements provide no obvious or robust generalizations, but they too may be useful when combined with other information.

The most useful correlates of helping as cues to its origin came from the study of the *demography* of communal compared to noncommunal species. Communal species tend to have lower rates of population increase, higher annual survival rates, and lower reproductive rates. The latter are typically achieved not by a reduction in clutch size but by a delay in the age at which breeding begins. The delay is often associated with a more conservative dispersal strategy, in which young may live one or more years in their natal territory without breeding.

The Three Main Questions

These observations suggested the first comprehensive ecological theory for the origin of helping. The key factor that united these elements of the puzzle was *territoriality*. In species with a relatively high survival rate for potential breeders, as in many communal birds, a surplus of nonbreeders can easily arise, thus creating conditions that favor group territoriality (Brown, 1969a; Fig. 3.2; Chapter 8). Under such conditions the best patches for breeding would be nearly continuously defended, and there would be more competitors than suitable territories. The best strategy for survival of a young bird until a breeding opportunity arose would be in many cases to remain at home, usually with its parents, and use the home territory as a base from which to make short dispersal forays until it was successful (Selander, 1964; Brown, 1969a, 1974, 1978a).

The point here is not to debate this particular theory. It is to identify three basic attributes of many communal social systems. These are delayed breeding, reduced dispersal, and helping. Helping is not necessarily linked to nonbreeding or nondispersal, and it is clearly a mistake to define or identify helpers as nonbreeders or nondispersers. Nevertheless, to use a popular phrase, delayed breeding and nondispersal "set the stage" for the evolution of helping in nuclear-family systems.

It is essential, therefore, to deal with the following questions both separately and in conjunction with each other:

(1) Why is *breeding* delayed in some populations and what factors determine age at first breeding?

(2) Why is *dispersal* delayed and or reduced, sometimes to the point of the bird never leaving the hatching site?

(3) Why *help*?

Note that question 3 does not necessarily include questions 1 and 2. An answer to questions 1 and 2 is *not* a sufficient answer to question 3.

Why Delay Breeding?

The principal alternative hypotheses for delayed breeding are: (1) territoriality leading to habitat saturation, (2) sex ratio, usually a shortage of females, (3) foraging success below threshold, (4) indirect selection, (5) parental suppression. Each of these may vary in its effect with yearly variations in environmental conditions.

The first, territoriality or habitat saturation, may be rejected in species that do not defend food resources (Pied Kingfisher). It is unlikely to be the main cause of delayed breeding when only males delay breeding. Removal of breeders should allow nonbreeders to take their places.

The second, sex ratio, may be rejected when both sexes delay breeding, as in Mexican Jays and many other species. Removal of some breeding males should allow some nonbreeders to breed. Monogamy is expected in most such cases.

The third, foraging skill, may usually be rejected when only one sex delays breeding. Removal of breeders should not result in their places being taken by nonbreeders. Demonstration of improvement in skill with age is needed.

The fourth, indirect selection, is unparsimonious as a cause of delayed breeding (not helping) when one of the first three hypotheses applies (Brown, 1978a; and later authors). If an excess of males occurs and only those males help, it would be unparsimonious to invoke indirect selection as the sole reason for the delay. The presence of defense of critical resources would also make this hypothesis unparsimonious. Removal of the breeders might or might not result in breeding by the nonbreeders. In general this is a difficult hypothesis to reject when breeding opportunities exist but are not taken.

The fifth, parental suppression, can work within the parent's area of dominance; however, it can hardly be invoked elsewhere. Since birds are highly mobile, parental suppression can be easily evaded.

Reviewing the empirical studies, it is regrettable that several authors have confounded helping with delayed breeding and that no author has seriously considered foraging skill. The problem is tractable using relatively simple methods, and the alternative hypotheses are falsifiable.

Two Kinds of Social System and Two Kinds of Kin Selection

Putting aside the question of whether or not kin selection is important, it is interesting that regular communal breeding can be divided into two types depending upon the type of kin selection involved (Fig. 10.3). In the more familiar type helpers are related to the young through their parents; kinship and kin selection are *indirect*. In the less familiar type helpers are related to some of the young through their own gametes; kinship and kin selection are *direct*. In the first type the system is based upon the *nuclear family*, and helpers are typically offspring of one or both breeders. In the second type, which I call *nest sharing*, breeders share a nest, and parenthood is probabilistic. Nest sharing may arise by *mate sharing*, in which males share one or more females at a single nest, or by *joint nesting*, in which two or more females lay their eggs in the same nest. The result is that breeders may care for young not their own because they have no reliable way to distinguish their own young from others. This new insight is the result of a decade of field studies and a modern clarification of kin selection (see Chapter 4).

Still, putting aside the question of the importance of kinship, it is noteworthy that on a purely descriptive level regular helping behavior is unknown in birds except in one of these two kinship systems, or a combination of them. Much has been made of the fact that in nest-sharing systems the helper-breeders are commonly unrelated *to each other*. The implication is that individuals can and do cooperate without being related to each other. This is supposed to be interpreted to mean that kinship is unimportant. Such statements ignore the more important fact that some nonzero probability of *close relatedness to a common pool of young* is characteristic of such systems. Indeed, individuals whose chances of being a parent appear slim to an observer tend to perform as predicted, that is, uncooperatively or even destructively.

Nest-sharing systems also shed light on nuclear-family systems. This is because they provide natural experiments on the importance of kinship. As more individuals of the same sex contribute their gametes, the average relatedness of an individual to the pooled young decreases from 1/2 to near zero. By contrast, in nuclear-family systems relatedness is insensitive to group size. Adding more siblings or half-sibs to the helper corps does not decrease their average relatedness to the young. Although group sizes are typically larger in nuclear-family systems than nest-sharing systems, so many other variables influence group size that this difference between the systems cannot safely be attributed to relatedness; however, the data do agree with the hypothesis.

We can use nest-sharing systems as a test of how low average relatedness can go and still justify "parental" behavior. In the joint-nesting anis, groups of four breeding pairs represent a rough limit to group size. In polyandrous species, sharing of a female by more than three males is extremely rare. Thus, an average relatedness of $r = 1/2 \times 1/4 = 1/8$ approximates the observed lower limit.

Obviously, a game theoretic approach is needed in which one variable is average relatedness to the young or a probability of being a parent. In Chapters 9 and 10 the roles of this and several other variables are explored.

Perhaps the most important result of the realization of this dichotomy is that we have some justification now for seeking two rather different types of ecological explanation of helping, one for nest-sharing systems and one for nuclear-family systems. Explanations for nuclear-family species need not apply to nest-sharing systems, and *vice versa.*

How Important Is Indirect Selection?

Recent discoveries have greatly strengthened the case for indirect selection. The most convincing evidence has come from studies on colonial species in which nonbreeding helpers have a free choice between helping kin or non-kin, or not helping at all. Alternative theories based on the difficulty of getting a territory can be rejected here. The strong preference for helping kin in Pied Kingfishers, European Bee-eaters, and White-fronted Bee-eaters (references in Chapter 8) has so far not been plausibly explained by any other hypothesis. There is even evidence for *altruism* in the Pied Kingfisher. The choice of helping to feed kin appears to lower the helper's chances for reproduction in the future, compared to other possible choices.

Strong evidence for indirect selection has also been presented recently for certain group-territorial species, namely, the Stripe-backed Wren and Bicolored Wren (Chapter 13). These new studies join several older studies on group-territorial species for which a strong role for indirect selection is evident and cannot now be rejected.

Much of the argument against indirect selection can be shown to have been based on confusion of the issues. As shown in Table 14.5, three general theories can be identified with respect to the role of indirect selection in explaining delayed breeding, reduced dispersal, and helping. One could imagine that helpers *give up* opportunities to breed in order to stay with their parents and help them (theory no. 2). Alternatively, one could argue that helpers were *prevented* from breeding by habitat saturation or other biological factors, that they remain with their parents as the best place to wait until a breeding opportunity arises, and that they choose helping mainly in such circumstances (theory no. 3). I have argued repeatedly in the past that theory no. 2 can be rejected or is at least unlikely for most cases of helping, as have several authors (references in Chapter 14). Theory no. 1 does not require indirect selection at all, although it may be present. Instead it explains helping by various largely untested special hypotheses that apply to very few species, such as the augmentation hypothesis or social bonding hypothesis.

Critics of indirect selection have directed their attention almost entirely at theory no. 2. They have virtually ignored theory no. 3, even though, in my judgment, it is strongly supported by data and cannot be rejected in a wide variety of studies on monogamous, nuclear-family species.

It is now time to consider theory no. 3 more carefully. Critics should also search for plausible alternative hypotheses to explain kin-preference by helpers in colonial species. Experiments should be devised that can lead to rejection of rival working hypotheses. Although it is important to be sure that theory no. 2 can be rejected, there never has been any enthusiasm for it among students of communal birds. Nobody has ever endorsed or advocated it. Its status seems to have changed from "straw man" to "dead horse." To continue beating it seems unlikely to accomplish much. Skeptics of indirect selection would do better to focus on theory no. 3 than theory no. 2.

Conflict within Groups

According to Hamilton's rule, genetic conflict of interest should diminish as the degree of genetic relatedness between individuals increases. Empirical findings on communal birds conform with this expectation. Conflict is most intense among sexual or reproductive rivals, and these are usually not closely related. Conflict is least apparent between nonbreeders and breeders in nuclear-family systems, which are based on parents and their offspring. Specifically, the nest robbery and egg destruction that have been found in Mexican Jays, Groove-billed Anis, and Acorn Woodpeckers tend to pit rival breeding females in the same unit against each other (Chapter 16). In contrast, such behavior is rarely observed in nuclear families, though they are not without more subtle rivalry.

One type of potential conflict is of special interest because of the attention it has received, namely, parent-offspring conflict (Chapter 15). The case of the eusocial insects suggests that parental suppression ("manipulation") of offspring might have played a role in avian systems, and Vehrencamp has modeled a system in which parents induce their offspring to act altruistically (Fig. 5.7). It has been shown mathematically, however, that helping is unlikely to have originated by parental manipulation (Chapter 15; Brown and Pimm, 1985).

A key parameter for understanding the parent-offspring relationship is the *variance in reproductive potential* among potential breeders and among members of a brood. The various conditions that favor delayed breeding tend to divide populations into breeders and nonbreeders, thus providing a natural source of variance in reproductive potential. Parents under some circumstances can utilize this variance to obtain help without suppressing breeding in their offspring (variance utilization). The fact that an offspring is equally related to its full sibs and its own offspring tends to minimize parent-offspring conflict in this situation.

An alternative is for the parents to somehow suppress breeding in their offspring (variance enhancement by suppression). This, however, does extract a cost from the parents, resulting in its being unprofitable to them unless the benefit to them from such suppression is very large. An origin of helping by this means is, therefore, less likely than by variance utilization.

Another possibility for the parents is to facilitate the attainment of breeding status by some of their young by allowing the young to remain longer on the parental territory before leaving or even allowing them to breed there. *Parental facilitation*, as this strategy is called, may yield a net profit for the parents even if a small reduction in the number of their own offspring is involved (Fig. 7.4), although such a reduction is not required.

Conflict for various ecologically important privileges may also be predicted, quite apart from considerations of relatedness, using energy-based models of group size (Chapter 8). These models (Fig. 8.7) predict that dominants may modulate their degree of suppression of subordinates to retain them and that subordinates may make sacrifices in order to be retained by dominants.

Alloparenting in Nuclear-family Systems

We have considered the environmental and social conditions that "set the stage" for helping in nuclear-family systems, namely, those that favor delayed breeding and delayed dispersal. We are now able to raise the question of the adaptive advantages of helping. We have recognized two types of communal kinship systems, one associated with probabilistic, direct relatedness to the young; the other, with indirect relatedness. We have considered the first type briefly above (and Chapters 9 and 10). We now consider the second type, in which the helpers are usually nonbreeding offspring of the breeders.

Breeders. Both breeders and nonbreeders should benefit from the system. Breeders may greatly increase their reproductive success by having helpers in some species; and in none with regular helping can we be sure that breeders do not benefit (Chapter 11). In some species breeding without helpers rarely or never occurs, leading us to wonder if it is even possible. Breeders are the focus for all the helpers in the group. Hence, breeder benefits are the sum of effects of each of their helpers. On an individual basis, therefore, the total augmentation of fitness to a breeder may be large, while the share of this from any one helper may be small.

Nonbreeders. The other key element in the helper-breeder relationship in nuclear families is the nonbreeding helper. Given that an individual is a nonbreeder in its parent's social unit, what are the payoffs to its own inclusive fitness to be obtained by behaving alloparentally as opposed to not behaving alloparentally? This is a critical question. It is the question to which phase 2 of my theory was addressed. That this is not the principal question addressed by Woolfenden and Fitzpatrick (1978, 1984) was pointed out early (Brown, 1978b), but confusion has persisted (see Chapter 14).

It may be useful first to list the major relevant hypotheses and then to attempt to falsify (reject or disprove) each. The null hypothesis, that helping is not adaptive, is considered in Chapter 2. We consider here the following hypotheses:

(1) indirect selection;
(2) augmentation, a form of generational mutualism, sometimes called "reciprocity";
(3) social bonding;
(4) pairing or mating.

Indirect Selection. The case for indirect selection is strong; and the case against it, weak in many species. The hypothesis that indirect selection is involved in the evolution of alloparental care (or of any other behavior) can be rejected by either of two kinds of evidence. (1) The hypothesis must be rejected if helpers are no more than randomly related to the young they care for. (2) The hypothesis must be rejected if no gain in fitness of the parent-recipients due to the helper occurs.

There has been a large amount of evidence gathered on these two points. In Chapter 12 it is shown that helpers in singular-breeding species are usually closely related to the young they feed in virtually every such species for which data are available. Probably in every species a few helpers are unrelated to the young, and these exceptions (roughly 10% in Green

Woodhoopoes) have been claimed to be sufficient basis for rejection of this hypothesis (Ligon and Ligon, 1978; Woolfenden and Fitzpatrick, 1984); however, the hypothesis does not require perfect accuracy in kin recognition. Even parent-offspring recognition is not perfect. In brief, we cannot reject the hypothesis using this kind of approach for group-territorial, singular-breeding species.

There is one exception among singular breeders which actually strengthens the case. In the Pied Kingfisher (Chapter 13), which is colonial and not group-territorial, related helpers can probably be distinguished from unrelated helpers by their behavior when joining the breeders. We can reject the hypothesis of indirect selection for the unrelated helpers, which join the breeders late, but not for the related ones, which are present earlier. Here, however, the evidence for indirect selection with the related helpers is strengthened because the related helpers devote more effort to helping than do the unrelated ones. They adjust their effort to their probable relatedness to the young. This approach appears to work for one colonial species with helpers and should be tried for others. It is, however, not expected to be suitable for rejecting the hypothesis in group-territorial species because helpers usually do not have such a choice within their territory. Therefore, accurate kin recognition would not be very useful and we would not expect it to be selected.

Augmentation. This hypothesis was proposed by Woolfenden and Fitzpatrick (1978) as an alternative to the hypothesis of indirect selection. Much of what they proposed in the same article dealt with delayed breeding and delayed dispersal, rather than alloparenting, and was similar to what others had proposed earlier (Selander, 1964; Brown, 1969a, 1974). It is, therefore, necessary to focus on those features which had not been previously proposed. Since Woolfenden and Fitzpatrick have not done this, it may be useful for me to restate their hypothesis in the terms employed in this book.

The hypothesis is that alloparental behavior has been selected primarily by the positive effect it has on F (probability of getting a suitable breeding territory and mate) of the helper in the following year mediated specifically by the increase, a, in group size, G, *due to the helper*. Thus, the increase in F due to other factors (the choice of nonbreeding and nondispersal) must be separated out before this hypothesis can be tested. In addition, the increase in group size due to the parents and other group members not included in a must be factored out. It must then be shown that $F = f(G)$ and more critically that $F = f(a)$.

In order for the augmentation hypothesis to work it is necessary that F be influenced by group size. This can occur if dispersal occurs in groups or by budding. Dispersal in groups is fairly common in a few species, but it is exceptional among communal breeders. Budding might occur in a variety of species but has been described only for male Scrub Jays. Dispersal by female Scrub Jays and females of most communal birds is typically by individuals entering established units and is not thought to be influenced importantly by source-group size. Therefore, the augmentation hypothesis can be rejected out of hand for female helpers in virtually all species. It can also be rejected for male helpers except in a few species. Even in Florida Scrub Jays it may not apply (Chapter 14).

In short, the applicability of the augmentation hypothesis as stated is extremely limited. Even in favorable cases, such as the Florida Scrub Jay, its failure to explain alloparenting in both sexes indicates a need for another hypothesis and casts doubt on its importance even in males.

Social Bonding. Assuming that living in an integrated social unit is beneficial to the individual, then mechanisms that keep the group together should also be beneficial. But what are these mechanisms? There has been no study of this, but almost any behavior that is done together by two or more members might be involved. Certainly allofeeding is a candidate. Could the helper be selected to feed a nestling purely as a device to gain that nestling's allegiance when older? It has been suggested that such allegiance might be valuable in dispersal or when the former helper breeds and needs helpers itself. In other words, the hypothesis is that individuals help purely as a means of recruiting future aid.

The bonding hypothesis can be rejected if the helper in its dispersal or future breeding does not utilize aid from the young it fed. This condition is met for female helpers in most species since they tend to disperse alone and breed in their new group, getting help sometimes from the new group members but rarely from former group members. In species with social dispersal this hypothesis cannot be so easily rejected.

There are so many potential means of social bonding, however, and so little is known about them that the case for social bonding through allofeeding in particular is weak and untested. Nevertheless, it is already clear that the hypothesis can be rejected for most species and cases of helping.

Future Pairing. In a variety of species males who have lost their female or young or who perhaps never had a female will begin to feed the young in an active nest not their own. This is sometimes resisted by the genetic father; but, either because he is secure in his paternity and needs help or because he cannot keep out the intruder-helper, the result is commonly a trio of feeders. This behavior is most often reported in species that do not normally have helpers and are not considered to have true communal social systems (Brown, 1983a).

Although this behavior by intruder-helpers seems maladaptive, it might benefit the helper if it raised its chances of mating in the next year, possibly even with the female whose nestlings he has fed. The hypothesis is that for these males alloparenting increases their probability of mating in the following year. The hypothesis resembles Zahavi's (1977) sexual selection hypothesis, which interprets helping as competition for mates.

A test of this hypothesis is available with the Pied Kingfisher. The hypothesis may be rejected if male helpers feeding the young of unrelated females have the same probability of mating in the next year as those feeding their mother's young. It is expected that males do not mate with their mothers. Reyer (1984) reported that *unrelated helpers had significantly higher rates of survival and pairing than related ones.* This supports the hypothesis.

Reciprocal Alloparenting in Plural-breeding Systems

Plural-breeding species are those in which two or more pairs, or females, breed in the same territory. They can be divided into joint-nesting species, in which all eggs are laid in one nest, and separate-nesting species, in which each female has her own nest. We have already discussed the joint-nesting species; so we limit out attention here to separate-nesting species. These species are thought to be derived from a singular-breeding condition, and in the Mexican Jay populations of each type occur. Therefore, we pass over the origins of helping and focus on a type of helping that cannot normally occur in singular-breeding species, namely, the mutual feeding of each others' fledglings by breeders in the same social unit. This occurs in joint-nesting species because the parents cannot identify their own young. In separate-nesting species, however, it is clear that each mother knows her own nest. Since reciprocal feeding begins in the late nestling stage, it cannot be attributed to inability to recognize one's own young.

The interesting feature of reciprocal allofeeding in Mexican Jays is that the hypothesis of indirect selection can be rejected on genealogical evidence (see Chapter 14). It is the only case of reciprocal aid-giving in communal birds for which indirect selection can be rejected. This does not mean, however, that a tit-for-tat strategy must be accepted. There may be a self-interest factor in reciprocal allofeeding that varies with the environment and may alter the payoff matrix away from a prisoner's dilemma. The situation has been modeled (Fig. 14.1), but a judgment based on present evidence would be premature.

Levels and Phases

Not only is no hypothesis adequate to explain all cases of avian helping behavior; no single hypothesis can adequately explain any one case of helping. This is because there may be several phases in the evolution of helping and because there are several aspects of the problem. For example, there are at least three separate and plausible alternative hypotheses to explain delayed breeding in communal birds (Chapter 5); yet none of these explains the evolution of helping behavior because for each one there are noncommunal birds that appear to delay breeding for the same reason (i.e., skill, territoriality, sex ratio, and their interactions with environment). This is why I consider delayed breeding as *setting the stage* for helping rather than explaining helping. Other authors have regarded theories for the evolution of delayed breeding, such as the various quite separate hypotheses that have been grouped together as an "ecological constraints theory," as an explanation of helping. Such hypotheses apply to only one phase of the process, delayed breeding; and, of course, they do not explain helping by breeders adequately either.

For the next phase it is necessary to consider both the donor and recipient of aid. For each, the costs and benefits of the relationship must be reckoned. I have provided theories for donor-helpers and recipient-parents. Energy is involved, and I have provided a theory of group formation based purely on the fitness correlates of energy (Chapter 8). Energy is not enough, however; the fitness of each party must be addressed in the most comprehensive possible way, using the theory of inclusive fitness.

Kin Selection and the Scientific Method

Kin selection has been a thorny issue. The basic concept and definition of kin selection were misunderstood by many sociobiologists for many years and still are by a few. Some biologists have presented the controversy over helping as a dichotomy between ecology and indirect fitness (e.g., Koenig and Pitelka, 1981), which it certainly is not (see Table 14.5 and the discussion of it). This misunderstanding implies a misconception of Hamilton's rule, which is an ecological model, and a misrepresentation of published theories, none of which can be classified as nonecological.

Early judgments against inclusive fitness theory have been reflected in the work of several ornithologists and are manifest in titles such as "Sociobiology is for the birds" (Ligon, 1981b) and "The selfish behavior of avian altruists" (Woolfenden, 1980; see also Woolfenden, 1981). A serious bias is evident in the frequently heard question, "If I have an explanation based on individual selection, why do I need kin selection?" Of course, kin selection is sometimes also individual selection (a semantic problem), but the main objection to such questions, in my view, is that they disregard the need for critical facts. They seem to me an unconscious attempt to evade the method of alternative falsifiable hypotheses. Is a hypothesis based upon direct selection but unsupported by facts inherently preferable to one that is supported by facts but which invokes both direct and indirect selection? A hypothesis should be rejected if the relevant data are inconsistent with it, and *only for that reason.* Of course, it is legitimate to inquire whether indirect selection is a *necessary* component of explanations of helping behavior; however, to avoid employing a double standard *one must ask the same question about direct selection.* Is direct selection necessary? These questions have not been equally popular. At the risk of being redundant I would like to emphasize that the "Is it necessary?" approach, when applied to indirect selection but not to direct selection, does not conform to the scientific method, and that the scientific method is the method of choice. The skepticism that has been shown toward indirect fitness is fine, but only if it is applied equally rigorously to the alternative hypotheses.

Our review of the facts does not reveal much empirical support for hypotheses that exclude indirect selection from explanations of regular helping by nonbreeders. Most of these can be rejected for most communal species on present evidence. On the other hand, indirect selection can probably be rejected only for a few cases, such as for late-arriving (unrelated) helpers in Pied Kingfishers.

I do not want to give the impression that relatedness is the only critical variable. Even for parental behavior by parents, relatedness is an insufficient explanation. Indirect kin selection is an ecological process, and relatedness is only one aspect of it. The same is true of kin selection for parental behavior. Hypotheses involving indirect selection are always ecological because natural selection is always ecological.

In order to understand the ecological aspect of indirect selection we need to examine the environmental factors that influence the costs and benefits that are incorporated in the modified version of Hamilton's rule that I have called the offspring rule. To do this requires detailed knowledge of the ecology of the species under investigation. It is only by the many painstaking, often long-term, behavioral-ecological studies that have been carried out in the last fifty years, and especially in the last decade, that we have arrived at our present state of understanding.

Where do we go from here? The views expressed in this book are not universally accepted. There would be little need for this book if they were. They have been and should continue to be contested. Regardless of how the future regards the issues discussed here, many new vistas in behavioral ecology have been opened by studies of communal birds. Long-term population studies, for example, seem to be on the threshold of providing new questions and answers that were inconceivable in the 1960s. New techniques have dealt with relevant genetic, energetic, and endocrine factors. New theory, such as game theory, needs further application. We have no predictive theory that relates community trophic structure to the types of social systems among its members. In all of these areas there is one simple, but beautiful approach that is available immediately—the well-controlled field experiment addressed to a critical issue.

Appendix

Reproduced below is my three-phase theory for the evolution of helping behavior in jays (Brown, 1974). This is followed by a brief review of tests of its predictions since its publication.

A Synopsis of the Hypothesized Origin and Maintenance of Communal Breeding and Altruism in Jays

"Three phases in the evolution of communal breeding and altruism can be recognized. (i) *K-selection Phase*: The initial phase in the origin of communal breeding from a non-communal species probably involves a period of *K*-selection in which competition for living space in an environment continuously at carrying capacity and with few suitable vacant territories causes selection for greater competitive ability. This phase is more likely to occur in non-migratory species inhabiting a stable, mature vegetation form. Since older, experienced individuals with mates and territories are probably better able to compete under such conditions than young ones, an older age structure and greater iteroparity are favored. With fewer territorial vacancies available, the optimal reproductive strategy changes from one of producing many young who disperse widely to find vacancies to one of producing fewer young who by virtue of experience gained in a year or more of immature apprenticeship have a better chance of competing successfully for a territory, remaining on it, and reproducing successfully. Reduced dispersal in the preceding phase sets the stage for the next phase. (ii) *Kin-selection Phase*: Because of the close genetic relationship between parent and offspring, the young who actively help their parents to raise young during their apprenticeship are selected more strongly than those who do not help at all or who help distant relatives. If the probability of a newly maturing jay finding a territorial vacancy by dispersing becomes as low as the probability of the death of its parent of the same sex, then a strategy will be favored in which half of the young remain with their parents in a subordinate, helping condition but ready to breed should the parent disappear. As the chances of a dispersant finding a good territory decrease, the adaptive value of traits which aid in the retention of good territories by the family increases. Since this happens simultaneously with the prolonged retention of young in the family, selection further reinforces the creation of family-owned territories because large families are likely to dominate small ones. This makes possible the third phase. (iii) *Kin-group-selection Phase*: If this trend continues, flock size and territory size increase, and more than one breeding pair may be tolerated in a territory. Inbreeding becomes more intense, and as the members of a flock become more nearly genetically uniform, the flock or clan becomes a more important genetic unit in natural selection. Ultimately selection comes to favor the most successful flock genotypes. Individual genotypes are selected mainly as they function harmoniously in a flock. Agressiveness within the flock, selfishness and cheating might be selected against if they lowered flock productivity and competitive ability.

Some Testable Predictions of the Proposed Theory

At the risk of some redundancy the following predictions are offered in the hope of stating briefly certain points which may be tested in future field investigations. When other communally breeding birds are studied the theory predicts them to have the characteristics listed, when compared with related non-communal species.

Correlates of K-selection

(1) Delayed maturity in both sexes.
(2) Lowered reproductive rate and increased survival rate.
(3) Diminished dispersal. Dispersants are fewer in number and do not travel as far.
(4) Not a regular, long-distance migrant; rarely a short-distance migrant.
(5) Characteristically found in stable as opposed to transitional habitats.
(6) Prominent density-dependent mortality.

Correlates of kin-selection (in temporary groups and families)

(7) Helpers (altruists) are on the average closely related to recipients, and less so to non-recipients.
(8) Helpers are found in small groups with adults, all of whom are kept together by strong social bonds.
(9) Outsiders are strongly repelled, especially by larger flocks (territoriality).

Correlates of kin-group-selection (in groups which last for many years)

(10) Territories differ significantly in productivity. The rank order of flocks in respect to production of young will tend to remain the same from year to year.
(11) Adaptations for retaining good territories in the same genetic lineage will be favored. The territory occupied by a flock will show little change in location, areas, and boundaries over intervals of many years. Larger flocks will show less change than smaller ones.
(12) Genetically the future occupants of a territory are descended from the present occupants, especially for larger flocks. Genetic differences between flocks (and between demes created by local topography) will tend to remain the same from year to year.
(13) Adaptations that reduce and formalize intra-flock conflict will be strengthened; intra-flock aggression should be inconspicuous.
(14) Because territories are defended by flocks rather than pairs, there should be virtually no vacant territories of good quality."

Commentary

This theory earned considerable approval in the 1970s among those workers who were interested in the application of inclusive fitness theory to behavioral phenomena in nature (see, for example, E. O. Wilson, 1975). The paper was reprinted in a collection of readings in sociobiology (Hunt, 1980) and has been frequently cited. On the other hand, the parts that invoked kin selection have been controversial.

The theory was formulated in 1972, well before most of the data discussed in this book were available. Subsequent studies, therefore, particularly those on jays, can serve as tests of the theory.

In general, the theory has stood the test of time and data remarkably well, at least in my opinion. For example, the theories of Koenig and Pitelka (1981), Emlen (1982a), and Woolfenden and Fitzpatrick (1984) have many similarities to those parts of the theory (as expressed in Brown, 1969a, 1974, 1978a) that do not involve "kin selection" (e.g., habitat saturation by territorial behavior; ecological constraints; individual advantage of delayed breeding; territorial inheritance).

The theory applies to nuclear-family systems and to plural-breeding species derived from them. It was not intended to apply to nest-sharing systems, such as those in the Groove-billed Ani and Acorn Woodpecker. Therefore, I compare it only to studies on nuclear-family systems and extended-family systems. It is convenient to discuss the predictions in order by section and number. All references to the Florida Scrub Jay are to Woolfenden and Fitzpatrick (1984), whose data are particularly useful as a test of the theory.

Correlates of K-selection. This term was used because the demographic correlates of communal breeding resemble the demographic syndrome of *K*-selection. The theory is, therefore, *based on demography*, not on any other connotations of the term *K*-selection, such as habitat stability.

(1) Delayed breeding is found in the Florida Scrub Jay and other species of jays with helpers (Table 2.2). In all these species the helpers are of both sexes. This makes a shortage of females unlikely as a cause of delayed breeding.

(2) Since species of jay in which helping occurs have a relatively low reproductive rate because of delayed breeding, they must also have a relatively high survival rate in order to maintain their populations. Data for jays in Table 3.2 show relatively high rates of survival in long-term studies, but detailed intrageneric comparisons of survival and reproductive rates are rare. One such study comparing two sympatric, congeneric species of shrikes confirmed the predictions of a higher survival rate and lower per capita reproductive rate in the communal species (Zack and Ligon, 1985a).

(3) After the prediction of reduced dispersal was made, it was confirmed in the Florida Scrub Jay and Mexican Jay (Figs. 7.1, 7.2). Less extensive data on other jays and other communal breeders also strikingly confirm this prediction (Chapters 6 and 7). This pattern contrasts with the pattern in noncommunal populations of the Scrub Jay and Steller's Jay, where the young leave the parents in their first summer (personal observation of banded birds; Atwood, 1980b).

(4) Migration is unknown in any communally breeding populations of jays. Elevational movements and short-distance movements relatively far from the breeding territories are, however, documented for noncommunal jays, such as western Scrub and Steller's jays (Westcott, 1969; Brown, 1963b). Blue Jays are well known for long-distance migrations in the eastern United States. With regard to other communal birds the prediction has held well. None are conspicuous for migration, and the vast majority is permanently resident on group territories.

(5) The prediction of habitat stability has raised semantic problems. The full prediction was as follows: "Their relatively old age structure and low dispersal suggest that [communal breeders] are adapted for long-continued competitive conditions at populations near their carrying capacities (*K*-selected) rather than for colonizing new or repetitively available habitats (*r*-selected). Consequently, we should expect to find communal breeding among species inhabiting stable, climax vegetation forms more frequently than among species characteristic of transient environments" (Brown, 1974:73). Stability, therefore, referred to *vegetation*, not rainfall.

In a general way the prediction has been supported. Communal breeders tend to be found in climax vegetation rather than in successional stages: for example, Florida Scrub Jay in sand scrub, Tufted Jay in heavily wooded barrancas, Beechey Jay in mature thorn forest. Exceptions occur among forest-edge inhabitants (Hardy et al., 1981), but in these cases the habitats are highly predictable with respect to the lifetime of an adult bird.

Stability of annual rainfall was not included in the prediction. For species inhabiting arid areas of unpredictable rainfall, the point to emphasize is that from the viewpoint of the bird these *habitats* are predictably the best when weather conditions permit breeding. They may be saturated at different levels in different years (see Chapter 5). In any case, the prediction was based on demographic correlates of communal breeding and is, therefore, not a good test of the essential parts of the theory.

(6) The prediction of density-dependent mortality has not been adequately tested. Stability in the density of actual breeders relative to the density of potential breeders in the Florida Scrub Jay and Mexican Jay (Brown, 1986) suggests density dependence.

Correlates of kin selection. (7) The prediction of close relatedness between nonbreeding helpers and their recipients is critical. It has been consistently upheld for jays in nuclear families. It has held also for many other species, as documented in Chapter 12.

(8) The prediction of small group sizes was made with implicit comparison to breeding colonies, winter flocks of migrants, and any other social groups of birds. This prediction has been verified, although helping is more common among colonial species than anticipated. Groups of jays and of most communal breeders are relatively small, about the size of a nuclear family but sometimes as large as an extended family (Chapter 12). The prediction of strong social bonds is, I suspect, virtually untestable, since social bonds are hard to define. However, the individuals

of some communally breeding species roost in the same nest and stick together so closely as to earn the colloquial name "happy family bird" in Australia.

(9) This prediction may be divided into two parts: (a) group territoriality; (b) larger groups have an advantage in defense of territories. Group territoriality is widespread among communal birds (Table 2.2). With the single exception of the Pinyon Jay it appears to be universal among communal jays. The advantage of larger groups in territorial defense has been demonstrated in the field for the Florida Scrub Jay and Pukeko (Craig, 1979).

Correlates of kin-group selection (in groups which last for many years). The concept of kin-group selection was not precisely defined in this paper, but it was based on the idea that the members of a territorial group were in a real sense a unit of selection not only because they cooperated in territorial defense but also because they were related to each other. Both of these ideas have been tested and confirmed. The closest to a mathematical model of kin-group selection is D. S. Wilson's (1975, 1977, 1980) model of trait-group selection. The essential parts of this model have been confirmed in several ways. Individuals in such groups cooperate in territorial defense, antipredator behavior, and other behaviors, and they are related (Chapter 12). The trait-group selection model is examined further in Chapter 12.

(10) The prediction that territories differ in production of young is confirmed for jays in Figure 7.2. These data are for a period of eleven years, during which the relative productivity was on average consistent. To the extent that production of young depends on vegetation, consistency among years is to be expected, since climax vegetation changes little from one year to the next. That territories of communally breeding birds vary in the vegetational correlates of group size has been shown in several studies (Chapter 11). Correlations between vegetation and reproductive success are also known. If territories are very similar in quality, they would not be expected to maintain the same rank order in different years. See also Chapter 7.

(11) The constancy of territorial boundaries and of number of territories in the decade since this prediction was made is impressive in the Mexican Jay (Figs. 7.2 and 17.1). In the Florida Scrub Jay, which has smaller units and, therefore, less stability of territorial boundaries, constancy in boundaries is less marked; but the number of breeding territories is remarkably stable (as judged by stability in the number of breeding pairs).

That good territories tend to be kept in the genetic lineage of the owners is confirmed in the Mexican Jay by Figures 7.2 and 12.6 and in the Florida Scrub Jay by budding. Species in which males inherit the entire natal territory, such as the Stripe-backed Wren, strikingly confirm this prediction. This prediction has been expanded into a theory of parental facilitation, which is presented and documented in Chapter 7.

(12) That future occupants of a territory should be descended from present ones follows from the preceding prediction and is confirmed, at least for the species mentioned, by the data just mentioned. Whether genetic differences between flocks remain the same from year to year is difficult to test. Although dispersal probably

prevents distinctive flock genotypes from persisting for long, territorial inheritance should tend to slow the genetic dilution caused by dispersal. See also Chapters 7 and 12.

(13) Intraflock conflict and aggression have been found to be strikingly reduced in Mexican Jays and many other communally breeding species. These findings, which are summarized in Chapters 15 and 16, are fully consistent with this prediction.

(14) The prediction that there should be few or no vacant, suitable territories may be rephrased as a prediction of habitat saturation. Much evidence confirms this prediction in Florida Scrub Jays and many other communally breeding birds (Chapter 5).

Inbreeding. I did not include inbreeding as a prediction, and it is not necessary for the theory. I did, however, include an estimate of inbreeding in Mexican Jays which proved to be much too high. With more data I discovered the error and argued on theoretical grounds against inbreeding as an important factor (Brown, 1978; Johnson and Brown, 1980).

Alternative Explanations. My goal in the paper was to integrate a genetic model (Hamilton's rule) with a realistic ecological setting (habitat saturation) for one group of birds, the jays. Both concepts were familiar in 1974, but their synergistic relationship had not been appreciated. I did not present an alternative theory or explanation, but I always felt that they should be developed and tested. Perhaps my theory stimulated these processes.

The ensuing decade of field work showed that the theory fits the data for jays—especially the Florida Scrub Jay—very well. Among species outside the intended taxonomic boundaries of the theory, many species resembled jays and fit the theory (e.g., babblers), while many did not (e.g., Acorn Woodpeckers and anis). As this book shows, several rather different lines of theory had to be developed for the latter species.

Modernization. The 1974 theory was only a beginning. It was a simple verbal model for a narrowly defined group of birds, and it did not consider many details that subsequent research has shown to be potentially significant. Communal societies, in which cooperation, altruism, and selfishness occur, are likely to continue to interest both theorists and field workers. Consequently, I expect continued development and testing of theory.

This book has presented a variety of more recent mathematical models, each relevant to a different aspect of the problem. Most of these amplify sections of the 1974 theory. Reasons for delayed breeding are articulated in the skill hypothesis, which is presented as one of several alternatives (*K*-selection phase). Inclusive fitness theory has been developed to allow empirical tests using the expanded offspring rule (Kin-selection phase). Relationships among individuals within social units have been modeled in terms of energy, parent-offspring relationships, and gene frequencies (Kin-group selection phase). In addition, a game theoretic model is described. In short, much progress has been made in expanding the basic three-phase model, as well as in exploring new directions.

Annotated Glossary

This glossary is more than a series of definitions. In response to suggestions from others I have included here notes on history and usage of terms rather than disrupt the train of thought by placing them in the text. The notes deal with details that are of concern mainly to specialists.

The purpose of technical terms, such as those in this glossary, is to facilitate accurate and precise communication between scientists. To this end I follow these guidelines:

(1) Use the definition supplied by the *originator* of the term if possible. If a modification is necessary, state the reason for it.

(2) Avoid changes in meaning or usage for established terms, and correct errors of usage that have crept into the literature.

(3) Introduce new terms only for concepts that previously had no name. Examples: indirect selection, plural and singular breeding.

(4) Avoid using new terms for established, named concepts.

(5) Use precisely descriptive terms in preference to general or vague terms if a choice exists. For example, joint nesting is preferable to communal nesting.

(6) When a part is referred to rather than the whole, use the term that specifies this part and avoid terms specifying the whole. For example, indirect selection is a part of kin selection; therefore, kin selection should not be used when indirect selection is intended.

(7) When designating an empirical, observable phenomenon avoid using a term that implies a particular theoretical explanation of that phenomenon. For example, helping can in theory be altruism or cooperation; therefore, to refer to it as cooperative breeding is prejudicial.

(8) The meaning of words that are in wide usage in the English language should not be redefined more narrowly if this can be easily avoided. For such words I have followed the Oxford English Dictionary (OED). Exceptions are tolerated for established terms such as altruism and helping.

Allofeed. To feed young that are not the offspring of the feeder; a special case of alloparenting.

Alloparent. "An individual that assists the parents in care of the young" (E. O. Wilson, 1975). To alloparent is to help. See *helper*.

Allopreen. To preen an individual other than oneself.

Altruism. A donor or initiator of an act is altruistic if the act causes a net increase in the direct fitness of the recipient and a net decrease in the direct fitness of the donor. This definition preserves the meaning used by Hamilton (1963, 1964). For further clarification see Chapter 4.

Attendant. A bird attending a nest, including parents and helpers (Dow, 1980). See *helper*.

Auxiliary. A helper. See discussion under *helper*.

Byproduct mutualism. A mutualism in which benefit to neighbors is an incidental byproduct of actions taken by the donor for its own benefit. Reciprocation is not required to maintain byproduct mutualism, as it is for score-keeping mutualisms.

Collateral kin. Kin who lie on different branches of a genealogy below a bifurcation, as opposed to lineal kin, who lie on the same line of descent or branch. Parents and offspring are lineal kin. Siblings are collateral kin of each other. See Figure 4.1.

Communal breeding. A system of breeding that is characterized by the normal presence of helpers at some or all nests. Often this results in the presence of three or more birds attending the young at a given nest, though this need not be true when a parent is missing. Brood capture, in which a parent takes over the care of young not its own by forcing the real parents to abandon them, which happens in some waterfowl, is excluded by tradition though not by the above definition. Cases in which individuals care for young not their own as a result of deception through intraspecific or interspecific brood parasitism are also excluded by tradition. In these cases the actual parents (the parasites) do not care for their own young. In this way brood parasitism differs from communal (cooperative) breeding.

The term cooperative breeding was used as a synonym for communal breeding by a few authors in the early 70s, including myself (Brown, 1970; Fry, 1972; Swainson, 1970), and several recent authors (e.g., Gaston, 1978; Emlen, 1978, and others), particularly in North America. Neither term may be said to have superseded the other. In a penultimate draft of the bibliography of this book there were 47 titles using communal (excluding those referring to joint nesting) and 60 using cooperative or co-operative. Authors will differ in their personal preferences and I do not expect that general usage will change much as a result of the arguments given below. Nevertheless, I have made a choice for my own usage and I owe the reader an explanation. Personally, I prefer to use the term communal breeding for the following reasons:

(1) Communal breeding was used in most of the literature devoted to the subject during the late 60s and early 70s (e.g., Brown, 1972, 1974; Craig, 1976; Dow, 1970, 1975; Hardy and Raitt, 1974; Ligon and Ligon, 1978, 1979; Ligon, 1981b; Rowley, 1968, 1977; Vernon, 1976; Zahavi, 1974). Therefore, its claim to be established *early* is well founded. My choice of communal in the early 1970s was influenced by Rowley (1968). For the history of these terms before 1968 see Brown (1978a).

(2) Communal breeding is widely used today. It is used in many *recent* papers (Dow, 1980; Dow and Gill, 1984; Craig, 1980; Ligon 1981b; Stacey, 1979a,b; and all of my own papers since 1972).

(3) A useful guideline in science is to avoid naming objective phenomena in terms of a particular hypothesis used to explain them, especially when viable alternative hypotheses exist (guideline #7). To violate this is to prejudice what should be a neutral endeavor. Controversy persists over whether

helpers should be interpreted as falling in the altruism quadrant of Figure 4.2, the mutualism quadrant (also termed cooperation), or the selfishness quadrant. For this stated reason Brown (1978a), Rowley (1981a), and Dow (1982) have chosen the less biased alternative, which is communal breeding. This argument does not depend on the name chosen for the cooperation quadrant. Even if this quadrant is designated mutualism (as in most of this book) and even if cooperative is used in the general dictionary sense, use of the term cooperative breeding still biases the issue. It still favors the mutualism quadrant over others. In a controversial field where this is a hot issue can this be justified? Individuals in communally breeding units have been shown to act selfishly on some occasions and altruistically on others; they are not always cooperative (by any definition of cooperation).

(4) Cooperation is conspicuous in some *noncommunal* social systems, such as breeding colonies that lack helpers.

For further discussion, see *group living, joint nesting.*

Cooperative breeding. See *communal breeding.*

Deme. A local population in which choice of mate is random with respect to most loci. This implies that isolation by distance or topography does not occur within the deme.

Direct fitness. Inclusive fitness minus indirect fitness. Direct fitness is synonymous with classical fitness and with individual fitness. Direct fitness has a personal component and a kinship component. The latter is responsible for the evolution of parental care, which, as emphasized by Hamilton (1964:1), is easily explained by classical individual fitness. For further clarification see Chapter 4.

Direct selection. Natural selection (including sexual selection) based on direct fitness. See *direct fitness.*

Dispersal. Movements of animals from a source, such as a roost, or a birthplace (Johnston, 1961; Berndt and Sternberg, 1961; Greenwood, 1980; Greenwood and Harvey, 1982). Excludes migration. In this book dispersal refers to movements from hatching site to site of first breeding (natal dispersal), whether successful or not. Further dispersal may occur later (breeding dispersal), but it is not discussed here.

ESS. See Box 14.1.

Evolutionarily stable strategy. See Box 14.1.

Extended family. A family containing kin beyond the nuclear family.

Floater. A nonbreeder that is a potential breeder and is not bound to a particular territory.

Flock. A number of animals traveling in company, commonly of the same or similar species. (Slightly modified from OED.)

Generational mutualism. A mutualistic relationship between members of different generations, as between parents and offspring. See Chapter 15.

Group. "An assemblage of persons, animals, or material things, standing near together, so as to form a collective unity. . . . A number of persons or things regarded as forming a unity on account of any kind of mutual or common relation, or classed together on account of a certain degree of similarity" (OED). Following guideline #8, I have not employed the specialized usage for group that was recommended by Rowley et al. (1979). See *group living.* For the designation of communal groups, rather than use a term that is in wide usage with a much broader meaning, I have found the term "primary social unit" (Brown and Orians, 1970), shortened to "unit" or "social unit" to be satisfactory. The term commune (Brown, 1978a) is also useful in this context.

Group-living. A species may be termed group-living if at any time in the year it normally lives in groups. Group-living, therefore, encompasses not only the communally breeding species but *also* species that breed colonially without helpers, those that breed in isolated pairs but form groups when not breeding, and any species that for whatever reason forms social groups on occasion (Alexander, 1974). In keeping with guidelines #6 and #8 this term should *not* be used as a substitute for communal breeding, as proposed by Zahavi (1976), and Rowley et al. (1979).

Group territory. A territory held by a group of *three* or more individuals. Group territories are normally but not invariably held all or most of the year. In most cases they are all-purpose territories, although some colonial bee-eaters are exceptions (Hegner et al., 1982).

Group selection. Selection at the level of groups of two or more individuals. See Chapter 12 for a discussion of kin selection as group selection.

Habitat saturation. The condition that occurs when that point is reached on the continuum of competitor pressure where all *suitable* territories are occupied and defended. Competitor pressure refers to the number of individuals seeking territories. If territories are essentially incompressible, saturation is a clearly defined point. For example, the transition from Level 1 to Level 2 in Figure 5.5 occurs at the point where the rich habitat is saturated. Similarly, the transition from Level 2 to Level 3 occurs when all available territories in all suitable habitats (in this case 2) are occupied and a class of nonowners and potential helpers is created (N_F), i.e., all suitable habitats are saturated. Sharing a territory may also be used as an indication of saturation. If territories are compressible and/or a continuous gradient of suitability of habitat occurs, saturation may be defined as the point at which nonbreeding or sharing of territories becomes preferable to taking the best unoccupied territory. The latter is illustrated in Figures 8.1 and 8.7 and modeled in Chapter 8. Defense costs are expected to rise sharply above the saturation point (Fig. 8.2).

Habitat saturation is either present or absent. It is a point, not a gradient. It seems reasonable to refer to a percentage saturation when below saturation, but not when above. In the latter case, the expression $N_F = a/(1 - s)$ in Figure 3.2 may be used to estimate or express the degree of competitor pressure. See Chapter 5 for further discussion.

Hamilton's Rule. See Chapter 4.

Helper. An individual that performs parent-like behavior toward young that are not genetically its own offspring, typically in company with the young's real parents (thus excluding brood parasitism and brood capture). Helpers may be breeders or nonbreeders. They may or may not benefit the young or their parents. Helpers may be altruistic, cooperative, or selfish. Note that breeding status and conferral of benefit or harm to recipient or helper are irrelevant to the definition. See quotations from Skutch in Chapters 1 and 2.

Some authors seem to understand the term in different ways, leading them to employ various substitutes, such as attendant, associate, auxiliary, secondary, supernumerary, and alloparent. Is there really a need for this proliferation of terms? As a step in furthering the precision and stability of terms used in behavioral ecology I briefly discuss the origin and utility of some of these terms and concepts.

As is well appreciated, the concept of a helper was introduced by Skutch in 1935 in a landmark paper on what he called helpers at the nest, or simply helpers. This paper and Skutch's later (1961a) review by definition and example established the concept of a helper. A helper was simply an individual that performed parent-like behavior toward young or potential young (as in nest building) that were not its own. Other kinds of behavior that might seem "helpful" such as flocking and alarm calling in winter flocks were excluded. Skutch's concept and term have persisted to the present. They have the merits of being objective, descriptive, and established, being used in nearly all papers on the subject, even by those authors who introduce substitutes. Furthermore, they direct credit where credit is due, namely, to the originator of the concept—not to the originator of a synonym.

Why then are any other terms needed? Usually they are not. There may, however, be a need for a term to designate individuals in a social unit other than the parents, without regard to whether or not these extra individuals perform parent-like behavior. Various terms have been suggested that might be useful for this purpose; but, unfortunately, instead of using them in this way many authors have used them as synonyms of helper. This practice destroys their usefulness, since no discrimination between the new term and helper is actually made. These terms have obscure origins, but some of them and their users are as follows: *auxiliary* (Armstrong, 1947; Parry, 1973); *supernumerary* (Armstrong, 1947; Rowley, 1965); *secondary* (Dow, 1980). Such terms might still prove useful if authors would define and use them differently from helper, retaining helper in its original sense. My own preference is to use simple descriptive terms instead, such as "extra bird" or "nonbreeder" or "additional breeder," as the situation requires.

Do we prejudge the issue of benefit by using helper? As defined and used by Skutch there is no issue involved to be prejudged. Helper is an objective and descriptive term. Neither "helping" nor "parental care" means that the behavior must be beneficial in every instance. Both helping and parental care can be detrimental in some situations. Certainly Skutch never required that benefit be shown in order to call behavior "helping." The conferral of benefit is simply not required by the definition. Therefore, prejudgment of an issue is not implied by use of helper. If we were to abandon helping and alloparenting for this reason, then we should also abandon parental care for the same reason to be consistent.

Can we avoid the implication of benefit by choosing a substitute term? As long as the substitute term is defined in the same way—as "parent-like behavior"—then the implication of potential benefit is likely to remain. Words like alloparent, attendant, and auxiliary certainly imply *potential* benefit, regardless of how defined. Furthermore, some terms have been used as synonyms of helper so frequently that they already have the same implications, regardless of how an author may use them (auxiliary, for example).

In conclusion, I suggest that the term helper be used only in cases in which parent-like behavior has been observed or in which this implication is intended. When this is not the case, other words that have not been used as helper-substitutes may be used. I fail to find persuasive the arguments invoked to discontinue this long-established term on the basis of prejudgment of issues or anthropocentrism.

Home range. The area in which an animal normally lives, exclusive of migrations, emigrations, dispersal movements, or unusual erratic wanderings. Home range is defined and estimated operationally and without reference to particular types of behavior (such as territoriality) and without reference to the home ranges of other individuals. Use of the term home range implies neither the presence nor absence of territoriality or of other individuals. Although this is self-evident from the definition, the term has been misused in these ways.

Inclusive fitness. The sum of direct fitness and indirect fitness. See glossary entries for these terms and Chapter 4.

Indirect fitness. The component of inclusive fitness that arises from effects on nondescendent kin and other nondescendent co-gene-carriers. See discussion in Chapter 4. Misunderstanding of the meaning and importance of indirect fitness and other terms reflecting the dichotomy between direct and indirect persists. The following analysis of the history of usage in this area is intended to dispel confusion by revealing its historical origins.

The main purposes in drawing the distinction between direct and indirect components of inclusive fitness are (1) to enable the critical issues to be recognized and (2) to facilitate precise framing of questions for empirical studies in a way that would enable evaluation of the essential difference made by inclusive-fitness theory.

Most of us have wondered why this essential distinction was not made earlier, and many have falsely jumped to the conclusion that it was. Scrutiny of the seminal works on inclusive fitness may shed some light on why this distinction was in fact not made. Hamilton (1964:3) in a mathematical model spoke of "the fitness a° of an individual" as "the sum of his basic unit, the effect σa of his personal genotype and the total e° of effects on kin due to his neighbors, which will depend on their genotypes:

$$a^{\bullet} = 1 + \sigma a + e^{\circ}."$$

West Eberhard (1975), possibly trying to follow this logic, divided inclusive fitness into "classical fitness" and a "kinship component." The jump from Hamilton's equation above to West Eberhard's verbal model may have been a cause of confusion because, although the expressions were outwardly similar, the assumptions in the two discussions were very different. Hamilton, in order to develop a simple mathematical model, had explicitly assumed a semelparous (one-time only) mode of reproduction with nonoverlapping generations, thus excluding parental care of offspring and other effects on descendent kin. Effects on kin of the same generation were included, but effects on other generations were not. West Eberhard did not make these assumptions and intended her discussion to apply to all generations. Although in Hamilton's model it is correct to regard the kinship term as an extension of classical theory (because only effects on nondescendent kin are included), in West Eberhard's case it is not. The difference is in the assumptions. As explained in Chapter 4, the modern view is distinguished from the classical by the inclusion of effects on *nondescendent kin*, not all kin.

Although West Eberhard assigned the *production* of offspring correctly to classical fitness, she did not discuss social effects on them, which fall in the kinship component according to Hamilton and Maynard Smith. It appears that West Eberhard's kinship component refers to social effects on *all* kin. For example, she later referred to the second term as "helping offspring or other relatives," thus explicitly combining effects on descendent and nondescendent kin (West Eberhard, 1981:3). Thus, although her paper was a valuable contribution, it did not identify the essential differences between the modern and classical theories. Her designation of the nonclassical component simply as "kinship component" did not describe its essential feature and, in view of Maynard Smith's (1964) definition of kin selection, could only lead to confusion.

Kin selection has sometimes been regarded as a synonym of indirect selection, but this too is incorrect. Maynard Smith (1964) included effects on both descendent and nondescendent kin under the term kin selection (see quote in Chapter 4). Therefore, *all kin* are included under kin selection, whereas only nondescendent kin are included in indirect selection.

In short, neither Hamilton nor Maynard Smith provided a verbal term for the new component of inclusive fitness. West Eberhard seemed to provide one; but she did not correctly identify its defining features (effects on nondescendent kin). Her term "kinship component" suggested interchangeability with "kinship selection," which includes social effects on all kin.

The direct-indirect dichotomy may seem arbitrary when viewed solely in terms of coefficients of relatedness, a view expressed by Krebs (1980:206); but it is far from arbitrary. It corresponds to a *fundamental biological dichotomy* (Table 4.1): direct effects are attributed to individuals who physically transmit their genes through their *own gametes*, usually to the affected individual; indirect effects are not mediated in this way and do not require the production or transmission of gametes by the individual. See Brown (1980).

The practical and historical implications of the apparent failure to recognize at an earlier date the importance of the direct-indirect dichotomy are extensive.

No study can claim to have demonstrated that inclusive fitness theory extends our understanding of the behavior of a species simply by demonstrating a kinship effect on behavior. It is necessary that the effect be via nondescendent kin, and effects on descendent kin must be excluded.

Studies in which the "essential difference" was ignored are common. Nearly all the literature on sociality in ground squirrels is preoccupied with kin selection and nepotism (aid to kin) and never mentions the required distinction between descendent and nondescendent kin. The confusion that resulted is nowhere better revealed than in the debate between Shields (1980) and Sherman (1980). All of that could have been avoided by identifying the critical issue and framing the questions properly in the beginning. Failure to do so has marred an entire generation of studies of inclusive fitness in mammals.

Ornithologists are not immune either. A study on helping behavior in Bell Miners (Clarke, 1984) demonstrated a correlation of effort with kinship but failed to do so when parental effort was excluded. Of course, kinship is important for parents; but since that is not the issue it should not be allowed to influence the analysis. When the analysis was confined to indirect effects, the result was not significant (see Chapter 13). Unfortunately, these examples are used uncritically in recent books as examples of studies on "kin selection" (e.g., Krebs and Davies, 1981; Trivers, 1985).

The direct-indirect dichotomy was introduced orally at a symposium on natural selection and social behavior held in 1978 at Ann Arbor, Michigan (Brown and Brown, 1981). The first publication of the dichotomy was a result of a Dahlem Symposium held in 1980 (Brown, 1980).

Indirect selection. See *indirect fitness.*

Interdemic selection. Differential success of demes because of their genetic properties. Success is achieved by reduced extinction in some models (Wynne-Edwards, 1962) and by greater colonization in others (Wright, 1945). See *deme.*

Intrademic selection. Selection within a deme, thus excluding interdemic selection.

Joint nesting. A system of nesting in which two or more females of the same species normally lay their eggs in the same nest. In communally breeding species the joint-nesting females typically share the care of the young (thus excluding egg parasitism). The first use of this term appears to have been by Skutch (1959:302) in describing Groove-billed Anis. Later I used it quite independently (Brown, 1978a:126). Since then it has been widely used. Unlike my earlier use of this term, which implied monogamy (Brown, 1978a), the present usage is independent of the mating system.

Vehrencamp (1977, 1978) chose to refer to joint nesting as communal nesting. Her usage was confusing because the term communal breeding had been widely used for a different meaning altogether and still is (see *communal breeding*). Joint nesting is preferred because (1) it is more precisely descriptive, (2) it is in wide use, and (3) it is not easily confounded with a completely different concept (i.e., communal breeding as used in this book and by many authors since the 1960s). Known joint-nesting species are listed in Table 2.2.

Kin selection. Selection that is dependent on social effects on kin. See quotation from Maynard Smith (1964) in Chapter 4 and *indirect selection.*

Lineal kin. See *collateral kin* and Figure 4.1.

Mutualism. A relationship between two or more individuals or species in which both parties benefit. Normally benefits are measured by their net effects on direct fitness. I regard cooperation as a synonym of mutualism. A graphic definition of cooperation and mutualism is provided in Figure 4.3A. I had hoped that we could reserve mutualism for interspecific relationships, its traditional and established context; so I proposed earlier that the upper right quadrant in the fitness-space diagram of Figure 4.3A be designated cooperation (Brown, 1975a:197), since Hamilton (1964) had assigned no name to it. West Eberhard (1975), however, used mutualism for this purpose. Both usages are common and legitimate.

Nuclear family. A family consisting of the mother, father, and their offspring but not including grandparents, grandoffspring, or other kin.

Offspring rule. See Chapter 4.

Parental facilitation. Aid given by parents to their offspring that improves the ability of the latter to achieve breeding status. Parental facilitation is not extended parental care in the usual sense, since the normal types of parental care, such as feeding young, are not involved. This subject is developed in Chapter 7 and the concept is applied to dispersal in anis in Chapter 10.

Parental manipulation. "Parental manipulation of progeny refers to parents adjusting or manipulating their parental investment, particularly by reducing the reproduction (inclusive fitness) of certain progeny in the interests of increasing their own inclusive fitness via other offspring. . . . I use the term 'manipulation' here in the dictionary sense of 'adroit or skillful management; fraudulent or deceptive treatment'." Quotation from Alexander (1974). The idea of variance enhancement is very similar to parental manipulation, differing from it mainly by emphasizing variance and de-emphasizing "adroit or skillful management; fraudulent or deceptive treatment." These subjects are discussed further in Chapter 15.

Plural breeding. A social system in which two or more females (monogamous or not) breed in the same communal social unit. The term does not apply to colonial species unless communal units exist within the colony. See *singular breeding.*

Polyandry. A mating system characterized by a behavioral bond between one breeding female and two or more males breeding with her. See Table 9.1.

Polygynandry. A mating system characterized by behavioral bonds between two or more breeding females and two or more males breeding with them. See Table 9.1.

Prisoner's Dilemma. This is a symmetrical game between two players that is additionally defined by its payoff matrix. See Chapter 14.

Reciprocal. "Of the nature of, pertaining to, a return made for something; given, felt, shown, etc., in return; correspondent" (OED).

Reciprocation, Reciprocity. "The state or condition of being reciprocal; a state or relationship in which there is mutual action, influence, giving and taking, correspondence, etc., between two parties or things" (OED). As used by Trivers (1971), who introduced the words to sociobiology, an exchange of favors is implied in which (1) the donor gives up or risks little but benefits the recipient more, in terms of inclusive fitness, and (2) the donor monitors the recipient to penalize failure to return favors by cessation of donor aid-giving to the "cheater." In the best known recent paper on the subject (Axelrod and Hamilton, 1981) reciprocation was associated most prominently with the Tit-for-tat strategy. In both of these influential papers reciprocity is associated with (1) a mutualistic relationship between reciprocators, (2) some kind of score-keeping used as a hedge against nonreciprocators. There has been a tendency in recent years to invoke reciprocity as an evolutionary explanation for a wide variety of cases of suspected mutualism (e.g., Ligon, 1983). Because some such cases may not satisfy the two criteria above, I have chosen in this book to use the generic term mutualism (or cooperation) and to subdivide it according to the second criterion, giving us byproduct mutualism and score-keeping mutualism as explained in Chapter 14. There has been some debate on exactly what reciprocity really means (Waltz, 1981). The usage adopted here obviates confusion of this sort and has the additional merit of being more precisely descriptive.

Relatedness. See Box 4.1.

Risk-sensitive behavior. Behavior in which options are chosen with respect to the variance in the reward or punishment. The demonstration of risk sensitivity requires that the mean be ruled out as the cause of the choice. This is done by making the means of the options equal and the variances unequal or by coupling the desired and chosen variance with the undesired mean. See Chapters 9 and 14.

Score-keeping mutualism. A type of mutualism that depends upon repeated exchange of favors and in which each participant adjusts its behavior in response to partial or total nonreciprocation by the other. Two of a family of such strategies for the prisoner's dilemma are Tit-for-tat and Judge. Further discussion will be found in Chapter 14. See also *mutualism* and *reciprocity*.

Separate nesting. A system of parental care in plural-breeding birds in which each breeding female has her own nest, as opposed to joint nesting. Examples in Table 2.2.

Singular breeding. A communal social system in which no more than one female breeds in each social unit, as opposed to plural breeding. Examples are given in Table 2.2.

Social unit. Groups of two or more individuals that associate together and behave in some respects as a unit, especially in group territoriality.

Supernumerary. A member of a social unit in addition to the breeding pair (Armstrong, 1947; Rowley, 1965). See discussion under *helper*.

Territory. A fixed area from which rival intruders are excluded by some combination of advertisement, threat, and attack (Brown, 1975).

Tit-for-tat. "This strategy is simply one of cooperating on the first move and then doing whatever the other player did on the preceding move. Thus TIT FOR TAT is a strategy of cooperation based on reciprocity." From Axelrod and Hamilton (1981). The game is a prisoner's dilemma.

Variance enhancement. Increase in the variance of reproductive potential among members of a parent's brood. The implication is that more brood members are thereby brought to lie above or below the threshold for breeding, thus increasing the proportion of the brood becoming breeders or potential helpers respectively. See Figure 15.3.

Variance utilization. Utilization by parents of aid from those of their offspring that are unable to breed for reasons not imposed by the parents. The implication is that offspring vary in their reproductive potential, some being above the threshold for attempting breeding and some, below. See Figure 15.3.

Literature Cited

Ainley, D. G., R. E. LeResche, and W. J. L. Sladen. 1983. *Breeding Biology of the Adelie Penguin.* Berkeley: University of California Press.

Alexander, R. D. 1974. The evolution of social behavior. *Ann. Rev. Ecol, Syst.* 5: 325–383.

Alexander, R. D., and P. W., Sherman. 1977. Local mate competition and parental investment in social insects. *Science* 196: 494–500.

Allan, T. A. 1979. Parental behavior of a replacement male Dark-eyed Junco. *Auk* 96: 630–631.

Alvarez, H. 1975. The social system of the Green Jay in Colombia. *Living Bird* 14: 5–44.

Amadon, D. 1944. Results of the Archbold Expedition. No. 50. A preliminary life history study of the Florida Jay, *Cyanocitta c. coerulescens. American Museum Novitates* 1252: 1–22.

Andersson, M. 1980. Nomadism and site tenacity as alternative reproductive tactics in birds. *J. Anim. Ecol.* 49: 175–184.

Andersson, M. 1984. Brood parasitism within species. In: Barnard, C. J., ed. *Producers and Scroungers: Strategies of Exploitation*, pp. 195—228. London: Croom Helm.

Anonymous. 1976. *Reader's Digest Complete Book of Australian Birds.* Sydney: Reader's Digest Serv. Pty. Ltd.

Armstrong, E. A. 1947. *Bird Display and Behaviour.* London: Lindsay Drummond.

Ashmole, P. 1963. The regulation of numbers of tropical oceanic birds. *Ibis* 103b: 458–473.

Atwood, J. L. 1980a. Social interactions in the Santa Cruz Island Scrub Jay. *Condor* 82: 440–448.

Atwood, J. L. 1980b. Breeding biology of the Santa Cruz Island Scrub Jay. In: Power, D. M., ed. *The California Islands: Proceedings of a Multidisciplinary Symposium*, pp. 675–688. Santa Barabara, Calif.: Santa Barbara Mus. Nat. History.

Austad, S. N., and K. N. Rabenold. 1985. Reproductive enhancement by helpers and an experimental examination of its mechanism in the Biciolored Wren: a facultatively communal breeder. *Behav. Ecol. Sociobiol.* 17:19–27.

Axelrod, R., and W. D. Hamilton. 1981. The evolution of cooperation. *Science* 211: 1390–1396.

Baker, M. C., C. S. Belcher, L. C. Deutsch, G. L. Sherman, and D. B. Thompson. 1981. Foraging success in junco flocks and the effects of social hierarchy. *Anim. Behav.* 29: 137–142.

Balda, R. P., and J. H. Balda. 1978. The care of young Piñon Jays (*Gymnorhinus cyanocephalus*) and their integration into the flocks. *J. Ornithol.* 119: 146–171.

Balda, R. P., and G. C. Bateman. 1971. Flocking and annual cycle of the Pinyon Jay, *Gymnorhinus cyanocephalus. Condor* 73: 287–302.

Balda, R. P., and J. L. Brown. 1977. Observations on the behaviour of Hall's Babbler. *Emu* 77: 111–117.

Baldwin, M. 1974. Studies of the Apostle Bird at Inverell. Part I: General behaviour. *Sunbird* 5: 77–88.

Baldwin, M. 1975. Studies of the Apostle Bird at Inverell. Part II: Breeding behaviour. *Sunbird* 6: 1–7.

Baldwin. P. H., and W. F. Hunter. 1963. Nesting and nest visitors of the Vaux's Swift in Montana. *Auk* 80: 81–85.

Ball, G. 1982. Barn Swallow fledgling successfully elicits feeding at a non-parental nest. *Wilson Bull.* 94: 362–363.

Barkan, C. P. L., J. L. Craig, J. L. Brown, A. M. Stewart, and S. D. Strahl. 1981. Dominance in social units of communal Mexican Jays. *Amer. Zool.* 21: 948.

Barkan, C. P. L., J. L. Craig, S. D. Strahl, A. M. Stewart, and J. L. Brown. 1986. Social Dominance in communal Mexican Jays *Aphelocoma ultramarina. Anim. Behav.*: 34: 175–187.

Beason, R. C., and L. L. Trout. 1984. Cooperative breeding in the Bobolink. *Wilson Bull.* 96: 709–710.

Bednarz, J. C. 1985. The breeding system of the Harris' Hawk. *Meeting of American Ornithologists' Union* 103: Abstract #59.

Bekoff, M., T. J. Daniels, and J. L. Gittleman. 1984. Life history patterns and the comparative social ecology of carnivores. *Ann. Rev. Ecol. Syst.* 15: 191–232.

Bell, H. L. 1982a. Social organization and feeding of the rufous babbler *Pomatostomus isidori. Emu* 82: 7–11.

Bell, H. L. 1982b. Co-operative breeding by the White-browed Scrub wren *Sericornis frontalis. Emu* 82: 315–316.

Bell, H. L. 1983a. A bird community of lowland rainforest in New Guinea. *Emu* 82: 256–275.

Bell, H. L. 1983b. Resource partitioning between three syntopic thornbills *Acanthizidae: Acanthiza* Vigors and Horsfield. Ph.D. thesis, University of New England, *Armidale, N.S.W.*

Bell, H. L. 1984. A note on communal breeding and dispersal of young of the Hooded Robin *Petroica cucullata. Emu* 84(3): 243–244.

Bell. H. L. 1985. The social organization and foraging behaviour of three syntopic thornbills *Acanthiza* spp. In: Keast, J. A., H. F. Recher, H. A. Ford, and D. Saunders, eds. *Birds of Eucalypt Forests and Woodlands: Ecology, Behaviour and Conservation.* Sydney: Surrey-Beattie.

Benson, C. W. 1946. Notes on the birds of southern Abyssinia. *Ibis* 88: 444–461.

Berndt, R., and H. Sternberg. 1968. Terms, studies and experiments on the problems of bird dispersion. *Ibis* 110: 256–269.

Bertram, B. C. R. 1979. Ostriches recognize their own eggs and discard others. *Nature* 279: 233–234.

Bertram, B. C. R. 1980a. Vigilance and group size in ostriches. *Anim. Behav.* 28: 278–286.

Bertram, B. C. R. 1980b. Breeding system and strategies of ostriches. *Proc. Internat. Ornithol. Congr.* 17: 890–894.

Birkhead, M. E. 1981. The social behaviour of the Dunnock *Prunella modularis*. *Ibis* 123: 75–84.
Bock, C. E., and J. H. Bock. 1974. Geographical ecology of the acorn woodpecker: diversity versus abundance of acorns. *Amer. Nat.* 108: 694–698.
Bock, W. J., and J. Farrand, Jr. 1980. The number of species and genera of recent birds: a contribution to comparative systematics. *American Museum Novitates* 2703: 1–29.
Borowski, R. 1978. Social inhibition of maturation in natural populations of *Xiphophorus variatus* (Pisces: Poeciliidae). *Science* 201: 933–935.
Brackbill, H. 1944. Juvenile cardinal helping at a nest. *Wilson Bull.* 56: 50.
Brackbill, H. 1970. Tufted titmouse breeding behavior. *Auk* 87: 522–536.
Britton, P. L., and H. A. Britton. 1977. The nest and eggs of the Chestnut-fronted Shrike *Prionops scopifrons*. *Scopus* 1: 86.
Brockelman, W. Y. 1975. Competition, the fitness of offspring, and optimal clutch size. *Amer. Nat.* 109: 677–699.
Brooke, R. K. 1975. Cooperative breeding, duetting, allopreening and swimming in the black crake. *Ostrich* 46: 190–191.
Brooker, M. G. 1969. The nesting of the chestnut-breasted quail-thrush in southwestern Queensland. *Emu* 69: 47.
Brown, C. R. 1984. Laying eggs in a neighbor's nest: benefit and cost of colonial nesting in swallows. *Science* 224: 518–519.
Brown, J. L. 1963a. Social organization and behavior of the Mexican jay. *Condor* 65: 126–153.
Brown, J. L. 1963b. Aggressiveness, dominance and social organization in the Steller jay. *Condor* 65: 460–484.
Brown, J. L. 1964. The evolution of diversity in avian territorial systems. *Wilson Bull.* 76: 160–169.
Brown, J. L. 1966. Types of group selection. *Nature* 211: 870.
Brown, J. L. 1969a. Territorial behavior and population regulation in birds. *Wilson Bull.* 81: 293–329.
Brown, J. L. 1969b. The buffer effect and productivity in tit populations. *Amer. Nat.* 103: 347–354.
Brown, J. L. 1970. Cooperative breeding and altruistic behavior in the Mexican jay. *Aphelocoma ultramarina. Anim. Behav.* 18: 366–378.
Brown, J. L. 1972. Communal feeding of nestlings in the Mexican jay (*Aphelocoma ultramarina*): interflock comparisons. *Anim Behav.* 20: 395–402.
Brown, J. L. 1974. Alternate routes to sociality in jays—with a theory for the evolution of altruism and communal breeding. *Amer. Zool.* 14: 63–80.
Brown, J. L. 1975a. *The Evolution of Behavior*. New York: Norton.
Brown, J. L. 1975b. Helpers among Arabian babblers, *Turdoides squamiceps*. *Ibis* 117: 243–244.
Brown, J. L. 1978a. Avian communal breeding systems. *Ann. Rev. Ecol. Syst.* 9: 123–155.
Brown, J. L. 1978b. Avian heirs of territory. *BioScience* 28: 750–752.
Brown, J. L. 1979. Another interpretation of commmunal breeding in green woodhoopoes. *Nature* 280: 174.

Brown, J. L. 1980. Fitness in complex avian social systems. In: Markl, H., ed. *Evolution of Social Behavior: Hypotheses and Empirical Tests*, pp. 115–128. Weinheim: Verlag Chemie.

Brown, J. L. 1981. On socio-ornithology. *Auk* 98: 417–418.

Brown, J. L. 1982a. Optimal group size in territorial animals. *J. Theor. Biol.* 95: 793–810.

Brown, J. L. 1982b. Behavioral ecology for sophomores. *Evolution* 36: 421–422.

Brown, J. L. 1983a. Cooperation—a biologist's dilemma. In: Rosenblatt, J. S., ed. *Advances in Behavior*. New York: Academic Press.

Brown, J. L. 1983b. Communal harvesting of a transient food resource in the Mexican Jay. *Wilson Bull.* 95: 286–287.

Brown, J. L. 1985. The evolution of helping behavior—an ontogenetic and comparative perspective. In: Gollin, E. S., ed. *The Evolution of Adaptive Skills*: Comparative and Ontogenetic Approaches, pp. 137–171. Hillsdale, N.J.: L. Erlbaum Assoc.

Brown, J. L. 1986. Cooperative breeding and the regulation of numbers. *Proc. Internat. Ornithol. Congr.* 18: 774–782.

Brown, J. L., and R. P. Balda. 1977. The relationship of habitat quality to group size in Hall's babbler (*Pomatostomus halli*). *Condor* 79: 312–320.

Brown, J. L., and E. R. Brown. 1980. Reciprocal aid-giving in a communal bird. *Z. Tierpsychol.* 53: 313–324.

Brown, J. L., and E. R. Brown. 1981a. Extended family system in a communal bird. *Science* 211: 959–960.

Brown, J. L., and E. R. Brown. 1981b. Kin selection and individual selection in babblers. In: Alexander, R. D., and D. W. Tinkle, eds. *Natural Selection and Social Behavior: Recent Research and New Theory* pp. 244–256. New York: Chiron Press.

Brown, J. L., and E. R. Brown. 1984. Parental facilitation: parent-offspring relations in communally breeding birds. *Behav. Ecol. Sociobiol.* 14: 203–209.

Brown, J. L., and E. R. Brown. 1985. Ecological correlates of group size in a communally breeding bird. *Condor* 87: 309–315.

Brown J. L., E. R. Brown, and S. D. Brown. 1982a. Morphological variation in a population of grey-crowned babblers: correlations with variables affecting social behavior. *Behav. Ecol. Sociobiol.* 10: 281–287.

Brown, J. L., E. R. Brown, S. D. Brown, and D. D. Dow. 1982b. Helpers: effects of experimental removal on reproductive success. *Science* 215: 421–422.

Brown, J. L., D. D. Dow, E. R. Brown, and S. D. Brown. 1978. Effects of helpers on feeding of nestlings in the grey-crowned babbler (*Pomatostomus temporalis*). *Behav. Ecol. Sociobiol.* 4: 43–59.

Brown, J. L., D. D. Dow, E. R. Brown, and S. D. Brown. 1983. Socioecology of the grey-crowned babbler: population structure, unit size, and vegetation correlates. *Behav. Ecol. Sociobiol.* 13: 115–124.

Brown, J. L., and G. H. Orians. 1970. Spacing patterns in mobile animals. *Ann. Rev. Ecol. Syst.* 1: 239–262.

Brown, J. L., and S. L. Pimm. 1985. The origin of helping: the role of variability in reproductive potential. *J. Theor. Biol.* 112: 465–477.

Brown, R. J., and M. N. Brown. 1980. Co-operative breeding in robins of the genus *Eopsaltria. Emu* 80: 89.

Brown, R. J., and M. N. Brown. 1982. Learning behaviour at the nest of the co-operatively breeding Yellow-rumped Thornbill *Acanthiza chrysorrhoa. Emu* 82: 111–112.

Buckley, F. G., and P. A. Buckley. 1974. Comparative feeding ecology of wintering adult and juvenile Royal Terns (Aves: Laridae, Sterninae). *Ecology* 55: 1053–1063.

Burger, J. 1981. Feeding competition between Laughing gulls and Herring gulls at a sanitary landfill. *Condor* 83: 328–335.

Burkitt, J. P. 1924. A study of the robin by means of marked birds. *Brit. Birds* 17: 294–303.

Butler, R. W., N. A. M Verbeek, and H. Richardson. 1984. The breeding biology of the Northwestern Crow. *Wilson Bull.* 96: 408–418.

Butts, W. K. 1930. A study of the chickadee and white-breasted nuthatch by means of marked individuals. Part I: Methods of marking birds. *Bird-Banding* 1: 149–168.

Bygott, J. D., B. C. R. Bertram, and J. P. Hanby. 1979. Male lions in large coalitions gain reproductive advantages. *Nature* 282: 839–841.

Caraco, T. 1979a. Time budgeting and group size: a theory. *Ecology* 60: 611–617.

Caraco, T. 1979b. Time budgeting and group size: a test of theory. *Ecology* 60: 618–627.

Caraco, T. 1979c. Ecological response of animal group size frequencies. In: Ord, J. K., G. P. Patil, and C. Taillie, eds. *Statistical Distributions in Ecological Work*, pp. 371–386. Burtonsville, Md.: Intern. Coop. Publ. House.

Caraco, T., and J. L. Brown. 1986. A game between communal breeders: when is food-sharing stable? *J. Theor. Biol.* 118: 379–393.

Caraco, T., S. Martindale, and H. R. Pulliam. 1980a. Avian flocking in the presence of a predator. *Nature* 285: 400–401.

Caraco, T., S. Martindale, and H. R. Pulliam. 1980b. Avian time budgets and distance to cover. *Auk* 97: 872–875.

Caraco, T., S. Martindale, and T. S. Whittam. 1980c. An empirical demonstration of risk-sensitive foraging preferences. *Anim. Behav.* 28: 820–830.

Carpenter, F. L., and R. E. MacMillen. 1976. Threshold model of feeding territoriality and test with a Hawaiian honeycreeper. *Science* 194: 639–642.

Carpenter, F. L., D. C. Paton, and M. A. Hixon. 1983. Weight gain and adjustment of feeding territory size in migrant hummingbirds. *Proc. Natl. Acad. Sci. USA* 80: 7259–7263.

Carrick, R. 1963. Ecological significance of territory in the Australian magpie, *Gymnorhina tibicen. Proc. Internat. Ornithol. Congr.* 13: 740–753.

Carrick. R. 1972. Population ecology of the Australian black-backed magpie, royal penguin, and silver gull. *U. S. Dept. Inter. Wildl. Res. Rep.* 2: 41–99.

Caughley, G. 1977. *Analysis of Vertebrate Populations.* Chichester: John Wiley and Sons.

Chamberlin, T. C. 1897. Studies for students. The method of multiple working hypotheses. *J. Geol.* 5: 837–848.

Chapman, G., and I. Rowley. 1978. Cooperative breeding by red-winged wrens (*Malurus elegans*). *Western Australian Naturalist* 14: 74.

Charnov. E. L. 1981. Kin selection and helpers at the nest: effects of paternity and biparental care. *Anim. Behav.* 29: 631–632.

Charnov, E. L., and J. R. Krebs. 1975. The evolution of alarm calls: altruism or manipulation?. *Amer. Nat.* 109: 107–112.

Clarke, M. F. 1984. Co-operative breeding by the Australian Bell Miner *Manorina melanophrys* Latham: a test of kin selection theory. *Behav. Ecol. Sociobiol.* 14: 137–146.

Clunie, F. 1973. Nest helpers at a White-breasted Woodswallow nest. *Notornis* 20: 378–380.

Clunie, F. 1976. Behaviour and nesting of Fijian White-breasted Woodswallows. *Notornis* 23: 61–75.

Cole, L. C. 1954. The population consequences of life history phenomena, *Q. Rev. Biol.* 29: 103–137.

Collias, E. C., and N. E. Collias. 1980b. Individual and sex differences in behaviour of the sociable weaver *Philetairus socius*. In: Johnson, D. N., ed. *Proc. Pan-African Ornithol. Congr.* 1976, 4: 243–251.

Collias, N. E., and E. C. Collias. 1964. Evolution of nest-building in the weaver-birds (Ploceidae). *Univ. of Calif. Publ. Zool.* 73: 1–239.

Collias, N. E., and E. C. Collias. 1977. Weaverbird nest aggression and evolution of the compound nest. *Auk* 94: 50–64.

Collias, N. E., and E. C. Collias. 1978a. Cooperative breeding in the White-browed Sparrow Weaver. *Auk* 95: 472–484.

Collias, N. E., and E. C. Collias. 1978b. Survival and intercolony movement of the White-browed Sparrow Weaver (*Plocepasser mahali*) over a two-year period. *Scopus* 3: 75–76.

Collias, N. E., and E. C. Collias. 1978c. Group territory, dominance, co-operative breeding in birds, and a new factor. *Anim. Behav.* 26: 121–122.

Collias, N. E., and E. C. Collias. 1980a. Behavior of the Grey-capped social weaver (*Pseudonigrita arnaudi*) in Kenya. *Auk* 97: 213–226.

Colquhoun, M. K. 1942. Notes on the social behavior of blue tits. *Brit. Birds* 35: 234–240.

Cooke, F., C. D. MacInnes, and J. P. Prevett. 1975. Gene flow between breeding populations of lesser snow geese. *Auk* 92: 493–510.

Counsilman, J. J. 1977a. Groups of the grey-crowned babbler in central southern Queensland. *Babbler* 1: 14–22.

Counsilman, J. J. 1977b. A comparison of two populations of the grey-crowned babbler (part 1). *Bird Behav.* 1: 43–82.

Counsilman, J. J. 1979. Notes on the breeding biology of the grey-crowned babbler. *Bird Behav.* 1(3): 114–124.

Counsilman, J. J. 1980. A comparison of two populations of the grey-crowned babbler (part 2). *Bird Behav.* 2: 1–109.

Counsilman, J. J., and B. King. 1977. Ageing and sexing the grey-crowned babbler (*Pomatostomus temporalis*). *Babbler* 1: 23–41.

Craig, A. 1983. Co-operative breeding in two African starlings, Sturnidae. *Ibis* 125: 114–115.
Craig, J. L. 1976. An interterritorial hierarchy: an advantage for a subordinate in a communal territory. *Z. Tierpsychol.* 42: 200–205.
Craig, J. L. 1977. The behaviour of a communal gallinule, the pukeko (*Porphyrio porphyrio melanotus*). *New Zeal. J. Zool.* 4: 413–433.
Craig, J. L. 1979. Habitat variation in the social organization of a communal gallinule, the Pukeko. *Porphyrio porphyrio melanotus. Behav. Ecol. Sociobiol.* 5: 331–358.
Craig, J. L. 1980a. Breeding success of a communal gallinule. *Behav. Ecol. Sociobiol.* 6: 289–295.
Craig, J. L. 1980b. Pair and group breeding behaviour of a communal Gallinule, Pukeko, *Porphyrio p. melanotus. Anim. Behav.* 28: 593–603.
Craig, J. L. 1984. Are communal pukeko caught in the prisoner's dilemma?. *Behav. Ecol. Sociobiol.* 14: 147–150.
Craig, J. L., and I. G. Jamieson. 1985. The relationship between presumed gamete contribution and parental investment in a communally breeding bird. *Behav. Ecol. Sociobiol.* 17: 207–211.
Craig. J. L., A. M. Stewart, and J. L. Brown. 1982. Subordinates must wait. *Z. Tierpsychol.* 60: 275–280.
Craig, R. 1983. Subfertility and the evolution of eusociality by kin selection. *J. Theor. Biol.* 100: 379–397.
Crome, F. H. J., and H. E. Brown. 1979. Notes on social organization and breeding of the Orange-footed Scrub-fowl *Megapodius reinwardt. Emu* 79: 111–119.
Crook, J. H. 1958, Études sur le comportement social de *Bubalornis a. albirostris* (Vieillot). *Alauda* 26: 161–195.
Crook, J. H. 1960. Nest form and construction in certain West African weaver-birds, *Ibis* 102: 1–25.
Crook, J. H. 1963. A comparative analysis of nest structure in the weaver birds. *Ibis* 105 : 238–262.
Crook, J. H. 1964. The evolution of social organisation and visual communication in the weaver birds (Ploceinae). *Behaviour Supplement* X: 1–178.
Crook, J. H. 1965. The adaptive significance of avian social organizations. *Symp. Zool. Soc. London* 14: 181–218.
Crook, J. R., and W. M. Shields. 1986. Barn swallow nest attendance: helping or harrassment? *Anim. Behav.*: in press.
Crossin, R. S. 1967. The breeding biology of the tufted jay. *Proc. West. Found. Vert. Zool.* 1: 265–297.
Crow, J. F., and M. Kimura. 1970. *An Introduction to Population Genetics Theory.* New York: Harper and Row.
Cullen, J. M. 1957. Plumage, age, and mortality in the Arctic tern. *Bird Study* 4: 197–207.
Curio, E. 1983. Why do young birds reproduce less well?. *Ibis* 125: 400–403.
Daly, M., and M. Wilson. 1978. *Sex, Evolution and Behavior.* North Scituate, Mass.: Duxbury Press.

Davies, N. B. 1978. Ecological questions about territorial behavior. In: Krebs, J. R., and N. B. Davies, eds. *Behavioural Ecology: An Evolutionary Approach*, pp. 317–350. Sunderland, Mass.: Sinauer.

Davies, N. B. 1983. Polyandry, cloaca-pecking and sperm competition in dunnocks. *Nature* 302: 334–336.

Davies, N. B. 1985. Cooperation and conflict among Dunnocks, *Prunella modularis*, in a variable mating system. *Anim. Behav.* 33: 628–648.

Davies, N. B. 1986. Reproductive success of Dunnocks, *Prunella modularis*, in a variable mating system. I. Factors influencing provisioning rate, nestling weight and fledging success. *J. Anim. Ecol.* 55: 123–138.

Davies, N. B., and A. I. Houston. 1981. Owners and satellites: The economics of territory defense in the pied wagtail, *Motacilla alba*. *J. Anim. Ecol.* 50: 157–180.

Davies, N. B., and A. I. Houston. 1986. Reproductive success of Dunnocks, *Prunella modularis*, in a variable mating system. II. Conflicts of interest among breeding adults. *J. Anim. Ecol.* 55: 139–154.

Davies, N. B., and A. Lundberg. 1984. Food distribution and a variable mating system in the Dunnock, *Prunella modularis*. *J. Anim. Ecol.* 53: 895–912.

Davies, S. J. J. F., and H. J. Frith. 1964. Some comments on the taxonomic position of the Magpie Goose *Anseranas semipalmata*. *Emu* 63: 265–272.

Davis, D. E. 1940a. Social nesting habits of the smooth-billed ani. *Auk* 57: 179–218.

Davis, D. E. 1940b. Social nesting habits of *Guira guira*. *Auk* 57: 472–484.

Davis, D. E. 1941. Social nesting habits of *Crotophaga major*. *Auk* 58: 179–183.

Davis, D. E. 1942. The phylogeny of social nesting in the Crotophaginae. *Q. Rev. Biol.* 17:115–134.

Davis, M. F. 1978. A helper at a tufted titmouse nest. *Auk* 95: 767.

Dawson, J. W., and R. W. Mannan. 1985. Composition and behavior of breeding groups of Harris' Hawk in Arizona. *Meeting of American Ornithologists' Union* 103: Abstract #60.

Decoux, J. P. 1982. Les particularités demographiques et sociecologiques du Coliou strié dans le Nord-est du Gabon. *Rev. Ecol.* (*Terre. Vie*) 36: 37–78.

Dexter, R. W. 1952. Extra-parental cooperation in the nesting of Chimney Swifts. *Wilson Bull.* 64: 133–139.

Dexter, R. W. 1981. Nesting success of chimney swifts related to age and the number of adults at the nest, and the subsequent fate of the visitors. *J. Field Ornithol.* 52: 228–232.

Diamond, A. W. 1980. Seasonality, population structure and breeding ecology of the Seychelles Brush Warbler *Acrocephalus sechellensis*. In Johnson, D. N., ed. *Proc. Pan-African Ornithol. Congr.* 1976: 4: 253–266.

Douthwaite, R. J. 1973. Pied kingfisher *Ceryle rudis* populations. *Ostrich* 44: 89–94.

Douthwaite, R. J. 1978. Breeding biology of the pied kingfisher, *Ceryle rudis*, on Lake Victoria. *J. East African Nat. Hist. Soc.* 31: 1–12.

Dow, D. D. 1970. Communal behaviour of nesting noisy miners. *Emu* 70: 131–134.

Dow, D. D. 1971. Australia's Noisy Miner. *Fauna* 3: 24–31.

Dow, D. D. 1973. Sex ratios and oral flange characteristics of selected genera of Australian honeyeaters in museum collections. *Emu* 73: 41–50.

Dow, D. D. 1975. Displays of the honeyeater, *Manorina melanocephala*. *Z. Tierpsychol.* 38: 70–96.

Dow, D. D. 1977. Indiscriminate interspecific aggression leading to almost sole occupancy of space by a single species of bird. *Emu* 77: 115–121.

Dow, D. D. 1978a. Reproductive behavior of the Noisy Miner, a communally breeding honeyeater. *Living Bird* 16: 163–185.

Dow, D. D. 1978b. Breeding biology and development of the young of *Manorina melanocephala*, a communally breeding honeyeater. *Emu* 78: 207–222.

Dow, D. D. 1979a. The influence of nests on the social behaviour of males in *Manorina melanocephala*, a communally breeding honeyeater. *Emu* 79: 71–83.

Dow, D. D. 1979b. Agonistic and spacing behaviour of the noisy miner *Manorina melanocephala*, a communally breeding honeyeater. *Ibis* 121: 423–436.

Dow, D. D. 1980a. Communally breeding Australian birds with an analysis of distributional and environmental factors. *Emu* 80: 121–140.

Dow, D. D. 1980b. Systems and strategies of communal breeding in Australian birds. *Proc. Internat. Ornithol. Congr.* 17: 875–881.

Dow, D. D. 1982. Communal breeding. *Roy. Austr. Ornithol. Union Newsletter* 54: 6–7.

Dow, D. D., and B. J. Gill. 1984. Environmental versus social factors as determinants of growth in nestlings of a communally breeding bird. *Oecologia* 63: 370–375.

Dow, D. D., and B. R. King. 1984. Communal building of brood and roost nests by the Grey-crowned Babbler *Pomatostomus temporalis*. *Emu* 84: 193–199.

Duebert, H. F. 1968. Two female mallards incubating on one nest. *Wilson Bull.* 80: 102.

Dunn, E. K. 1972. Effect of age on the fishing ability of Sandwich Terns *Sterna sandvicensis*. *Ibis* 114: 360–366.

Dyer, M., and C. H. Fry. 1980. The origin and role of helpers in bee-eaters. *Proc. Internat. Ornithol. Congr.* 17: 862–868.

Dyrcz, A. 1977. Nest-helpers in the Alpine Accentor *Prunella collaris*. *Ibis* 119: 215.

Earle, R. A. 1983. Aspects of the breeding biology of the Whitebrowed Sparrow-weaver *Plocepasser mahali* (Aves: Ploceidae). *Navorsinge van die Nasionale Museum* 4: 177–191.

Eisenmann, E. 1946. Acorn storing by *Balanosphyra formicivora* in Panama. *Auk* 63: 250.

Ellis, D. H., and W. H. Whaley. 1979. Two winter breeding records for the Harris' Hawk. *Auk* 96: 413.

Emlen, S. T. 1978. Cooperative breeding. In: Krebs, J. R., and N. B. Davies, eds. *Behavioural Ecology: An Evolutionary Approach.* Sunderland, Mass.: Sinauer.

Emlen, S. T. 1981. Altruism, kinship, and reciprocity in the white-fronted bee-eater. In: Alexander, R. D., and D. W. Tinkle, eds. *Natural Selection and Social Behavior: Recent Research and New Theory*, pp. 217–230. New York: Chiron Press.

Emlen, S. T. 1982a. The evolution of helping. I. An ecological constraints model. *Amer. Nat.* 119: 29–39.

Emlen, S. T. 1982b. The evolution of helping. II. The role of behavioral conflict. *Amer. Nat.* 119: 40–53.

Emlen, S. T. 1983. Erratum. *Amer. Nat.* 121: 755.

Emlen, S. T. 1984. Cooperative breeding in birds and mammals. In: Krebs, J. R., and N. B. Davies, eds. *Behavioural Ecology: An Evolutionary Approach*, 2nd ed., pp. 305–339. Sunderland, Mass.: Sinauer.

Emlen, S. T., and N. J. Demong. 1980. Bee-eaters: an alternative route to co-operative breeding?. *Proc. Internat. Ornithol. Congr.* 17: 895–901.

Emlen, S. T., J. M. Emlen, and S. A. Levin. 1986. Sex-ratio selection in species with helpers-at-the-nest. *Amer. Nat.* 127: 1–8.

Emlen, S. T., and L. W. Oring. 1977. Ecology, sexual selection, and the evolution of mating systems. *Science* 197: 215–223.

Emlen, S. T., and S. L. Vehrencamp. 1983. Cooperative breeding strategies among birds. In: Brush, A. H., and G. A. Clark, eds. *Perspectives in Ornithology*, pp. 93–120. Cambridge: Cambridge University Press.

Emlen, S. T., and S. L. Vehrencamp. 1985. Cooperative breeding strategies among birds. In: Holldobler, B., and M. Lindauer, eds. *Experimental Behavioral Ecology*, pp. 359–374. Stuttgart–New York: G. Fischer Verlag.

Emlen, S. T., and P. H. Wrege. 1986. Forced copulation and intraspecific parasitism: two costs of social living in the White-fronted Bee-eater. *Z. Tierpsychol.*: 71: 2–29.

Ervin, S. 1979. Bushtit helpers: accident or altruism. *Bird Behav.* 1(3): 93–97.

Faaborg, J., T. De Vries, C. B. Patterson, and C. R. Griffin. 1980. Preliminary observations on the occurrence and evolution of polyandry in the Galapagos hawk (*Buteo galapagoensis*). *Auk* 97: 581–590.

Faaborg, J., and C. B. Patterson. 1981. The characteristics and occurrence of co-operative polyandry. *Ibis* 123: 477–484.

Filewood, L. W. C., K. Hough, I. C. Morris, and D. E. Peters. 1978. Helpers at the nest of the Rainbow Bee-eater. *Emu* 78: 43–44.

Fjeldsa, J. 1973. Territory and the regulation of population density and recruitment in the Horned Grebe (*Podiceps auritus auritus* Boje 1922). *Vidensk. Medd. Dansk Naturhist. Foren.* 136: 117–189.

Ford, J. 1963. Breeding behavior of the Yellow-tailed Thornbill in south Western Australia. *Emu* 63: 185–200.

Fraga, R. M. 1972. Cooperative breeding and a case of successive polyandry in the Bay-winged Cowbird. *Auk* 89: 447–449.

Fretwell, S. D., and H. L. Lucas. 1970. On territorial behaviour and other factors influencing habitat distribution in birds. *Acta Biotheor.* 19: 16–36.

Frith, H. J., and S. J. J. F. Davies. 1961. Ecology of the Magpie Goose, *Anseranas semipalmata* Latham (Anatidae). *CSIRO Wildl. Res.* 6: 91–141.

Fry, C. H. 1972. The social organisation of bee-eaters (Meropidae) and co-operative breeding in hot-climate birds. *Ibis* 114: 1–14.

Fry, C. H. 1977. The evolutionary significance of co-operative breeding in birds. In: Stonehouse, B., and C. Perrins, eds. *Evolutionary Ecology*, pp. 127–135. Baltimore: University Park Press.

Fry, C. H. 1980. Survival and longevity among tropical land-birds. In: Johnson, D. N., ed. *Proc. Pan-African Ornithol. Congr. 1976* 4: 333–343.

Garnett, S. T. 1980. The social organization of the Dusky Moorhen, *Gallinula tenebrosa* (Aves: Rallidae). *Aust. Wildl. Res.* 7: 103–112.

Gaston, A. J. 1973. The ecology and behaviour of the long-tailed tit. *Ibis.* 115: 330–351.

Gaston, A. J. 1976. Brood parasitism by the pied crested cuckoo (*Clamator jacobinus*). *J. Anim. Ecol.* 45: 331–345.

Gaston, A. J. 1977a. Social behaviour within groups of jungle babblers (*Turdoides striatus*). *Anim. Behav.* 25: 828–848.

Gaston, A. J. 1977b. Notes on the Striated Babbler *Turdoides earlei* (Blyth) near Delhi. *J. Bombay Nat. Hist. Soc.* 75: 219–220.

Gaston, A. J. 1978a. Ecology of the common babbler *Turdoides caudatus. Ibis* 120: 415–432.

Gaston, A. J. 1978b. Demography of the jungle babbler *Turdoides striatus. J. Anim. Ecol.* 47: 845–870.

Gaston, A. J. 1978c. The evolution of group territorial behavior and cooperative breeding. *Amer. Nat.* 112: 1091–1100.

Gaston, A. J. 1978d. Social behaviour of the Yellow-eyed Babbler *Chrysomma sinensis. Ibis* 120: 361–364.

Gauthreaux, S. A., Jr. 1978. The ecological significance of behavioral dominance. *Perspectives in Ethology* 3: 17–54.

Gavrilov, E. I. 1963. The biology of the Eastern Spanish Sparrow *Passer hispaniolensis transcaspicus* Tschusi, in Kazakhstan. *J. Bombay Nat. Hist. Soc.* 60: 301–317.

Geist, V. 1977. A comparison of social adaptations in relation to ecology in gallinaceous bird and ungulate societies. *Ann. Rev. Ecol. Syst.* 8: 193–207.

Gill, F. B., and L. L. Wolf. 1975. Economics of feeding territoriality in the Golden-winged Sunbird. *Ecology* 56: 333–345.

Gittleman, J. L. 1985. Functions of communal care in mammals. In: Greenwood, P. J., and M. Slatkin, eds. *Evolution–Essays in Honour of John Maynard Smith*, pp. 187–205. Cambridge: Cambridge University Press.

Glas, P. 1960. Factors governing density in the chaffinch (*Fringilla coelebs*) in different types of wood. *Arch. Neerl. Zool.* 13: 466–472.

Gould, S. J. 1977. *Ontogeny and Phylogeny.* Cambridge, Mass. Harvard University Press.

Gowaty, P. A. 1981. An extension of the Orians-Verner-Willson model to account for mating systems besides polygyny. *Amer. Nat.* 118: 851–859.

Gowaty, P. A., and A. A. Karlin. 1984. Multiple maternity and paternity in apparently monogamous eastern bluebirds (*Sialia sialis*). *Behav. Ecol. Sociobiol.* 15: 91–95.

Gowaty, P. A., and M. R. Lennartz. 1985. Sex ratios of nestling and fledgling red-cockaded woodpeckers (*Picoides borealis*) favor males. *Amer. Nat.* 126: 347–353.

Grafen, A. 1982. How not to measure inclusive fitness. *Nature* 298: 425–426.

Grafen, A. 1984. Natural selection, kin selection and group selection. In: Krebs, J. R., and N. B. Davies, eds. *Behavioural Ecology: An Evolutionary Approach*, 2nd ed., pp. 62–84. Sunderland, Mass.: Sinauer.

Grant, P. R., and N. Grant. 1979. Breeding and feeding of Galapagos mockingbirds, *Nesomimus parvulus*. *Auk* 96: 723–736.

Graves, J. A., and A. Whiten. 1980. Adoption of strange chicks by herring gulls. *Z. Tierpsychol.* 54: 267–278.

Greenwood, P. J. 1980. Mating systems, philopatry and dispersal in birds and mammals. *Anim. Behav.* 28: 1140–1162.

Greenwood, P. J., and P. H. Harvey. 1982. The natal and breeding dispersal of birds. *Ann. Rev. Ecol. Syst.* 13: 1–21.

Greenwood, P. J., et al., eds. 1985. *Evolution: Essays in Honour of John Maynard Smith*. Cambridge: Cambridge University Press.

Greig-Smith, P. W. 1976. Observations on the social behaviour of helmet-shrikes. *Bull. Nigerian Ornithol. Soc.* 12: 25–30.

Greig-Smith, P. W. 1979. Observations of nesting and group behaviour of Seychelles White-eyes *Zosterops modesta*. *Ibis* 121: 344–348.

Griffin, C. R. 1976. A preliminary comparison of Texas and Arizona Harris' Hawk (*Parabuteo unicinctus*) populations. *Raptor Research* 10: 50–54.

Grimes, L. G. 1976a. The occurrence of cooperative breeding behaviour in African birds. *Ostrich* 47: 1–15.

Grimes, L. G. 1976b. Co-operative breeding in African birds. In: Frith, H. J., and J. H. Calaby, eds. *Proc. 16th Internat. Ornithol. Congr.*, pp. 667–684. Canberra: Australian Acad. Sci.

Grimes, L. G. 1980. Observations of group behaviour and breeding biology of the yellow-billed shrike *Corvinella corvina*. *Ibis* 122: 166–192.

Gross, A. O. 1949. Nesting of the Mexican jay in the Santa Rita Mountains, Arizona. *Condor* 51: 241–249.

Gutierrez, R. J., and W. D. Koenig. 1978. Characteristics of storage trees used by acorn woodpeckers in two California woodlands. *J. Forestry* 76: 162–164.

Hamilton, W. D. 1963. The evolution of altruistic behaviour. *Amer. Nat.* 97: 354–356.

Hamilton, W. D. 1964. The genetical evolution of social behaviour. I. and II. *J. Theor. Biol.* 7: 1–52.

Hamilton, W. D. 1970. Selfish and spiteful behaviour in an evolutionary model. *Nature* 228: 1218–1220.

Hamilton, W. D. 1971. Geometry for the selfish herd. *J. Theor. Biol.* 31: 295–311.

Hamilton, W. D. 1975. Innate social aptitudes in man: an approach from evolutionary genetics. In: Fox, R., ed. *Biosocial Anthropology*, pp. 133–155. New York: Wiley.

Hamilton, W. J., III. 1959. Aggressive behavior in migrant pectoral sandpipers. *Condor* 61: 161–179.

Hannon, S. J., R. L. Mumme, W. D. Koenig, and F. A. Pitelka. 1985. Replacement of breeders and within-group conflict in the cooperatively breeding acorn woodpecker. *Behav. Ecol. Sociobiol.* 4: 303–312.

Hardy, J. W. 1961. Studies in behavior and phylogeny of certain New World jays (Garrulinae). *Univ. Kansas Sci. Bull.* 42: 13–149.
Hardy, J. W. 1973. Age and sex differences in the black-and-blue jays of Middle America. *Bird-Banding* 44: 81–90.
Hardy, J. W. 1974. Behavior and its evolution in neotropical jays (Cissilopha). *Bird-Banding* 45: 253–268.
Hardy, J. W. 1976. Comparative breeding behavior and ecology of the bushy-crested and Nelson San Blas jays. *Wilson Bull.* 88: 96–120.
Hardy, J. W., T. A. Webber, and R. F. Raitt. 1981. Communal social biology of the southern San Blas jay. *Bull. Florida State Mus.* 26: 203–263.
Harper, D. G. C. 1985. Interactions between adult robins and chicks belonging to other pairs. *Anim. Behav.* 33: 876–884.
Harris, J. E. K., and O. M. G. Newman. 1974. Communal nesting behaviour in the genus *Sericornis*. *Bull. Tasmanian Field. Nat. Club* 37: 4–5.
Harrison, C. J. O. 1969. Helpers at the nest in Australian passerine birds. *Emu* 69: 30–40.
Hatch, J. J. 1966. Collective territories in Galapagos Mockingbirds, with notes on other behavior. *Wilson Bull.* 78: 198–207.
Hegner, R. E., S. T. Emlen, and N. J. Demong. 1982. Spatial organization of the white-fronted bee-eater. *Nature* 298: 264–266.
Heinrich, B. 1979. *Bumblebee Economics.* Cambridge, Mass.: Harvard University Press.
Holmes, W. G., and P. W. Sherman. 1983. Kin recognition in animals. *Amer. Sci.* 71: 46–55.
Hosono, T. 1983. [A study of the life history of the Blue Magpie. II. Breeding helpers and nest-parasitism by cuckoos]. *J. Yamashina Inst. Ornithol.* 15: 63–71.
Houston, A. I., and N. B. Davies. 1985. The evolution of cooperation and life history in the Dunnock, *Prunella modularis*. In: Sibly, R. M., and R. H. Smith, eds. *Behavioural Ecology: The Ecological Consequences of Adaptive Behaviour*, pp. 471–487. Oxford: Blackwell.
Howe, R. W. and R. A. Noske. 1980. Cooperative feeding of fledglings by Crested Shrike-tits. *Emu* 80: 40.
Huels, T. R. 1981. Cooperative breeding in the golden-breasted starling *Cosmopsarus regius*. *Ibis* 123: 539–542.
Hunt, J. H. 1980. *Selected Readings in Sociobiology.* New York: McGraw-Hill.
Hunter, L. A. 1985. The effects of helpers in cooperatively breeding purple gallinules. *Behav. Ecol. Sociobiol.* 18: 147–153.
Hutchinson, G. E., and R. H. MacArthur. 1959. Appendix: On the theoretical significance of aggressive neglect in interspecific competition. *Amer. Nat.* 93: 133–134.
Immelmann, K. 1960. Behavioural observations on several species of Western Australian birds. *Emu* 60: 237–244.
Immelmann, K. 1961. Beiträge zur Biologie und Ethologie australischer Honigfresser (Meliphagidae). *J. Ornithol.* 102: 164–207.

Immelmann, K. 1966. Beobachtungen an Schwalbenstaren. *J. Ornithol.* 107: 37–69.

Jarvis, J. U. M. 1981. Eusociality in a mammal: cooperative breeding in naked mole-rat colonies. *Science* 212: 571–573.

Jeyasingh, D. E. J. 1976. Fecal feeding in the white-headed babbler *Turdoides affinis. J. Bombay Nat. Hist. Soc.* 73: 218.

Johnsgard, P. A. 1961. Breeding biology of the Magpie Goose. *Annual Report of the Wildfowl Trust* 12: 92–103.

Johnsingh, A. J. T., and K. Paramanandham. 1982. Group care of white headed babblers *Turdoides affinis* for a pied crested cuckoo *Clamator jacobinus* chick. *Ibis* 124: 179–183.

Johnson, M. S., and J. L. Brown. 1980. Genetic variation among trait groups and apparent absence of close inbreeding in Grey-crowned Babblers. *Behav. Ecol. Sociobiol.* 7: 93–98.

Johnston, R. F. 1961. Population movements of birds. *Condor* 63: 386–389.

Joste, N., J. D. Ligon, and P. B. Stacey. 1985. Shared paternity in the acorn woodpecker (*Melanerpes formicivorus*). *Behav. Ecol. Sociobiol.* 17: 39–41.

Joste, N. E., W. D. Koenig, R. L. Mumme, and F. A. Pitelka. 1982. Intragroup dynamics of a cooperative breeder: an analysis of reproductive roles in the acorn woodpecker. *Behav. Ecol. Sociobiol.* 11: 195–201.

Kemp. A. C. 1978. A review of the hornbills: biology and radiation. *Living Bird* 17: 105–136.

Kemp. A. C., and M. I. Kemp. 1974. Observations on the Buffalo Weaver. *Bokmakierie* 26: 55–58.

Kemp, A. C., and M. I. Kemp. 1980. The biology of the southern ground hornbill *Bucorvus leadbeateri* (Vigors) (Aves: Bucerotidae). *Annals of the Transvaal Museum* 32: 65–100.

Kepler, A. K. 1977. Comparative study of todies (Todidae) with emphasis on the Puerto Rican today, *Todus mexicanus. Publ. Nuttall Ornithol. Club* 16: 1–190.

Kilham, L. 1956. Breeding and other habits of casqued hornbills (*Bycanistes subcylindricus*). *Smiths. Misc. Coll.* 131(9): 1–45.

Kilham, L. 1977. Altruism in nesting yellow-bellied sapsucker. *Auk* 94: 613–614.

Kilham, L. 1984. Cooperative breeding of American Crows. *J. Field Ornithol.* 55: 349–356.

Kiltie, R. A., and J. W. Fitzpatrick. 1984. Reproduction and social organization of the Black-capped Donacobius (*Donacobius atricapillus*) in southeastern Peru. *Auk* 101: 804–811.

King. B. R. 1980. Social organization and behaviour of the grey-crowned babbler *Pomatosomus temporalis. Emu* 80: 59–76.

Kinnaird, M. F., and P. R. Grant. 1982. Cooperative breeding in the Galapagos Mockingbird *Nesomimus parvulus. Behav. Ecol. Sociobiol.* 10: 65–73.

Klomp, H. 1972. Regulation of the size of bird populations by means of territorial behavior. *Netherlands J. Zool.* 22: 456–488.

Kluyver, H. M., and L. Tinbergen. 1953. Territory and regulation of density in titmice. *Arch. Neerl. Zool.* 10: 265–286.

Koenig, W. D. 1981a. Space competition in the acorn woodpecker: power struggles in a cooperative breeder. *Anim. Behav.* 29: 396–427.

Koenig, W. D. 1981b. Reproductive success, group size, and the evolution of co-operative breeding in the acorn woodpecker. *Amer. Nat.* 117: 421–443.

Koenig, W. D., R. L. Mumme, and F. A. Pitelka. 1983. Female roles in co-operatively breeding Acorn Woodpeckers. In: Wasser, S. K., ed. *Social Behavior of Female Vertebrates*, pp. 235–261. New York: Academic Press.

Koenig, W. D., R. L. Mumme, and F. A. Pitelka. 1984. The breeding system of the Acorn Woodpecker in Central coastal California. *Z. Tierpsychol.* 65: 289–308.

Koenig, W. D., and F. A. Pitelka. 1979. Relatedness and inbreeding avoidance: counterploys in the communally nesting acorn woodpecker. *Science* 206: 1103–1105.

Koenig, W. D., and F. A. Pitelka. 1981. Ecological factors and kin selection in the evolution of cooperative breeding in birds. In: Alexander, R. D., and D. W. Tinkle, eds. *Natural Selection and Social Behavior: Recent Research and New Theory*. New York: Chiron Press.

Koenig, W. D., and P. L. Williams. 1979. Notes on the status of acorn woodpeckers, in central Mexico. *Condor* 81: 317–318.

Koester, F. 1971. Zum Nistverhalten des Ani, *Crotophaga ani. Bonn. Zool. Beitr.* 22: 4–27.

Krebs, J. R. 1980. Measuring fitness in social systems. In: Markl, H., ed. *Evolution of Social Behavior: Hypotheses and Empirical Tests*, pp. 205–218. Weinheim: Verlag Chemie.

Krebs J. R., and N. B. Davies. 1981. *An Introduction to Behavioural Ecology*. Oxford: Blackwell.

Krekorian, C. O. 1978. Alloparental care in the Purple Gallinule. *Condor* 80: 382–390.

Lack, D. 1947. The significance of clutch size, parts I–II. *Ibis* 89: 302–352.

Lack, D. 1948a. The significance of clutch size, part III. *Ibis* 90: 25–45.

Lack, D. 1948b. Natural selection and family size in the starling. *Evolution* 2: 95–110.

Lack, D. 1953. *The Life of the Robin*. London: Penguin Books.

Lack, D. 1954a. The evolution of reproductive rates. In: Huxley, J., A. C. Hardy, and E. B. Ford, eds. *Evolution as a Process*, pp. 143–156. London: Allen and Unwin.

Lack, D. 1954b. *The Natural Regulation of Animal Numbers*. London: Oxford University Press.

Lack, D. 1966. *Population Studies of Birds*. Oxford: Clarendon Press.

Lack, D. 1968. *Ecological Adaptations for Breeding in Birds*. London: Methuen.

Lack, D., and E. Lack. 1958. The nesting of the Long-tailed Tit. *Bird Study* 5: 1–19.

Lane, R. 1978. Co-operative breeding in the Australian Little Grebe. *Sunbird* 9: 2.

Lawton, M. F., and C. F. Guindon. 1981. Flock composition, breeding success, and learning in the Brown Jay. *Condor* 82: 27–33.

Lawton, M. F., and R. O. Lawton. 1985. The breeding biology of the Brown Jay in Monteverde, Costa Rica. *Condor* 87: 192–204.

Leach, F. A. 1925. Communism in the California woodpecker. *Condor* 27: 12–19.

Leighton, M. 1982. "Fruit resources and patterns of feeding, spacing and grouping among sympatric Bornean hornbills (*Bucerotidae*)." Ph.D. thesis, University of California, Davis.

Lewis, D. M. 1981. Determinants of reproductive success of the white-browed sparrow weaver, *Plocepasser mahali. Behav. Ecol. Sociobiol.* 9: 83–93.

Lewis, D. M. 1982a. Dispersal in a population of white-browed sparrow weavers. *Condor* 84: 306–312.

Lewis, D. M. 1982b. Cooperative breeding in a population of White-browed Sparrow Weavers *Plocepasser mahali. Ibis* 124: 511–522.

Ligon, J. D. 1970. Behavior and breeding biology of the red-cockaded woodpecker. *Auk* 87: 255–278.

Ligon, J. D. 1981a. Demographic patterns and communal breeding in the green woodhoopoe, *Phoeniculus purpureus*. In: Alexander, R. D., and D. W. Tinkle, eds. *Natural Selection and Social Behavior: Recent Research and New Theory*, pp. 231–243. New York: Chiron Press.

Ligon, J. D. 1981b. Sociobiology is for the birds. *Auk* 98: 409–412.

Ligon, J. D. 1983. Cooperation and reciprocity in avian social systems. *Amer. Nat.* 121: 366–383.

Ligon, J. D. 1985. The Florida Scrub Jay. *Science* 227: 1573–1574.

Ligon, J. D., and S. L. Husar. 1974. Notes on the behavioral ecology of Couch's Mexican jay. *Auk* 91: 841–843.

Ligon, J. D., and S. H. Ligon, 1978. Communal breeding in green woodhoopoes as a case for reciprocity. *Nature* 280: 174.

Ligon, J. D., and S. H. Ligon. 1979. The communal social system of the green woodhoopoe in Kenya. *Living Bird* 17: 159–197.

Ligon, J. D., and S. H. Ligon. 1983. Reciprocity in the Green Woodhoopoe (*Phoeniculus purpureus*). *Anim. Behav.* 31: 480–489.

Macdonald, D. W. 1983. The ecology of carnivore social behaviour. *Nature* 301: 379–384.

Mack, K. J. 1967. Apostle birds in the Murray Lands. *South Aust. Ornithol.* 25: 139–145.

Mackness, B. S. 1976. A helper at the nest of the Brown Warbler. *Sunbird* 7: 79.

MacLean, A. A. E. 1982. "A comparative study of the improvement of foraging success with age in three species of gulls on the Niagara river: relationships to deferred breeding." Ph.D. thesis, University of Rochester.

MacLean, A. A. E. 1986. Age-specific foraging ability and the evolution of deferred breeding in three species of gulls. *Wilson Bull.*: 98: 267–279.

Maclean, G. L. 1973. The sociable weaver, Parts 1–5. *Ostrich* 44: 176–261.

MacRoberts, M. H., and B. R. MacRoberts. 1976. Social organization and behavior of the acorn woodpecker in central coastal California. *Ornithol. Monogr.* 21: 1–115.

Mader, W. J. 1975a. Extra adults at Harris' hawk nests. *Condor* 77: 482–485.

Mader, W. J. 1975b. Biology of the Harris' hawk in southern Arizona. *Living Bird* 14: 59–85.

Mader, W. J. 1977. Harris' hawks lay three clutches of eggs in one year. *Auk* 94: 370–371.
Mader, W. J. 1978. A comparative nesting study of Red-tailed hawks and Harris' hawks in southern Arizona. *Auk* 95: 327–337.
Mader, W. J. 1979. Breeding behavior of a polyandrous trio of Harris' hawks in southern Arizona. *Auk* 96: 776–788.
Malcolm, J. M. 1971. Two female trumpeter swans share a nest. *Murrelet* 52: 24–25.
Matessi. C., and S. D. Jayakar. 1973. A model for the evolution of altruistic behavior. *Genetics* 74: S174.
Matessi, C., and S. D. Jayakar. 1976. Conditions for the evolution of altruism under Darwinian selection. *Theor. Popul. Biol.* 9: 360–387.
Maynard Smith, J. 1964. Group selection and kin selection. *Nature* 201: 1145–1147.
Maynard Smith, J. 1976. Group selection. *Q. Rev. Biol.* 51: 277–283.
Maynard Smith, J. 1982. *Evolution and the Theory of Games.* New York: Cambridge University Press.
Maynard Smith, J., and G. R. Price. 1973. The logic of animal conflict. *Nature* 246: 15–18.
Maynard Smith, J., and M. G. Ridpath. 1972. Wife sharing in the Tasmanian native hen, *Tribonyx mortierii*: a case of kin selection?. *Amer. Nat.* 106: 447–452.
McGillivray, W. B. 1980. Communal nesting in the House Sparrow. *J. Field Ornithol.* 51: 371–372.
McKaye, K. R. 1981. Natural selection and the evolution of interspecific brood care in fishes. In: Alexander, R. D., and D. W. Tinkle, eds. *Natural Selection and Social Behavior: Recent Research and New Theory*, pp. 173–183. New York: Chiron Press.
Michael, E. 1926. Acorn storing methods of the California and Lewis' Woodpeckers. *Condor* 28: 68–69.
Michener, C. D. 1974. *The Social Behavior of the Bees.* Cambridge, Mass.: Harvard University Press.
Michod, R. E. 1982. The theory of kin selection. *Ann. Rev. Ecol. Syst.* 13: 23–55.
Miskell, J. 1977. Co-operative feeding of young at the nest by Fischer's starling *Spreo fischeri. Scopus* 1: 87–88.
Moehlman, P. D. 1979. Jackal helpers and pup survival. *Nature* 277: 382–383.
Moffat, J. D. M., M. J. Whitmore, and E. M. Date. 1983. Communal breeding by Striped Honeyeaters. *Emu* 83: 202–203.
Monnert, R. J. 1983. L'aide à l'élevage chez le faucon pèlerin. *Alauda* 51: 241–250.
Moreau, R. E. 1947. Relations between number in brood, feeding rate and nestling period in nine species of birds in Tanganyika Territory. *J. Anim. Ecol.* 16: 205–209.
Morony, J. J., Jr., W. J. Bock, and J. Farrand, Jr. 1975. *Reference List of the Birds of the World.* New York: Amer. Mus. Nat. Hist.
Morrison, M. L., R. D. Slack, and E. Shanley, Jr. 1978. Age and foraging ability relationships of Olivaceous Cormorants. *Wilson Bull.* 90: 414–422.

Morton, S. R., and G. D. Parry. 1974. The auxiliary social system in kookaburras: a reappraisal of its adaptive significance. *Emu* 74: 196–198.

Mugaas, J. N., and J. R. King. 1981. Annual variation of daily energy expenditure by the Black-billed Magpie: a study of thermal and behavioral energetics. Studies in Avian Biology 5: 1–78.

Mumme, R. L., W. D. Koenig, and F. A. Pitelka. 1983a. Reproductive competition in the communal acorn woodpecker: sisters destroy each other's eggs. *Nature* 306: 583–584.

Mumme, R. L., W. D. Koenig, and F. A. Pitelka. 1983b. Mate guarding in the Acorn Woodpecker: within-group reproductive competition in a cooperative breeder. *Anim. Behav.* 31: 1094–1106.

Mumme, R. L., W. D. Koenig, R. M. Zink, and J. A. Marten. 1985. Genetic variation and parent age in a California population of Acorn Woodpeckers. *Auk* 102: 305–312.

Munro, J., and J. Bedard. 1977. Gull predation and creching behaviour in the common eider. *J. Anim. Ecol.* 46: 799–810.

Myers, G. R., and D. W. Waller. 1977. Helpers at the nest in Barn Swallows. *Auk* 94: 596.

Myers, J. P., P. G. Connors, and F. A. Pitelka. 1979. Territoriality in non-breeding shorebirds. *Shorebirds in Marine Environments. Studies in Avian Biology.* 2: 231–246.

Nakamura, T. 1972. Home range structure of a population of *Aegithalos caudatus. Misc. Reports Yamashina Inst. Ornithol.* 6: 424–488.

Nakamura, T. 1973. Population size and distribution of breeding birds in Kawatabi grassland. In: Numata, M., ed. *Ecological Studies of Japanese Grassland.* Internat. Biol. Program.

Nias, R. C. 1984. Territory quality and group size in the superb blue fairy-wren *Malurus cyaneus. Emu* 84: 178–180.

Nice, M. M. 1943. *Studies in the Life History of the Song Sparrow, 2.* New York: Dover, 1964 ed.

Norris, R. A. 1958. Comparative biosystematics and life history of the nuthatches *Sitta pygmaea* and *Sitta pusilla. Univ. Calif. Publ. Zool.* 56: 119–300.

Noske, R. A. 1980a. Co-operative breeding and plumage variation in the Orange-winged (Varied) Sittella. *Corella* 4: 45–53.

Noske, R. A. 1980b. Cooperative breeding by tree creepers. *Emu* 80: 35–36.

Noske, R. A. 1983. Communal behaviour of brown-headed honeyeaters. *Emu* 83: 38–41.

Ohara, H., and S. Yamagishi. 1984. The first record of helping at the nest in the Short-tailed Bush Warbler *Cettia squameiceps. Ornithol. Society of Japan* 33: 39–41.

Ohara, H., and S. Yamagishi. 1985. A helper at the nest of the Short-tailed Bush Warbler *Cettia squameiceps. J. Yamashina Inst. Ornithol.* 17: 67–73.

Orians, G. H. 1961. The ecology of blackbird (Agelaius) social systems in birds and mammals. *Ecol. Monogr.* 31: 285–312.

Orians, G. H. 1969. Age and hunting success in the Brown Pelican. *Anim. Behav.* 17: 316–319.

Orians, G. H., L. Erickmann, and J. C. Schultz. 1977a. Nesting and other habits of the Bolivian Blackbird (*Oreospar bolivianus*). *Condor* 79: 250–256.

Orians, G. H., C. E. Orians, and K. J. Orians. 1977b. Helpers at the nest in some Argentine blackbirds. In: Stonehouse, B., and C. Perrins, eds. *Evolutionary Ecology*. Baltimore: University Park Press.

Oring, L. W. 1982. Avian mating systems. In: Farner, D., J. King, and K. Parkes, eds. *Avian Biology*, pp. 1–92. New York: Academic Press.

Oster, G. F., and E. O. Wilson. 1978. *Caste and Ecology in the Social Insects*. Princeton, N.J.: Princeton University Press.

Owens, D. D., and M. J. Owens. 1984. Helping behaviour in brown hyenas. *Nature* 308: 843–845.

Parker, G. A., and N. Knowlton. 1980. The evolution of territory size—some ESS models. *J. Theor. Biol.* 84: 445–476.

Parker, G. A., and M. R. Macnair. 1978. Models of parent-offspring conflict. *Anim. Behav.* 26: 97–110.

Parker, J. W., and M. Ports. 1982. Helping at the nest by yearling Mississippi Kites. *Raptor Research* 16: 14–17.

Parry, V. 1973. The auxiliary social system and its effect on territory and breeding in kookaburras. *Emu* 73: 81–100.

Patterson, I. J. 1982. *The Shelduck: A Study in Behavioural Ecology*. Cambridge: Cambridge University Press.

Payne, R. B. 1977. The ecology of brood parasitism in birds. *Ann. Rev. Ecol. Syst.* 8: 1–28.

Payne, R. B., L. L. Payne, and I. Rowley. 1985. Splendid wren *Malurus splendens* response to cuckoos: an experimental test of social organization in a communal bird. *Behaviour* 94: 108–127.

Pierce, B. A., and H. M. Smith. 1978. Neoteny or paedogenesis. *J. Herpetology* 13: 119–121.

Platt, J. R. 1964. Strong inference. *Science* 146: 347–353.

Price, G. R. 1972. Extension of covariance selection mathematics. *Ann. Hum.* Genet. 35: 485–490.

Price, T., S. Millington, and P. Grant. 1983. Helping at the nest in Darwin's finches as misdirected parental care. *Auk* 100: 192–194.

Pulliam, H. R. 1973. On the advantages of flocking. *J. Theor Biol.* 38: 419–422.

Pulliam, H. R. 1976. The principles of optimal behavior and the theory of communities. In: Bateson, P. P. G., and P. H. Klopfer, eds. *Perspectives in Ethology*, pp. 311–332. New York: Plenum Press.

Pulliam, H. R. 1981. On experimental tests of sociobiological theory. *Auk* 98: 406–412.

Pulliam, H. R., and T. Caraco. 1984. Living in groups: is there an optimal group size? In: Krebs, J. R., and N. B. Davies, eds. *Behavioural Ecology: An Evolutionary Approach*, 2nd ed., pp. 122–147. Sunderland, mass.: Sinauer.

Pulliam, H. R., and G. C. Millikan. 1982. Social behavior in the non-reproductive season. In: Farner, D., J. King, and K. Parkes, eds. *Avian Biology*. New York: Academic Press.

Pulliam, H. R., G. H. Pyke, and T. Caraco. 1982. The scanning behavior of juncos: A game-theoretical approach. *J. Theor. Biol.* 95: 89–103.

Rabenold, K. N. 1984. Cooperative enhancement of reproductive success in tropical wren societies. *Ecology* 65: 871–885.

Rabenold, K. N. 1985. Cooperation in breeding by nonreproductive wrens: kinship, reciprocity, and demography. *Behav. Ecol. Sociobiol.* 17: 1–17.

Rabenold, K. N., and C. R. Christensen. 1979. Effects of aggregation on feeding and survival in a communal wren. *Behav. Ecol. Sociobiol.* 6: 39–44.

Radke, E. L., and J. Klimosewki. 1977. Late fledging date for Harris' Hawk. *Wilson Bull.* 89: 469.

Raitt, R. J. and J. W. Hardy. 1976. Behavioral ecology of the Yucatan jay. *Wilson Bull.* 88: 529–554.

Raitt, R. J., and J. W. Hardy. 1979. Social behavior, habitat, and food of the Beechey jay. *Wilson Bull.* 91: 1–15.

Raitt, R. J., S. R. Winterstein, and J. W. Hardy. 1984. Structure and dynamics of communal groups in the Beechey Jay. *Wilson Bull.* 96: 206–227.

Rappole, J. H., and D. W. Warner. 1976. Relationships between behavior, physiology and weather in avian transients at a migration stopover site. *Oecologia* 26: 193–212.

Rappole, J. H., D. W. Warner, and M. Ramos. 1977. Territoriality and population structure in a small passerine community. *Am. Midl. Nat.* 97: 110–119.

Real, L. A. 1981. Uncertainty and plant-pollinator interactions: the foraging behavior of bees and wasps on artificial flowers. *Ecology* 62: 20–26.

Recher, H. F. 1977. Ecology of co-existing White-cheeked and New Holland Honeyeaters. *Emu* 77: 136–142.

Recher, H. F., and J. A. Recher. 1969. Some aspects of the ecology of migrant shorebirds. II. Aggression. *Wilson Bull.* 81: 140–154.

Reyer, H.-U. 1980a. Flexible helper structure as an ecological adaptation in the pied kingfisher (*Ceryle rudis rudis* L.). *Behav. Ecol. Sociobiol.* 6: 219–227.

Reyer, H.-U. 1980b. Sexual dimorphism and co-operative breeding in the Striped Kingfisher. *Ostrich* 51: 117–118.

Reyer, H.-U. 1982. Soziale Strategien und Irhe Evolution. *Naturw. Runds.* 35: 6–17.

Reyer, H.-U. 1984. Investment and relatedness: a cost/benefit analysis of breeding and helping in the pied kingfisher (*Ceryle rudis*). *Anim. Behav.* 32: 1163–1178.

Reyer, H.-U. 1986. Breeder-helper-interactions in the Pied Kingfisher reflect the costs and benefits of cooperative breeding. *Behaviour*: 96: 277–303.

Reyer, H.-U., and K. Westerterp. 1985. Parental energy expenditure: a proximate cause of helper recruitment in the pied kingfisher (*Ceryle rudis*). *Behav. Ecol. Sociobiol.* 17: 363–369.

Reynolds, J. F. 1965. Behavior of the Black-lored Babbler. *E. Afr. Wildlife Jour.* 3: 130.

Ricklefs, R. E. 1975. The evolution of co-operative breeding in birds. *Ibis* 117: 531–534

Ricklefs, R. E. 1980. Watch-dog behaviour observed at the nest of a cooperative breeding bird, the rufous-margined flycatcher *Myiozetetes cayanensis*. *Ibis* 122: 116–118.

Ricklefs. R. E. 1983. Comparative avian demography. In: Johnston, R. F., ed. *Current Ornithology*, pp. 1–32. London: Plenum Press.

Ridpath, M. G. 1972. The Tasmanian native hen, *Tribonyx mortierii*. I–III. *CSIRO Wildl. Res.* 17: 1–118.

Riedman, M. L. 1982. The evolution of alloparental care and adoption in mammals and birds. *Q. Rev. Biol.* 57: 405–435.

Ritter, W. E. 1938. *The California Woodpecker and I*. Berkeley: University of California Press.

Roberts, R. C. 1979. Habitat and resource relationships in acorn woodpeckers. *Condor* 81: 1–8.

Robinson, A. 1956. The annual reproductory cycle of the magpie, *Gymnorhinus dorsalis* Campbell, in south-western Australia. *Emu* 56: 233–336.

Rohwer. S., and P. W. Ewald. 1981. The cost of dominance and advantage of subordination in a badge signaling system. *Evolution* 35: 441–454.

Rood, J. P., 1978. Dwarf mongoose helpers at the den. *Z. Tierpsychol.* 48: 277–287.

Rowan, M. K. 1967. A study of the colies of southern Africa. *Ostrich* 6: 63–115.

Rowley, I. 1965a. The life history of the superb blue wren, *Malurus cyaneus. Emu* 64: 251–297.

Rowley, I. 1965b. White-winged choughs. *Austral. Nat. Hist.* 15: 81–85.

Rowley, I. 1968. Communal species of Australian birds. *Bonn. Zool. Beitr.* 19: 362–368.

Rowley, I. 1976. Co-operative breeding in Australian birds. In: Frith, H. J., and J. H. Calaby, eds. *Proc. 16th Internat. Ornithol. Congr.*, pp. 657–666. Canberra: Australian Acad. Sci.

Rowley, I. 1977. Communal activities among white-winged choughs *Corcorax melanorhamphus. Ibis* 120: 1–20.

Rowley, I. 1981a. The communal way of life in the Splendid Wren, *Malurus splendens. Z. Tierpsychol.* 55: 228–267.

Rowley, I. 1981b. A relic population of blue-breasted wrens, *Malurus pulcherinus* in the central wheatbelt. *Western Australian Naturalist* 15: 1–8.

Rowley, I., S. T. Emlen, A. J. Gaston, and G. E. Woolfenden. 1979. A definition of 'group'. *Ibis* 121: 231.

Sabine, W. S. 1949. Dominance in winter flocks of juncos and tree sparrows. *Physiol. Zool.* 22: 68–85.

Sabine, W. S. 1959. The winter society of the Oregon junco: intolerance, dominance and the pecking order. *Condor* 61: 110–135.

Sappington, J. N. 1977. Breeding biology of house sparrows in north Mississippi. *Wilson Bull.* 89: 300–309.

Sauer, F. G., and E. M. Sauer. 1959. Polygamie beim Sud-Afrikanischen Strauss (*Struthio camelus australis*). *Bonn. Zool. Beitr.* 10: 266–285.

Sauer, F. G., and E. M. Sauer. 1966. The behavior and ecology of the South African Ostrich. *Living Bird* 5: 45–75.

Schreiber, R. W., and S. N. Young. 1974. Evidence for learning to feed in Laughing gulls. *Fla. Field Nat.* 2: 16–17.

Searcy, W. A. 1978. Foraging success in three age classes of Glaucous-winged Gulls. *Auk* 95: 586–588.

Selander, R. K. 1959. Polymorphism in Mexican Brown jays. *Auk* 76: 385–417.
Selander, R. K. 1964. Speciation in wrens of the genus *Campylorhynchus*. *Univ. Calif. Publ. Zool.* 74: 1–224.
Selander, R. K., and D. R. Giller. 1959. The avifauna of the Barranca de Oblatos, Jalisco, Mexico. *Condor* 61: 210–222.
Sherman, P. W. 1980. The meaning of nepotism. *Amer. Nat.* 116: 604–606.
Shields, W. M. 1980. Ground squirrel alarm calls: nepotism or parental care? *Amer. Nat.* 116: 599–603.
Short, L. L. 1970. Notes on the habits of some Argentine and Peruvian woodpeckers (Aves Picidae). *American Museum Novitates* 2413: 1–37.
Short, L. L., and J. F. M. Horne. 1980. Ground barbets of East Africa. *Living Bird* 18: 179–186.
Shy, M. M. 1982. Interspecific feeding among birds: a review. *J. Field Ornithol.* 53: 370–393.
Sick, H. 1959. Notes on the biology of two Brazilian swifts, *Chaetura andrei* and *Chaetura cinereiventris*. *Auk* 76: 471–477.
Skutch, A. F. 1935. Helpers at the nest. *Auk* 52: 257–273.
Skutch, A. F. 1943. The family life of Central American woodpeckers. *Sci. Monthly* 56: 358–364.
Skutch, A. F. 1953. The White-throated Magpie-jay. *Wilson Bull.* 65: 68–74.
Skutch, A. F. 1954. Life histories of Central American birds. *Pacific Coast Avifauna* 31: 1–448.
Skutch, A. F. 1958. Roosting and nesting of Aracari toucans. *Condor* 60: 201–219.
Skutch, A. F. 1959. Life history of the groove-billed ani. *Auk* 76: 281–317.
Skutch, A. F. 1960. Life histories of Central American birds. II. *Pacific Coast Avifauna* 34: 1–593.
Skutch, A. F. 1961a. Helpers among birds. *Condor* 63: 198–226.
Skutch, A. F. 1961b. The nest as a dormitory. *Ibis* 103a: 50–70.
Skutch, A. F. 1967. Adaptive limitation of the reproductive rate of birds. *Ibis* 109: 579–599.
Skutch, A. F. 1969a. A study of the rufous-fronted thornbird and associated birds. *Wilson Bulll.* 81: 5–43.
Skutch, A. F. 1969b. Life histories of Central American birds. III. *Pacific Coast Avifauna* 35:,
Skutch, A. F. 1972. Studies of tropical American birds. *Publ. Nuttall Ornithol. Club* 10: 1–228.
Smith, A. J., and B. I. Robertson. 1978. Social organization of Bell Miners. *Emu.* 78: 169–178.
Smith, C. C. 1968. The adaptive nature of social organization in the genus of tree squirrels *Tamiasciurus*. *Ecol. Monogr.* 38: 31–63.
Smith, C. C., and S. D. Fretwell. 1974. The optimal balance between size and number of offspring. *Amer. Nat.* 108: 499–506.
Smith, S. M. 1978. The underworld in a territorial sparrow: Adaptive strategy for floaters. *Amer. Nat.* 112: 571–582.
Snow, B., and D. Snow. 1982. Territory and social organization in a population of Dunnocks, *Prunella modularis*. *J. Yamashina Inst. Ornithol.* 14: 281–292.

Snow, D. W. 1971. Observations on the purple-throated fruit-crow in Guyana. *Living Bird* 10: 5–17.

Snow, D. W., and C. T. Collins. 1962. Social breeding behavior of the Mexican tanager. *Condor* 64: 161.

Snow, D. W., and A. Lill. 1974. Longevity records for some neotropical land birds. *Condor* 76: 262–267.

Stacey, P. B. 1979a. Habitat saturation and communal breeding in the acorn woodpecker. *Anim. Behav.* 27: 1153–1166.

Stacey, P. B. 1979b. Kinship, promiscuity and communal breeding in the acorn woodpecker. *Behav. Ecol. Sociobiol.* 6: 53–66.

Stacey, P. B. 1981. Foraging behavior of the Acorn Woodpecker in Belize, Central America. *Condor* 83: 336–339.

Stacey, P. B. 1982. Female promiscuity and male reproductive success in social birds and mammals. *Amer. Nat.* 120: 51–64.

Stacey, P. B., and C. E. Bock. 1978. Social plasticity in the acorn woodpecker. *Science* 202: 1298–1300.

Stacey, P. B., and T. C. Edwards, Jr. 1983. Possible cases of infanticide by immigrant females in a group-breeding bird. *Auk* 100: 731–733.

Stacey, P. B., and R. Jansma. 1977. Storage of pinyon nuts by the Acorn Woodpecker in New Mexico. *Wilson Bull.* 89: 150–151.

Stacey, P. B., and W. D. Koenig. 1984. Cooperative breeding in the acorn woodpecker. *Sci. Amer.* (August): 114–121.

Stallcup, J. A., and G. E. Woolfenden. 1978. Family status and contribution to breeding by Florida Scrub Jays. *Anim. Behav.* 26: 1144–1156.

Stearns, S. C. 1976. Life-history tactics: a review of the ideas. *Q. Rev. Biol.* 51: 3–47.

Strahl, S. D. 1985. "The behavior and socio-ecology of the Hoatzin, Opisthocomus hoatzin, in the llanos of Venezuela." Ph.D. thesis, State University of New York, Albany.

Swainson, G. W. 1970. Co-operative rearing in the bell miner. *Emu* 70: 183–188.

Tarbell, A. T. 1983. A yearling helper with a tufted titmouse brood. *J. Field Ornithol.* 54: 89.

Tarboton, W. R. 1981. Cooperative breeding and group territoriality in the Black Tit. *Ostrich* 52: 216–225.

Thangamani, A., K. Paramanandham, and A. J. T. Johnsingh. 1981. Helpers among the Black Drongo (*Dicrurus adsimilis*). *J. Bombay Nat. Hist. Soc.* 78: 602–603.

Thomas, B. T. 1979. Behavior and breeding of the White-bearded Flycatcher (*Conopias inornata*). *Auk* 96: 767–775.

Thomas, B. T. 1983. The plain-fronted thornbird: nest construction, material choice, and nest defense behavior. *Wilson Bull.* 95: 106–117.

Tidemann, S. C. 1980. Notes on breeding and social behaviour of the white-winged fairy-wren. *Emu* 80: 157–161.

Trail, P. W. 1980. Ecological correlates of social organization in a communally breeding bird, the acorn woodpecker, *Melanerpes formicivorus*. *Behav. Ecol. Sociobiol.* 7: 83–92.

Trail, P. W., S. D. Strahl, and J. L. Brown. 1981. Infanticide and selfish behavior in a communally breeding bird, the Mexican jay (*Aphelocoma ultramarina*). *Amer. Nat.* 118: 72–82.

Trivers, R. L. 1971. The evolution of reciprocal altruism. *Q. Rev. Biol.* 46: 35–57.

Trivers, R. L. 1974. Parent-offspring conflict. *Amer. Zool.* 14: 249–264.

Trivers, R. L. 1985. *Social Evolution.* Menlo Park, Calif. Benjamin /Cummings.

Trivers, R. L., and H. Hare. 1976. Haplodiploidy and the evolution of the social insects. *Science* 191: 249–263.

Vehrencamp, S. L. 1977. Relative fecundity and parental effort in communally nesting anis, *Crotophaga sulcirostris. Science* 197: 403–405.

Vehrencamp, S. L. 1978. The adaptive significance of communal nesting in groove-billed anis (*Crotophaga sulcirostris*). *Behav. Ecol. Sociobiol.* 4: 1–33.

Vehrencamp, S. L. 1979. The roles of individual, kin, and group selection in the evolution of sociality. In: Marler, P., and J. G. Vandenbergh, eds. *Handbook of Behavioral Neurobiology* Vol. 3. New York: Plenum Press.

Vehrencamp, S. L. 1980. To skew or not to skew. *Proc. Internat. Ornithol. Congr.* 17: 869–874.

Vehrencamp, S. L. 1983. A model for the evolution of despotic versus egalitarian societies. *Anim. Behav.* 31: 667–682.

Vehrencamp, S. L., B. S. Bowen, and R. R. Koford. 1986a. The effect of breeding unit size on fitness components in Groove-billed Anis. In: Clutton-Brock, T. H., ed. *Reproductive Success.* Chicago: University of Chicago Press.

Vehrencamp, S. L., B. S. Bowen, and R. R. Koford. 1986b. Breeding roles and pairing patterns within communal groups of Groove-billed Anis. *Anim. Behav.*: 35: 347—366.

Verbeek, N. A. M., and R. W. Butler. 1981. Cooperative breeding of the north-western crow *Corvus caurinus. Ibis* 123: 183–189.

Vernon, C. J. 1976. Communal feeding of nestlings by the arrowmarked babbler. *Ostrich* 47: 134–136.

Vernon, F. J., and C. J. Vernon. 1978. Notes on the social behaviour of the Dusky-faced Warbler. *Ostrich* 49: 92–93.

Vine, I. 1971. Risk of visual detection and pursuit by a predator and the selective advantage of flocking behaviour *J. Theor. Biol.* 30: 405–422.

Vries, T., de. 1973. *The Galapogos Hawk.* Amsterdam: Free University Press.

Vries, T., de. 1975. The breeding biology of the Galapagos hawk, *Buteo galapagoensis. LeGerfaut* 65: 29–57.

Wade, M. J. 1978. A critical review of the models of group selection. *Q. Rev. Biol.* 53: 101–114.

Wagner, H. O. 1955. Bruthelfer unter den Vögeln. *Veröff Überseemus. Bremen Reihe A* 2: 327–330.

Walters, J. R., and B. F. Walters. 1980. Cooperative breeding by southern lapwings *Vanellus chilensis. Ibis* 122: 505–509.

Walters, J. R. 1982. Parental behavior in lapwings (Charadriidae) and its relationships with clutch sizes and mating systems. *Evolution* 36: 1030–1040.

Waltz, E. C. 1981. "Reciprocal altruism" and "spite" in gulls: a comment. *Amer. Nat.* 118: 588–592.

Waser, P. M. 1981. Sociality or territorial defense. The influence of resource. renewal. *Behav. Ecol. Sociobiol.* 8: 231–237.

Watson, A. 1985. Social class, socially-induced loss, recruitment and breeding of red grouse. *Oecologia* 67: 493–498.

Weatherhead, P. J., and R. J. Robertson. 1980. Altruism in the savannah sparrow. *Behav. Ecol. Sociobiol.* 6: 185–186.

Welty, J. C. 1975. *The Life of Birds.* Philadelphia: Saunders.

Westcott, P. W. 1969. Relationships among three species of jays wintering in southeastern Arizona. *Condor* 71: 353–359.

West Eberhard, M. J. 1975. The evolution of social behavior by kin selection. *Q. Rev. Biol.* 50: 1–33.

West Eberhard, M. J. 1981. Intragroup selection and the evolution of insect societies. In: Alexander, R. D., and D. W. Tinkle, eds. *Natural Selection and Social Behavior: Recent Research New Theory*, pp. 3–17. New York: Chiron Press.

Wetmore, A., and B. H. Swales. 1931. The birds of Haiti and the Dominican Republic. *Bull. U. S. Nat. Mus.* 155: 1–483.

Wettin, P. 1984. Simultaneous polyandry in the Purple Swamphen. *Emu* 84: 111–113.

White, S. R. 1950. Domestic co-operation among Chestnut-tailed Thornbills. *Western Australian Naturalist* 2: 93–94.

Whitney, G. 1976. Genetic substrates for the initial evolution of human sociality. I. Sex chromosome mechanisms. *Amer. Nat.* 110: 867–875.

Wiley, R. H., and K. N. Rabenold. 1984. The evolution of cooperative breeding by delayed reciprocity and queuing for favorable social position, *Evolution*: 38: 609–621.

Wiley, R. H., and M. S. Wiley. 1977. Recognition of neighbor's duets by striped-backed wrens *Campylorhynchus nuchalis. Behaviour* 62: 10–34.

Wilkinson, R. 1978. Co-operative breeding in the Chestnut-bellied Starling *Spreo pulcher. Bull, Nigerian Ornithol. Soc.* 14(46): 71–72.

Wilkinson, R. 1982. Social organization and communal breeding in the chestnut-bellied starling (*Spreo pulcher*). *Anim. Behav.* 30: 1118–1128.

Wilkinson, R., and A. E. Brown. 1984. Effect of helpers on the feeding rates of nestlings in the chestnut-bellied starling *Spreo pulcher. J. Anim. Ecol.* 53: 1–10.

Williams, G. C. 1966a. *Adaptation and Natural Selection.* Princeton, N.J.: Princeton University Press.

Williams, G. C. 1966b. Natural selection, the cost of reproduction, and a refinement of Lack's principle. *Amer. Nat.* 100: 687–690.

Wilson, D. S. 1975. A theory of group selection. *Proc. Natl. Acad. Sci. USA* 72: 143–146.

Wilson, D. S. 1977. Structured demes and the evolution of group-advantageous traits. *Amer. Nat.* 111: 157—185.

Wilson, D. S. 1980. *The Natural Selection of Populations and Communities.* Menlo Park, Calif.: Benjamin/Cummings.

Wilson, D. S. 1983. The group selection controversy: history and current status. *Ann. Rev. Ecol. Syst.* 14: 159–187.

Wilson, E. O. 1971. *The Insect Societies.* Cambridge, Mass.: Harvard University Press.

Wilson, E. O. 1975. *Sociobiology.* Cambridge, Mass.: Harvard University Press.

Winterstein, S. R. 1985. "Ecology and sociobiology of Black-throated Magpie-jay." Ph.D. thesis, New Mexico State University.

Winterstein, S. R., and R. J. Raitt. 1983. Nestling growth and development and the breeding ecology of the Beechey jay. *Wilson Bull.* 95: 256–268.

Woodall, P. F. 1975. On the life history of the Bronze Mannikin. *Ostrich* 46: 55–86.

Woolfenden, G. E. 1975. Florida scrub jay helpers at the nest. *Auk* 92: 1–15.

Woolfenden, G. E. 1976. Co-operative breeding in American birds. In: Frith, H. J., and J. H. Calaby, eds. *Proc. 16th Internat. Ornithol. Congr.*, pp. 674–684. Canberra: Australian Acad. Sci.

Woolfenden, G. E. 1978. Growth and survival of young Florida Scrub Jays. *Wilson Bull.* 90: 1–18.

Woolfenden, G. E. 1980. The selfish behavior of avian altruists. *Proc. Internat. Ornithol. Congress* 17: 886–889.

Woolfenden, G. E. 1981. Selfish behavior by Florida scrub jay helpers. In: Alexander, R. D., and D. W. Tinkle, eds. *Natural Selection and Social Behavior: Recent Research and New Theory*, pp. 257–260. New York: Chiron Press.

Woolfenden, G. E., and J. W. Fitzpatrick. 1977. Dominance in the Florida scrub jay. *Condor* 79: 1–12.

Woolfenden, G. E., and J. W. Fitzpatrick. 1978. The inheritance of territory in group-breeding birds. *BioScience* 28: 104–108.

Woolfenden, G. E., and J. W. Fitzpatrick. 1984. *The Florida Scrub Jay: Demography of a Cooperative-breeding Bird.* Princeton, N.J.: Princeton University Press.

Wright, S. 1945. Tempo and mode in evolution: a critical review. *Ecology* 26: 415–419.

Wynne-Edwards, V. C. 1962. *Animal Dispersion in Relation to Social Behavior.* New York: Hafner.

Yamashina, M. 1938. A sociable breeding habit among Timaliine birds. *Proc. Internat. Ornithol. Congr.* 9: 453–456.

Yom-Tov, Y. 1980. Intraspecific nest parasitism in birds. *Biol. Rev.* 55: 93–108.

Yom-Tov, Y., G. M. Dunnet, and A. Anderson. 1974. Intraspecific nest parasitism in the starling *Sturnus vulgaris*.. *Ibis* 116: 87–90.

Young, E. C. 1978. Behavioural ecology of Lonnbergi skuas in relation to environment of the Chatham islands. *New Zeal. J. Zool.* 5: 401–416.

Zack, S., and J. D. Ligon. 1985a. Cooperative breeding in Lanius shrikes. I. Habitat and demography of two sympatric species. *Auk* 102: 754–765.

Zack, S., and J. D. Ligon. 1985b. Cooperative breeding in Lanius shrikes. II. Maintenance of group-living in a nonsaturated habitat. *Auk* 102: 766–773.

Zahavi, A. 1974. Communal nesting by the Arabian babbler. *Ibis* 116: 84–87.

Zahavi, A. 1976. Co-operative nesting in Eurasian birds. In: Frith, H. J., and J. H.

Calaby, eds. *Proc. 16th Internat. Ornithol. Congr.*, pp. 685–693. Canberra: Australian Acad. Sci.

Zahavi, A. 1977. Reliability in communication systems and the evolution of altruism. In: Stonehouse, B., and C. Perrins, eds. *Evolutionary Ecology*, pp. 253–259. Baltimore: University Park Press.

Zahavi, A. 1981. Natural selection, sexual selection and the selection of signals. In: Scudder, G. G. E., and J. L. Reveal, eds. *Evolution Today–Proceedings, ICSEB II*, pp. 133–138.

Author Index

Taxonomic Index

Subject Index

Library of Congress Cataloging-in-Publication Data

Brown, Jerram L.
Helping and communal breeding in birds.

(Monographs in behavior and ecology)
Bibliography: p.
Includes indexes.
1. Birds—Behavior. 2. Birds—Ecology. 3. Birds—Evolution. 4. Birds—Reproduction. I. Title. II. Series.
QL698.3.B76 1987 598.2'51 86-18669
ISBN 0-691-08447-5 (alk. paper)
ISBN 0-691-08448-3 (pbk.)